Mathematics of Planet Earth

Volume 10

This series provides a variety of well-written books of a variety of levels and styles, highlighting the fundamental role played by mathematics in a huge range of planetary contexts on a global scale. Climate, ecology, sustainability, public health, diseases and epidemics, management of resources and risk analysis are important elements. The mathematical sciences play a key role in these and many other processes relevant to Planet Earth, both as a fundamental discipline and as a key component of cross-disciplinary research. This creates the need, both in education and research, for books that are introductory to and abreast of these developments.

Springer's MoPE series will provide a variety of such books, including monographs, textbooks, contributed volumes and briefs suitable for users of mathematics, mathematicians doing research in related applications, and students interested in how mathematics interacts with the world around us. The series welcomes submissions on any topic of current relevance to the international Mathematics of Planet Earth effort, and particularly encourages surveys, tutorials and shorter communications in a lively tutorial style, offering a clear exposition of broad appeal.

Responsible Editors:
Martin Peters, Heidelberg (martin.peters@springer.com)
Robinson dos Santos, São Paulo (robinson.dossantos@springer.com)

Additional Editorial Contacts:
Donna Chernyk, New York (donna.chernyk@springer.com)
Masayuki Nakamura, Tokyo (masayuki.nakamura@springer.com)

Bertrand Chapron • Dan Crisan • Darryl Holm •
Etienne Mémin • Anna Radomska

Editors

Stochastic Transport in Upper Ocean Dynamics

STUOD 2021 Workshop, London, UK,
September 20–23

 Springer

Editors
Bertrand Chapron
Ifremer – Institut Français de Recherche
pour l'Exploitation de la Mer
Plouzané, France

Dan Crisan
Imperial College London
London, UK

Darryl Holm
Imperial College London
London, UK

Etienne Mémin
Campus Universitaire de Beaulieu
Inria – Institut National de Recherche en
Sciences et Technologies du Numérique
Rennes, France

Anna Radomska
Imperial College London
London, UK

This work was supported by Horizon 2020 Framework Programme (856408)

ISSN 2524-4264   ISSN 2524-4272   (electronic)
Mathematics of Planet Earth
ISBN 978-3-031-18990-6   ISBN 978-3-031-18988-3   (eBook)
https://doi.org/10.1007/978-3-031-18988-3

Mathematics Subject Classification: 60Hxx, 60H17, 70L10, 35R60, 37M05, 37-11, 35Qxx, 65Pxx, 00B25

This Springer imprint is published by the registered company Springer Nature Switzerland AG
The registered company address is: Gewerbestrasse 11, 6330 Cham, Switzerland

Preface

This volume contains the Proceedings of the 2nd Stochastic Transport in Upper Ocean Dynamics Workshop held on 20–23 September 2021. After the success of the first workshop, the STUOD Principal Investigators: Prof. Dan Crisan (ICL), Prof. Bertrand Chapron (IFREMER), Prof. Darryl Holm (ICL) and Prof. Etienne Mémin (INRIA) were delighted to be back with another educational and inspirational event. "Stochastic Transport in Upper Ocean Dynamics" (STUOD) project is supported by an ERC Synergy Grant, led by Imperial College London, National Institute for Research in Digital Science and Technology (INRIA) and the French Research Institute for Exploitation of the Sea (IFREMER). The project aims to deliver new capabilities for assessing variability and uncertainty in upper ocean dynamics and provide decision makers a means of quantifying the effects of local patterns of sea level rise, heat uptake, carbon storage and change of oxygen content and pH in the ocean. The project will make use of multimodal data and will enhance the scientific understanding of marine debris transport, tracking of oil spills and accumulation of plastic in the sea.

As in the previous year, the 2nd STUOD Annual Workshop 2021 focused on a range of fundamental topical areas, including:

1. Observations at high resolution of upper ocean properties such as temperature, salinity, topography, wind, waves and velocity
2. Large-scale numerical simulations
3. Data-based stochastic equations for upper ocean dynamics that quantify simulation error
4. Stochastic data assimilation to reduce uncertainty

Each chapter in the present volume illustrates one or several of these topical areas. Many chapters offer new mathematical frameworks that are intended to enhance future research in the STUOD project.

The event brought together 65 participants from 11 countries: UK 28, France 22, USA 1, Canada 1, Australia 1, Czech Republic 1, Germany 4, Italy 4, Ireland 1, South Africa 1 and Switzerland 1. Moreover, the workshop was well attended by early-career academics, post-graduate students, industry representatives (Watson-

Marlow Fluid Technology Group, OceanScope), senior members of the community and invited guests.

The scientific program of this 4-day hybrid event included invited presentations by STUOD Advisory Board Members: Prof Alberto Carrassi (University of Reading, NCEO), Prof Franco Flandoli (Scuola Normale Superiore) and Prof Sebastian Reich (University of Potsdam), Dr Eniko Székely (École Polytechnique Fédérale de Lausanne, Swiss Data Science Center), individual presentations by the STUOD Principal Investigators and post-doctoral Researchers, snapshot presentations and demos. The speakers included leading mid-career and senior researchers as well as early-career researchers. Moreover, the forum yielded opportunities for investigators at an early stage of their career to have discussions with established scientist, fostering potential future research collaborations, networking as well as inclusion and training of the next generation of researchers.

The photograph above shows some participants attending the event in person during a break between lectures.

Most of the lectures were video-recorded and may be viewed on the STUOD YouTube channel.

The following is a brief description of the 19 contributions included in the proceedings:

The submitted manuscripts include the paper by **Dan Crisan and Prince Romeo Mensah**, entitled "**Blow-up of Strong Solutions of the Thermal Quasi-Geostrophic Equation**". This paper concerns the system of coupled equations that governs the evolution of the buoyancy and potential vorticity of a fluid. This system has been shown in recent work of the authors and their collaborators to possess a local in time solution. In this paper, the authors give a characterization of the blow-up of solutions of the system in the spirit of the classical Beale–Kato–Majda blow-up criterion for the solution of the Euler equation.

The contribution of **Arnaud Debussche, Berenger Hug, and Etienne Mémin**, entitled "**Modelling Under Location Uncertainty: A Convergent Large-Scale Representation of the Navier-Stokes Equations**", introduces martingale solutions for 2D and 3D stochastic Navier-Stokes equations in the framework of the modelling under location uncertainty (LU). Such solutions are unique when the spatial dimension is 2D. The authors also prove that, if the noise intensity goes to zero, these solutions converge to a solution of the deterministic Navier-Stokes equation.

Evgueni Dinvay considers in the paper "**A Stochastic Benjamin-Bona-Mahony Type Equation**" a particular nonlinear dispersive stochastic equation recently introduced as a model describing surface water waves under location uncertainty. The corresponding noise term is introduced through a Hamiltonian formulation, which guarantees the energy conservation of the flow. The author shows that the initial-value problem has a unique solution.

Benjamin Dufée, Etienne Mémin, and Dan Crisan investigate in the paper "**Observation-Based Noise Calibration: An Efficient Dynamics for the Ensemble Kalman Filter**" the calibration of the stochastic noise in order to guide its realizations towards the observational data used for the assimilation. This is done in the context of the stochastic parametrization under location uncertainty (LU) and data assimilation. The new methodology is mathematically justified by the use of the Girsanov theorem and yields significant improvements in the experiments carried out on the surface quasi-geostrophic (SQG) model, when applied to ensemble Kalman filters. The test case studied in the paper shows improvements of the peak MSE from 85% to 93%.

The paper by **Camilla Fiorini, Pierre-Marie Boulvard, Long Li, and Etienne Mémin**, entitled "**A Two-Step Numerical Scheme in Time for Surface Quasi Geostrophic Equations Under Location Uncertainty**", considers the surface quasi-geostrophic (SQG) system under location uncertainty (LU) and proposes a Milstein-type scheme for these equations, which is then used in a multi-step method. The SQG system considered in the paper consists of one stochastic partial differential equation, which models the stochastic transport of the buoyancy, and a linear operator linking the velocity and the buoyancy. In the LU setting, the Euler-Maruyama scheme converges with weak order 1 and strong order 0.5. The authors develop higher order schemes in time, based on a Milstein-type scheme in a multi-step framework. They compare different kinds of Milstein schemes. The scheme with the best performance is then included in the two-step scheme. Finally, they show how their two-step scheme decreases the error in comparison to other multi-step schemes.

The contribution of **Franco Flandoli and Eliseo Luongo**, entitled "**The Dissipation Properties of Transport Noise**", presents in a compact way the latest results about the dissipation properties of transport noise in fluid mechanics. Motivated by the fact that transport noise is natural in a passive scalar equation for the heat diffusion and transport, the authors introduce several results about enhanced dissipation due to the noise. Rigorous statements are matched with numerical experiments to understand that the sufficient conditions stated are not yet optimal but give a first useful indication.

Daniel Goodair presents in the paper "**Existence and Uniqueness of Maximal Solutions to a 3D Navier-Stokes Equation with Stochastic Lie Transport**" a criterion for showing that an abstract SPDE possesses a unique maximal strong solution. This is then applied to a 3D stochastic Navier-Stokes equation. Inspired by the classical work of Kato and Lai, the author provides a comparable result in the stochastic case applicable to a variety of noise structures such as additive, multiplicative and transport. In particular, the criterion is designed to fit viscous fluid dynamics models with stochastic advection by lie transport. Its application to the incompressible Navier-Stokes equation matches the existence and uniqueness result of the deterministic theory.

Darryl D. Holm, Ruiao Hu, and Oliver D. Street present in "**Coupling of Waves to Sea Surface Currents Via Horizontal Density Gradients**" a set of mathematical models and numerical simulations motivated by satellite observations of horizontal sea surface fluid motions that show the close coordination between thermal fronts and the vertical motion of waves or, after an approximation, the slowly varying envelope of the rapidly oscillating waves. This coordination of fluid movements with wave envelopes occurs most dramatically when strong horizontal buoyancy gradients are present, e.g., at thermal fronts. The nonlinear models of this coordinated movement presented in the paper may provide future opportunities for the optimal design of satellite imagery that could simultaneously capture the dynamics of both waves and currents directly. The models derived in the paper appear first in their un-approximated form, then again with a slowly varying envelope (SVE) approximation using the WKB approach. The WKB wave-current-buoyancy interaction model derived by the authors for a free surface with horizontal buoyancy gradients indicates that the mechanism for these correlations is the ponderomotive force of the slowly varying envelope of rapidly oscillating waves acting on the surface currents via the horizontal buoyancy gradient. In this model, the buoyancy gradient appears explicitly in the WKB wave momentum, which in turn generates density-weighted potential vorticity whenever the buoyancy gradient is not aligned with the wave-envelope gradient.

The contribution of **Ruiao Hu and Stuart Patching**, entitled "**Variational Stochastic Parameterisations and Their Applications to Primitive Equation Models**", presents a numerical investigation into the stochastic parameterizations of the primitive equations (PE) using the stochastic advection by lie transport (SALT) and stochastic forcing by lie transport (SFLT) frameworks. These frameworks were chosen due to their structure-preserving introduction of stochasticity, which decomposes the transport velocity and fluid momentum into their drift and stochastic parts, respectively. In this paper, the authors develop a new calibration methodology to implement the momentum decomposition of SFLT, and they compare this methodology with the Lagrangian path methodology implemented for SALT. The resulting stochastic primitive equations are then integrated numerically using a modification of the FESOM2 code. For certain choices of the stochastic parameters, the authors show that SALT causes an increase in the eddy kinetic energy field and an improvement in the spatial spectrum. SFLT also shows improvements in these areas, though to a lesser extent. The SALT approach, however, produces an excessive downwards diffusion of temperature, compared to high-resolution deterministic simulations.

The paper by **Oana Lang and Wei Pan**, entitled "**A Pathwise Parameterisation for Stochastic Transport**", sets the stage for a new probabilistic approach to effectively calibrate in a pathwise manner a general class of stochastic nonlinear fluid dynamics models. The authors focus on a 2D Euler SALT equation, showing that the driving stochastic parameter can be calibrated in an optimal way to match a set of given data. Moreover, they show that this model is robust with respect to the stochastic parameters.

The work by **Long Li, Etienne Mémin, and Gilles Tissot**, entitled "**Stochastic Parameterization with Dynamic Mode Decomposition**", considers a physical stochastic parameterization to account for the effects of the unresolved small scale on the large-scale flow dynamics. This random model is based on a stochastic transport principle, which ensures a strong energy conservation. The dynamic mode decomposition (DMD) is performed on high-resolution data to learn a basis of the unresolved velocity field, on which the stochastic transport velocity is expressed. Time-harmonic property of DMD modes allows the authors to perform a clean separation between time-differentiable and time-decorrelated components. The corresponding random scheme is assessed on a quasi-geostrophic (QG) model.

The paper by **Alexander Lobbe**, entitled "**Deep Learning for the Benes Filter**", concerns the filtering problem, in other words, the optimal estimation of a hidden state given partial and noisy observations. Filtering is extensively studied in the theoretical and applied mathematical literature. One of the central challenges in filtering today is the numerical approximation of the optimal filter. The author presents a brief study of a new numerical method based on the mesh-free neural network representation of the density of the solution of the filtering problem achieved by deep learning. Based on the classical SPDE splitting method, the algorithm introduced includes a recursive normalization procedure to recover the normalized conditional distribution of the signal process. The present work uses the Benes model as a benchmark: within the analytically tractable setting of the Benes filter, the author discusses the role of nonlinearity in the filtering model equations for the choice of the domain of the neural network. Further, he presents the first study of the neural network method with an adaptive domain for the Benes model.

Data assimilation techniques are the state-of-the-art approaches in the reconstruction of a spatio-temporal geophysical state such as the atmosphere or the ocean. These methods rely on a numerical model that fills the spatial and temporal gaps in the observational network. Unfortunately, limitations regarding the uncertainty of the state estimate may arise when considering the restriction of the data assimilation problems to a small subset of observations, as encountered for instance in ocean surface reconstruction. These limitations motivated the exploration of reconstruction techniques that do not rely on numerical models. In this context, the increasing availability of geophysical observations and model simulations motivates the exploitation of machine learning tools to tackle the reconstruction of ocean surface variables. In the paper **"End-to-End Kalman Filter in a High Dimensional Linear Embedding of the Observations"**, by **Said Ouala, Pierre Tandeo, Bertrand Chapron, Fabrice Collard and Ronan Fablet**, the authors formulate sea surface spatio-temporal reconstruction problems as state space Bayesian smoothing problems with unknown augmented linear dynamics. The solution of the smoothing problem, given by the Kalman smoother, is written in a differentiable framework which allows, given some training data, to optimize the parameters of the state space model.

Large-scale weather can often be successfully described using a small amount of patterns. A statistical description of re-analysed pressure fields identifies these recurring patterns with clusters in state space, also called *regimes*. Recently, these weather regimes have been described through instantaneous, local indicators of dimension and persistence, borrowed from dynamical systems theory and extreme value theory. Using similar indicators and going further, **Paul Platzer, Bertrand Chapron, and Pierre Tandeo** focus in the paper **"Dynamical Properties of Weather Regime Transitions"** on weather regime transitions. They use sixty years of winter-time sea-level pressure reanalysis data centred on the North-Atlantic Ocean and western Europe. These experiments reveal regime-dependent behaviours of dimension and persistence near transitions, although in average one observes an increase of dimension and a decrease of persistence near transitions. The effect of transition on persistence is stronger and lasts longer than on dimension. The findings confirm the relevance of such dynamical indicators for the study of large-scale weather regimes and reveal their potential to be used for both the understanding and detection of weather regime transitions.

Standard maximum likelihood or Bayesian approaches to parameter estimation for stochastic differential equations are known not to be robust to perturbations in the continuous-in-time data. In the paper **"Frequentist Perspective on Robust Parameter Estimation Using the Ensemble Kalman Filter"**, **Sebastian Reich** gives a rather elementary explanation of this observation in the context of continuous-time parameter estimation using an ensemble Kalman filter. The author employs the frequentist perspective to shed new light on two robust estimation techniques; namely subsampling the data and rough path corrections. He also illustrates the findings through a simple numerical experiment.

The contribution of **Valentin Resseguier, Erwan Hascoet and Bertrand Chapron**, entitled **"Random Ocean Swell-Rays: A Stochastic Framework"**, concerns swell systems that radiate across ocean basins. Far from their sources, emerging surface waves have low steepness characteristics, with very slow amplitude variations. Swell propagation then closely follows principles of geometrical optics, that is, the eikonal approximation to the wave equation, with a constant wave period along geodesics, when following a wave packet at its group speed. The phase averaged evolution of quasi-linear wave fields is then dominated by interactions with underlying current and/or topography changes. Comparable to the propagation of light in a slowly varying medium, over many wavelengths, cumulative effects can lead to refraction. This opens the possibility of using surface swell waves as probes to estimate turbulence along their propagating path.

Louis Thiry, Long Li and Etienne Mémin present in the paper, entitled **"Modified (Hyper-) Viscosity for Coarse-Resolution Ocean Models"**, a simple parameterization for coarse-resolution ocean models. To replace computationally expensive high-resolution ocean models, the authors develop a computationally cheap parameterization for coarse-resolution models based solely on the modification of the viscosity term in advection equations. The parametrization is meant to reproduce the mean quantities like pressure, velocity or vorticity computed from a high-resolution reference solution or using observations. The authors test this new parameterization on a double-gyre quasi-geostrophic model in the eddy-permitting regime. The results show that the proposed scheme significantly improves the energy statistics and the intrinsic variability on the coarse mesh. This method will serve as a deterministic basis model for coarse-resolution stochastic parameterizations in future works.

Resolving numerically all the scale interactions of ocean dynamics in a high-resolution realistic configuration is today far beyond reach, and only large-scale representations can be afforded. **Francesco L. Tucciarone, Etienne Mémin and Long Li** study in the paper **"Primitive Equations Under Location Uncertainty: Analytical Description and Model Development"** a stochastic parameterization of the ocean primitive equations derived within the modelling under location uncertainty framework. Numerical assessments built with the NEMO core's code are provided for a double-gyres configuration.

The paper by **Yicun Zhen, Bertrand Chapron and Etienne Mémin**, entitled **"Bridging Koopman Operator and Time-Series Auto-Correlation Based Hilbert-Schmidt Operator"**, considers Hilbert-Schmidt operators associated with stationary continuous-time processes. A Hilbert space and a (time-shift) continuous one-parameter semigroup of isometries are introduced and analysed. Under some technical assumptions, the continuous one-parameter semigroup is shown to be equivalent, almost surely, to the classical Koopman one-parameter semigroup.

Finally, the STUOD Organizing Committee would like to acknowledge the financial and in-kind support received from several sources: the European Research Council (ERC) under the European Union's Horizon 2020 Research and Innovation Programme (ERC, Grant Agreement No 856408) – for providing funds to cover the travel expenses of the invited speakers, catering costs and administrative support; Imperial College London – for offering the conference venue.

STUOD Organizing Committee:
Prof. Bertrand Chapron (IFREMER)
Prof. Dan Crisan (ICL)
Prof. Darryl Holm (ICL)
Prof. Etienne Mémin (INRIA)
Dr Anna Radomska (ICL)
May 2022

Plouzané, France   Bertrand Chapron
London, UK   Dan Crisan
London, UK   Darryl Holm
Rennes, France   Etienne Mémin
London, UK   Anna Radomska
May 2022

Organization

Program Chairs

Dan Crisan	Imperial College London
Anna Radomska	Imperial College London
Etienne Mémin	Inria, France
Bertrand Chapron	Ifremer, France
Darryl Holm	Imperial College, United Kingdom

Contents

Blow-Up of Strong Solutions of the Thermal Quasi-Geostrophic Equation 1
Dan Crisan and Prince Romeo Mensah

Modeling Under Location Uncertainty: A Convergent Large-Scale Representation of the Navier-Stokes Equations 15
Arnaud Debussche, Berenger Hug, and Etienne Mémin

A Stochastic Benjamin-Bona-Mahony Type Equation 27
Evgueni Dinvay

Observation-Based Noise Calibration: An Efficient Dynamics for the Ensemble Kalman Filter 43
Benjamin Dufée, Etienne Mémin, and Dan Crisan

A Two-Step Numerical Scheme in Time for Surface Quasi Geostrophic Equations Under Location Uncertainty 57
Camilla Fiorini, Pierre-Marie Boulvard, Long Li, and Etienne Mémin

The Dissipation Properties of Transport Noise 69
Franco Flandoli and Eliseo Luongo

Existence and Uniqueness of Maximal Solutions to a 3D Navier-Stokes Equation with Stochastic Lie Transport 87
Daniel Goodair

Coupling of Waves to Sea Surface Currents Via Horizontal Density Gradients 109
Darryl D. Holm, Ruiao Hu, and Oliver D. Street

Variational Stochastic Parameterisations and Their Applications to Primitive Equation Models 135
Ruiao Hu and Stuart Patching

A Pathwise Parameterisation for Stochastic Transport...................... 159
Oana Lang and Wei Pan

Stochastic Parameterization with Dynamic Mode Decomposition 179
Long Li, Etienne Mémin, and Gilles Tissot

Deep Learning for the Benes Filter 195
Alexander Lobbe

**End-to-End Kalman Filter in a High Dimensional Linear
Embedding of the Observations** 211
Said Ouala, Pierre Tandeo, Bertrand Chapron, Fabrice Collard,
and Ronan Fablet

Dynamical Properties of Weather Regime Transitions 223
Paul Platzer, Bertrand Chapron, and Pierre Tandeo

**Frequentist Perspective on Robust Parameter Estimation Using
the Ensemble Kalman Filter** 237
Sebastian Reich

Random Ocean Swell-Rays: A Stochastic Framework 259
Valentin Resseguier, Erwan Hascoët, and Bertrand Chapron

Modified (Hyper-)Viscosity for Coarse-Resolution Ocean Models 273
Louis Thiry, Long Li, and Etienne Mémin

**Primitive Equations Under Location Uncertainty: Analytical
Description and Model Development** 287
Francesco L. Tucciarone, Etienne Mémin, and Long Li

**Bridging Koopman Operator and Time-Series Auto-Correlation
Based Hilbert–Schmidt Operator** 301
Yicun Zhen, Bertrand Chapron, and Etienne Mémin

Index.. 317

Blow-Up of Strong Solutions of the Thermal Quasi-Geostrophic Equation

Dan Crisan and Prince Romeo Mensah

Abstract The Thermal Quasi-Geostrophic (TQG) equation is a coupled system of equations that governs the evolution of the buoyancy and the potential vorticity of a fluid. It has a local in time solution as proved in Crisan et al. (Theoretical and computational analysis of the thermal quasi-geostrophic model. Preprint arXiv:2106.14850, 2021). In this paper, we give a criterion for the blow-up of solutions to the Thermal Quasi-Geostrophic equation, in the spirit of the classical Beale–Kato–Majda blow-up criterion (cf. Beale et al., Comm. Math. Phys. 94(1), 61–66, 1984) for the solution of the Euler equation.

Keywords Blow-up criterion · Thermal Quasi-Qeostrophic equation · Modified Helmholtz operator

1 Introduction

The Thermal Quasi-Geostrophic (TQG) equation is a coupled system of equations governed by the evolution of the buoyancy $b : (t, \mathbf{x}) \in [0, T] \times \mathbb{R}^2 \mapsto b(t, \mathbf{x}) \in \mathbb{R}$ and the potential vorticity $q : (t, \mathbf{x}) \in [0, T] \times \mathbb{R}^2 \mapsto q(t, \mathbf{x}) \in \mathbb{R}$ in the following way:

$$\partial_t b + (\mathbf{u} \cdot \nabla)b = 0, \tag{1}$$

$$\partial_t q + (\mathbf{u} \cdot \nabla)(q - b) = -(\mathbf{u}_h \cdot \nabla)b, \tag{2}$$

$$b(0, x) = b_0(x), \qquad q(0, x) = q_0(x), \tag{3}$$

D. Crisan · P. R. Mensah (✉)
Department of Mathematics, Imperial College, London, UK
e-mail: d.crisan@imperial.ac.uk; p.mensah@imperial.ac.uk

© The Author(s) 2023
B. Chapron et al. (eds.), *Stochastic Transport in Upper Ocean Dynamics*,
Mathematics of Planet Earth 10, https://doi.org/10.1007/978-3-031-18988-3_1

where

$$\mathbf{u} = \nabla^\perp \psi, \qquad \mathbf{u}_h = \frac{1}{2}\nabla^\perp h, \qquad q = (\Delta - 1)\psi + f. \tag{4}$$

Here, $\psi : (t, \mathbf{x}) \in [0, T] \times \mathbb{R}^2 \mapsto \psi(t, \mathbf{x}) \in \mathbb{R}$ is the streamfunction, $h : \mathbf{x} \in \mathbb{R}^2 \mapsto h(\mathbf{x}) \in \mathbb{R}$ is the spatial variation around a constant bathymetry profile and $f : \mathbf{x} \in \mathbb{R}^2 \mapsto f(\mathbf{x}) \in \mathbb{R}$ is the Coriolis parameter. Since we are working on the whole space, we can supplement our system with the far-field condition

$$\lim_{|\mathbf{x}| \to \infty} (b(\mathbf{x}), \mathbf{u}(\mathbf{x})) = 0.$$

Our given set of data is $(\mathbf{u}_h, f, b_0, q_0)$ with regularity class:

$$\mathbf{u}_h \in W^{3,2}_{\mathrm{div}}(\mathbb{R}^2; \mathbb{R}^2), \quad f \in W^{2,2}(\mathbb{R}^2), \quad b_0 \in W^{3,2}(\mathbb{R}^2), \quad q_0 \in W^{2,2}(\mathbb{R}^2). \tag{5}$$

The TQG equation models the dynamics of a submesoscale geophysical fluid in thermal geostrophic balance, for which the Rossby number, the Froude number and the stratification parameter are all of the same asymptotic order. For a historical overview, modelling and other issues pertaining to the TQG equation, we refer the reader to [4].

In the following, we are interested in *strong solutions* of the system (1)–(4) which can naturally be defined in terms of just b and q although the unknowns in the evolutionary Eqs. (1)–(2) are b, q and $\mathbf{u}$. This is because for a given f, one can recover the velocity $\mathbf{u}$ from the vorticity q by solving the equation

$$\mathbf{u} = \nabla^\perp (\Delta - 1)^{-1}(q - f)$$

derived from (4). Also note that a consequence of the equation $\mathbf{u} = \nabla^\perp \psi$ in (4) is that $\mathrm{div}\,\mathbf{u} = 0$. This means that the fluid is incompressible. With these information in hand, we now make precise, the notion of a strong solution.

Definition 1 (Local Strong Solution) Let $(\mathbf{u}_h, f, b_0, q_0)$ be of regularity class (5). For some $T > 0$, we call the triple (b, q, T) a *strong solution* to the system (1)–(4) if the following holds:

– The buoyancy b satisfies $b \in C([0, T]; W^{3,2}(\mathbb{R}^2))$ and the equation

$$b(t) = b_0 - \int_0^t \mathrm{div}(b\mathbf{u})\,\mathrm{d}\tau,$$

holds for all $t \in [0, T]$;
– the potential vorticity q satisfies $q \in C([0, T]; W^{2,2}(\mathbb{R}^2))$ and the equation

$$q(t) = q_0 - \int_0^t \left[\operatorname{div}((q - b)\mathbf{u}) + \operatorname{div}(b\mathbf{u}_h) \right] d\tau$$

holds for all $t \in [0, T]$.

Such local strong solutions exist on a maximal time interval. We define this as follows.

Definition 2 (Maximal Solution) Let $(\mathbf{u}_h, f, b_0, q_0)$ be of regularity class (5). For some $T > 0$, we call $(b, q, T_{\max})$ a *maximal solution* to the system (1)–(4) if:

- there exists an increasing sequence of time steps $(T_n)_{n \in \mathbb{N}}$ whose limit is $T_{\max} \in (0, \infty]$;
- for each $n \in \mathbb{N}$, the triple (b, q, T_n) is a local strong solution to the system (1)–(4) with initial condition (b_0, q_0);
- if $T_{\max} < \infty$, then

$$\limsup_{T_n \to T_{\max}} \|b(T_n)\|^2_{W^{3,2}(\mathbb{R}^2)} + \|q(T_n)\|^2_{W^{2,2}(\mathbb{R}^2)} = \infty. \tag{6}$$

We shall call $T_{\max} > 0$ the *maximal time*.

The existence of a unique local strong solution of (1)–(4) has recently been shown in [4, Theorem 2.10] on the torus. A unique maximal solutions also exist [4, Theorem 2.14] and the result also applies to the whole space [4, Remark 2.1]. We state the result here for completeness.

Theorem 1 *For $(\mathbf{u}_h, f, b_0, q_0)$ of regularity class (5), there exist a unique maximal solution (b, q, T) of the system (1)–(4).*

Before we state our main result, let us first present some notations used throughout this work.

1.1 Notations

In the following, we write $F \lesssim G$ if there exists a generic constant $c > 0$ (that may vary from line to line) such that $F \leq c\,G$. Functions mapping into $\mathbb{R}^2$ are **boldfaced** (for example the velocity $\mathbf{u}$) while those mapping into $\mathbb{R}$ are not (for example the buoyancy b and vorticity q). For $k \in \mathbb{N} \cup \{0\}$ and $p \in [1, \infty]$, $W^{k,p}(\mathbb{R}^2)$ is the usual Sobolev space of functions mapping into $\mathbb{R}$ with a natural modification for functions mapping into $\mathbb{R}^2$. For $p = 2$, $W^{k,2}(\mathbb{R}^2)$ is a Hilbert space with inner product $\langle u, v \rangle_{W^{k,2}(\mathbb{R}^2)} = \sum_{|\beta| \leq k} \langle \partial^\beta u, \partial^\beta v \rangle$, where $\langle \cdot, \cdot \rangle$ denotes the standard L^2-inner product. For general $s \in \mathbb{R}$, we use the norm

$$\|v\|_{W^{s,2}(\mathbb{R}^2)} \equiv \left(\int_{\mathbb{R}^2} \left(1 + |\xi|^2\right)^s |\hat{v}(\xi)|^2 \, d\xi \right)^{\frac{1}{2}} \tag{7}$$

defined in frequency space. Here, $\widehat{v}(\xi)$ denotes the Fourier coefficients of v. For simplicity, we write $\| \cdot \|_{s,2}$ for $\| \cdot \|_{W^{s,2}(\mathbb{R}^2)}$. When $k = s = 0$, we get the usual $L^2(\mathbb{R}^2)$ space whose norm we will simply denote by $\| \cdot \|_2$. A similar notation will be used for norms $\| \cdot \|_p$ of general $L^p(\mathbb{R}^2)$ spaces for any $p \in [1, \infty]$ as well as for the inner product $\langle \cdot, \cdot \rangle_{k,2} := \langle \cdot, \cdot \rangle_{W^{k,2}(\mathbb{R}^2)}$ when $k \in \mathbb{N}$. Additionally, $W^{k,p}_{\mathrm{div}}(\mathbb{R}^2)$ represents the space of divergence-free vector-valued functions in $W^{k,p}(\mathbb{R}^2)$.

With respect to differential operators, we let $\nabla_0 := (\partial_{x_1}, \partial_{x_2}, 0)^T$ and $\nabla_0^\perp := (-\partial_{x_2}, \partial_{x_1}, 0)$ be the three-dimensional extensions of the two-dimensional differential operators $\nabla = (\partial_{x_1}, \partial_{x_2})^T$ and $\nabla^\perp := (-\partial_{x_2}, \partial_{x_1})$ by zero respectively. The Laplacian $\Delta = \mathrm{div}\nabla = \partial_{x_1 x_1} + \partial_{x_2 x_2}$ remains two-dimensional.

1.2   Main Result

Our main result is to give a blow-up criterion, of Beale–Kato–Majda-type [2], for a strong solution (b, q, T) of (1)–(4). In particular, we show the following result.

Theorem 2 *Suppose that (b, q, T) is a local strong solution of (1)–(4). If*

$$\int_0^T \left(\|q(t)\|_\infty + \|\nabla b(t)\|_\infty \right) dt \equiv K < \infty, \tag{8}$$

then there exists a solution (b', q', T') with $T' > T$, such that $(b', q') = (b, q)$ on $[0, T]$. Moreover, for all $t \in [0, T]$,

$$\|b(t)\|_{3,2} + \|q(t)\|_{2,2} \leq \left[e + \|b_0\|_{3,2} + \|q_0\|_{2,2} \right]^{\exp(cK)} \exp[cT \exp(cK)].$$

An immediate consequence of the above theorem is the following:

Corollary 1 *Assume that (b, q, T) is a maximal solution. If $T < \infty$, then*

$$\int_0^T \left(\|q(t)\|_\infty + \|\nabla b(t)\|_\infty \right) dt = \infty$$

and in particular,

$$\sup_{t \uparrow T} \left(\|q(t)\|_\infty + \|\nabla b(t)\|_\infty \right) = \infty.$$

2   Blow-Up

We devote the entirety of this section to the proof of Theorem 2. In order to achieve our goal, we first derive a suitable exact solution for what is referred to as the modified Helmholtz equation. Some authors also call it the Screened Poisson equation [3] while others rather mistakenly call it the Helmholtz equation. Refer to [1] for the difference between the Helmholtz equation and modified Helmholtz equation.

2.1 Estimate for the 2D Modified Helmholtz Equation or the Screened Poisson Equation

In the following, we want to find an exact solution $\psi : \mathbb{R}^2 \to \mathbb{R}$ of

$$(\Delta - 1)\psi(\mathbf{x}) = w(\mathbf{x}), \qquad \lim_{|\mathbf{x}| \to \infty} \psi(\mathbf{x}) = 0 \qquad (9)$$

for a given function $w \in W^{2,2}(\mathbb{R}^2)$. The corresponding two-dimensional free space Green's function $G^{\text{free}}(\mathbf{x})$ for (9) must therefore solve

$$(\Delta - 1)G^{\text{free}}(\mathbf{x} - \mathbf{y}) = \delta(\mathbf{x} - \mathbf{y}), \qquad \lim_{|\mathbf{x}| \to \infty} G^{\text{free}}(\mathbf{x} - \mathbf{y})(\mathbf{x}) = 0 \qquad (10)$$

in the sense of distributions. Indeed, one can verify that the Green's function is given by

$$G^{\text{free}}(\mathbf{x} - \mathbf{y}) = \frac{1}{2\pi} K_0(|\mathbf{x} - \mathbf{y}|) \qquad (11)$$

see [1, Table 9.5], where

$$K_0(z) = \int_0^\infty \frac{e^{-\sqrt{z^2 + r^2}}}{\sqrt{z^2 + r^2}} \, dr$$

is the modified Bessel function of the second kind, see equation (8.432-9), page 917 of [5] with $\nu = 0$ and $x = 1$. However, since the integral above is an even function, it follows that

$$G^{\text{free}}(\mathbf{x} - \mathbf{y}) = \frac{i}{4} H_0^{(1)}(i|\mathbf{x} - \mathbf{y}|) = \frac{1}{4\pi} \int_{\mathbb{R}} \frac{e^{-\sqrt{|\mathbf{x} - \mathbf{y}|^2 + r^2}}}{\sqrt{|\mathbf{x} - \mathbf{y}|^2 + r^2}} \, dr \qquad (12)$$

which is the zeroth-order Hankel function of the first kind, see equation (11.117) in [1] and equation (8.421-9) of [5] on page 915. Therefore,

$$\psi(\mathbf{x}) = \frac{1}{4\pi} \int_{\mathbb{R}^3} \frac{e^{-|(\mathbf{x} - \mathbf{y}, -r)|}}{|(\mathbf{x} - \mathbf{y}, -r)|} w((\mathbf{y}, 0)) \, d\mathbf{y} dr \qquad (13)$$

$$=: \psi((\mathbf{x}, 0)) \qquad (14)$$

where we have used the identity $\sqrt{|\mathbf{x} - \mathbf{y}|^2 + r^2} = |(\mathbf{x}, 0) - (\mathbf{y}, r)| = |(\mathbf{x} - \mathbf{y}, -r)|$. We can therefore view the argument of the streamfunction ψ as a $3D$-vector with zero vertical component.

2.2 Log-Sobolev Estimate for Velocity Gradient

Our goal now is to find a suitable estimate for the Lipschitz norm of $\mathbf{u}$ that solves

$$\mathbf{u} = \nabla^{\perp}\psi, \qquad (\Delta - 1)\psi = w \tag{15}$$

where $w \in W^{2,2}(\mathbb{R}^2)$ is given. In particular, inspired by Beale et al. [2], we aim to show Proposition 1 below. This log-estimate is the crucial ingredient that allow us to obtain our blow-up criterion in terms of just the buoyancy gradient and the vorticity although preliminary estimate may have suggested estimating the velocity gradient as well.

Proposition 1 *For a given $w \in W^{2,2}(\mathbb{R}^2)$, any $\mathbf{u}$ solving (15) satisfies*

$$\|\mathbf{u}\|_{1,\infty} \lesssim 1 + (1 + 2\ln^+(\|w\|_{2,2}))\|w\|_{\infty} \tag{16}$$

where $\ln^+ a = \ln a$ if $a \geq 1$ and $\ln^+ a = 0$ otherwise.

Proof To show (16), we fix $L \in (0, 1]$ and for $\mathbf{z} \in \mathbb{R}^3$, we let $\zeta_L(\mathbf{z})$ be a smooth cut-off function satisfying

$$\zeta_L(\mathbf{z}) = \begin{cases} 1 & : |\mathbf{z}| < L, \\ 0 & : |\mathbf{z}| > 2L \end{cases}$$

and $|\partial\zeta_L(\mathbf{z})| \lesssim L^{-1}$ where $\partial := \nabla_0^{\perp}$ or ∇_0 as well as $|\nabla_0\nabla_0^{\perp}\zeta_L(\mathbf{z})| \lesssim L^{-2}$. This latter requirement ensures that the point of inflection of the graph of the cut-off, the portion that is constant, concave upwards and concave downwards are all captured. We now define the following

$$B_1 := \{(\mathbf{y}, r) \in \mathbb{R}^3 : |(\mathbf{x}, 0) - (\mathbf{y}, r)| = |(\mathbf{x} - \mathbf{y}, -r)| < 2L\},$$

$$B_2 := \{(\mathbf{y}, r) \in \mathbb{R}^3 : L \leq |(\mathbf{x} - \mathbf{y}, -r)| \leq 1\},$$

$$B_3 := \{(\mathbf{y}, r) \in \mathbb{R}^3 : |(\mathbf{x} - \mathbf{y}, -r)| > 1\},$$

so that by adding and subtracting ζ_L, we obtain

$$|\nabla\mathbf{u}(\mathbf{x})| = |\nabla_0\nabla_0^{\perp}\psi((\mathbf{x}, 0))| \leq |\nabla_0(\mathbf{u}_1((\mathbf{x}, 0)), 0)| + |\nabla_0(\mathbf{u}_2^1((\mathbf{x}, 0)), 0)|$$
$$+ |\nabla_0(\mathbf{u}_2^2((\mathbf{x}, 0)), 0)| + |\nabla_0(\mathbf{u}_2^3((\mathbf{x}, 0)), 0)|$$

$$+ |\nabla_0(\mathbf{u}_2^4((\mathbf{x}, 0)), 0)| + |\nabla_0(\mathbf{u}_3((\mathbf{x}, 0)), 0)|$$

$$=: |\nabla_0\mathbf{u}_1| + |\nabla_0\mathbf{u}_2^1| + |\nabla_0\mathbf{u}_2^2| + |\nabla_0\mathbf{u}_2^3|$$

$$+ |\nabla_0\mathbf{u}_2^4| + |\nabla_0\mathbf{u}_3|$$

where

$$\nabla_0\mathbf{u}_1 := \frac{1}{4\pi}\int_{B_1} \zeta_L((\mathbf{x} - \mathbf{y}, -r))\frac{e^{-|(\mathbf{x}-\mathbf{y},-r)|}}{|(\mathbf{x} - \mathbf{y}, -r)|}\nabla_0\nabla_0^\perp w((\mathbf{y}, 0))\, \mathbf{dy}dr,$$

$$\nabla_0\mathbf{u}_2^1 := \frac{1}{4\pi}\int_{B_2} [1 - \zeta_L((\mathbf{x} - \mathbf{y}, -r))]\nabla_0\nabla_0^\perp\left[\frac{e^{-|(\mathbf{x}-\mathbf{y},-r)|}}{|(\mathbf{x} - \mathbf{y}, -r)|}\right]w((\mathbf{y}, 0))\, \mathbf{dy}dr,$$

$$\nabla_0\mathbf{u}_2^2 := \frac{1}{4\pi}\int_{B_2} \nabla_0\nabla_0^\perp[1 - \zeta_L((\mathbf{x} - \mathbf{y}, -r))]\frac{e^{-|(\mathbf{x}-\mathbf{y},-r)|}}{|(\mathbf{x} - \mathbf{y}, -r)|}\, w((\mathbf{y}, 0))\, \mathbf{dy}dr,$$

$$\nabla_0\mathbf{u}_2^3 := \frac{1}{4\pi}\int_{B_2} \nabla_0^\perp[1 - \zeta_L((\mathbf{x} - \mathbf{y}, -r))]\nabla_0\left[\frac{e^{-|(\mathbf{x}-\mathbf{y},-r)|}}{|(\mathbf{x} - \mathbf{y}, -r)|}\right]w((\mathbf{y}, 0))\, \mathbf{dy}dr,$$

$$\nabla_0\mathbf{u}_2^4 := \frac{1}{4\pi}\int_{B_2} \nabla_0[1 - \zeta_L((\mathbf{x} - \mathbf{y}, -r))]\nabla_0^\perp\left[\frac{e^{-|(\mathbf{x}-\mathbf{y},-r)|}}{|(\mathbf{x} - \mathbf{y}, -r)|}\right]w((\mathbf{y}, 0))\, \mathbf{dy}dr,$$

$$\nabla_0\mathbf{u}_3 := \frac{1}{4\pi}\int_{B_3} \nabla_0\nabla_0^\perp\left[[1 - \zeta_L((\mathbf{x} - \mathbf{y}, -r))]\frac{e^{-|(\mathbf{x}-\mathbf{y},-r)|}}{|(\mathbf{x} - \mathbf{y}, -r)|}\right]w((\mathbf{y}, 0))\, \mathbf{dy}dr.$$

For $L \in (0, 1]$, we have that

$$|\nabla_0\mathbf{u}_1| \lesssim \left(\int_{B_1} \frac{e^{-2|(\mathbf{x},0)-(\mathbf{y},r)|}}{|(\mathbf{x}, 0) - (\mathbf{y}, r)|^2}\, \mathbf{dy}dr\right)^{\frac{1}{2}}\|\nabla_0\nabla_0^\perp w((\mathbf{y}, 0))\|_2$$

$$\lesssim \left(\int_0^{2L} \frac{e^{-2s}}{s^2} s^2 ds\right)^{\frac{1}{2}}\|w\|_{2,2} \lesssim \left(1 - e^{-4L}\right)^{\frac{1}{2}}\|w\|_{2,2} \lesssim \pi L^{\frac{1}{2}}\|w\|_{2,2}.$$

Now note that

$$\nabla_0\mathbf{u}_2^1 := \frac{1}{4\pi}\int_{B_2} [1 - \zeta_L((\mathbf{x} - \mathbf{y}, -r))]\left\{\frac{2(\mathbf{x} - \mathbf{y})^T(\mathbf{x} - \mathbf{y})^\perp}{(|\mathbf{x} - \mathbf{y}|^2 + r^2)^2} + \frac{3(\mathbf{x} - \mathbf{y})^T(\mathbf{x} - \mathbf{y})^\perp}{(|\mathbf{x} - \mathbf{y}|^2 + r^2)^{\frac{5}{2}}}\right.$$

$$- \frac{1}{|\mathbf{x} - \mathbf{y}|^2 + r^2}\begin{pmatrix} 0 & 1 & 0 \\ -1 & 0 & 0 \\ 0 & 0 & 0 \end{pmatrix} - \frac{1}{(|\mathbf{x} - \mathbf{y}|^2 + r^2)^{\frac{3}{2}}}\begin{pmatrix} 0 & 1 & 0 \\ -1 & 0 & 0 \\ 0 & 0 & 0 \end{pmatrix}$$

$$\left. + \frac{(\mathbf{x} - \mathbf{y})^T(\mathbf{x} - \mathbf{y})^\perp}{(|\mathbf{x} - \mathbf{y}|^2 + r^2)^{\frac{3}{2}}} + \frac{(\mathbf{x} - \mathbf{y})^T(\mathbf{x} - \mathbf{y})^\perp}{(|\mathbf{x} - \mathbf{y}|^2 + r^2)^2}\right\}e^{-|(\mathbf{x}-\mathbf{y},-r)|}w(\mathbf{y})\, dr\mathbf{dy}$$

$$=: \sum_{i=1}^{6} \mathbb{K}_i((\mathbf{x} - \mathbf{y}, -r)).$$

Clearly, $|\mathbf{x} - \mathbf{y}|^2 \le |\mathbf{x} - \mathbf{y}|^2 + r^2 = |(\mathbf{x} - \mathbf{y}, -r)|^2$ and for any $L \in (0, 1]$, the inequalities

$$(e^{-L} - e^{-1}) \le (1 - e^{-1}) \le (1 - e^{-1})(1 - \ln(L)) \lesssim (1 - \ln(L))$$

holds independent of L. Therefore, for $L \in (0, 1]$, it follows that

$$|\mathbb{K}_1((\mathbf{x} - \mathbf{y}, -r))| + |\mathbb{K}_3((\mathbf{x} - \mathbf{y}, -r)) + |\mathbb{K}_6((\mathbf{x} - \mathbf{y}, -r))|$$

$$\lesssim \|w\|_\infty \int_{B_2} \frac{e^{-|(\mathbf{x} - \mathbf{y}, -r)|}}{|(\mathbf{x} - \mathbf{y}, -r)|^2} \, dr dy$$

$$\lesssim \|w\|_\infty \int_L^1 \frac{e^{-s}}{s^2} s^2 \, ds$$

$$\lesssim \|w\|_\infty (1 - \ln(L)).$$

Again, we can use $|\mathbf{x} - \mathbf{y}|^2 \le |\mathbf{x} - \mathbf{y}|^2 + r^2$ and the fact that the inequalities

$$(e^{-L}(L + 1) - 2e^{-1}) \le (1 - 2e^{-1}) \le (1 - 2e^{-1})(1 - \ln(L)) \lesssim (1 - \ln(L))$$

holds independent of any $L \in (0, 1]$ to obtain

$$|\mathbb{K}_5((\mathbf{x} - \mathbf{y}, -r))| \lesssim \|w\|_\infty \int_{B_2} \frac{e^{-|(\mathbf{x} - \mathbf{y}, -r)|}}{|(\mathbf{x} - \mathbf{y}, -r)|} \, dr dy$$

$$\lesssim \|w\|_\infty \int_L^1 \frac{e^{-s}}{s} s^2 \, ds$$

$$\lesssim \|w\|_\infty (1 - \ln(L)).$$

Finally, for $\mathbb{K}_2$ and $\mathbb{K}_4$, we also obtain

$$|\mathbb{K}_2((\mathbf{x} - \mathbf{y}, -r))| + |\mathbb{K}_4((\mathbf{x} - \mathbf{y}, -r)) \lesssim \|w\|_\infty \int_{B_2} \frac{e^{-|(\mathbf{x} - \mathbf{y}, -r)|}}{|(\mathbf{x} - \mathbf{y}, -r)|^3} \, dr dy$$

$$\lesssim \|w\|_\infty \int_L^1 \frac{e^{-s}}{s^3} s^2 \, ds$$

$$\lesssim \|w\|_\infty e^{-L} \int_L^1 \frac{1}{s} \, ds$$

$$\lesssim \|w\|_\infty e^{-0} (-\ln(L)).$$

We have shown that

$$|\nabla_0 \mathbf{u}_2^1| \lesssim \|w\|_\infty (1 - \ln(L)) \tag{17}$$

for $L \in (0, 1]$. Also, the quantity $(1/L^2)[e^{-L}(L+1) - e^{-2L}(2L+1)]$ is uniformly bounded for any $L \in (0, 1]$ and as such,

$$|\nabla_0 \mathbf{u}_2^2| \lesssim \left(\int_L^{2L} \frac{e^{-s}}{sL^2} s^2 ds \right) \|w\|_\infty \lesssim \|w\|_\infty. \tag{18}$$

Next, we note that the estimate for $\nabla_0 \mathbf{u}_2^3$ and $\nabla_0 \mathbf{u}_2^4$ will be the same where in particular,

$$\nabla_0 \mathbf{u}_2^3 := \frac{-1}{4\pi} \int_{B_2} \nabla_0^\perp \left[1 - \zeta_L((\mathbf{x} - \mathbf{y}, -r)) \right] \left\{ \frac{(\mathbf{x} - \mathbf{y})^T}{|\mathbf{x} - \mathbf{y}|^2 + r^2} \right.$$
$$\left. + \frac{(\mathbf{x} - \mathbf{y})^T}{(|\mathbf{x} - \mathbf{y}|^2 + r^2)^{\frac{3}{2}}} \right\} e^{-|(\mathbf{x} - \mathbf{y}, -r)|} w(\mathbf{y}) \, dr dy$$
$$=: \mathbb{K}_7((\mathbf{x} - \mathbf{y}, -r)) + \mathbb{K}_8((\mathbf{x} - \mathbf{y}, -r)).$$

Since $|\mathbf{x} - \mathbf{y}| \leq 1$ holds on B_2, it follows from the condition $|\nabla_0^\perp \zeta_L(\mathbf{z})| \lesssim L^{-1}$ that

$$|\mathbb{K}_7((\mathbf{x} - \mathbf{y}, -r))| \lesssim \left(\int_L^{2L} \frac{e^{-s}}{s^2 L} s^2 \, ds \right) \|w\|_\infty \lesssim \|w\|_\infty$$

since $(1/L)[e^{-L} - e^{-2L}]$ is uniformly bounded in L. Similarly, we can use the fact that $|\mathbf{x} - \mathbf{y}| \leq \sqrt{|\mathbf{x} - \mathbf{y}|^2 + r^2}$ to obtain

$$|\mathbb{K}_8((\mathbf{x} - \mathbf{y}, -r))| \lesssim \left(\int_L^{2L} \frac{e^{-s}}{s^2 L} s^2 \, ds \right) \|w\|_\infty \lesssim \|w\|_\infty.$$

We can therefore conclude that,

$$|\nabla_0 \mathbf{u}_2^3| + |\nabla_0 \mathbf{u}_2^4| \lesssim \|w\|_\infty. \tag{19}$$

Similar to the estimate for $\nabla \mathbf{u}_2^1$, we have that

$$|\nabla_0 \mathbf{u}_3| \lesssim \|w\|_\infty. \tag{20}$$

It follows by summing up the various estimates above that

$$\|\nabla \mathbf{u}\|_\infty \lesssim L^{\frac{1}{2}} \|w\|_{2,2} + (1 - \ln(L)) \|w\|_\infty. \tag{21}$$

It remains to show that the estimate (21) also holds for $\mathbf{u}$. For this, we first recall that

$$\mathbf{u}(\mathbf{x}) = \frac{1}{4\pi} \int_{\mathbb{R}^3} \nabla_0^\perp \left[\frac{e^{-|(\mathbf{x} - \mathbf{y}, -r)|}}{|(\mathbf{x} - \mathbf{y}, -r)|} \right] w((\mathbf{y}, 0)) \, dy dr. \tag{22}$$

We now use the inequalities

$$|\mathbf{x} - \mathbf{y}| \le \left(|\mathbf{x} - \mathbf{y}|^2 + r^2\right)^{\frac{1}{2}} \tag{23}$$

and

$$\frac{1}{4\pi}\left|\nabla_0^\perp \frac{e^{-|(\mathbf{x}-\mathbf{y},-r)|}}{|(\mathbf{x}-\mathbf{y},-r)|}\right| \lesssim \left[\frac{|\mathbf{x}-\mathbf{y}|}{(|\mathbf{x}-\mathbf{y}|^2+r^2)^{\frac{3}{2}}} + \frac{|\mathbf{x}-\mathbf{y}|}{|\mathbf{x}-\mathbf{y}|^2+r^2}\right] e^{-|(\mathbf{x}-\mathbf{y},-r)|}$$

to obtain

$$\begin{aligned}
\|\mathbf{u}\|_\infty &\lesssim \|w\|_\infty \int_{\mathbb{R}^3}\left[\frac{1}{|\mathbf{x}-\mathbf{y}|^2+r^2} + \frac{1}{(|\mathbf{x}-\mathbf{y}|^2+r^2)^{\frac{1}{2}}}\right] e^{-|(\mathbf{x}-\mathbf{y},-r)|} \mathrm{d}\mathbf{y}\mathrm{d}r \\
&\lesssim \|w\|_\infty \int_0^\infty \frac{e^{-s}}{s^2} s^2 \mathrm{d}s + \|w\|_\infty \int_0^\infty \frac{e^{-s}}{s} s^2 \mathrm{d}s \\
&\lesssim \|w\|_\infty.
\end{aligned} \tag{24}$$

Therefore, it follows from (21) and (24) that

$$\|\mathbf{u}\|_{1,\infty} \lesssim L^{\frac{1}{2}}\|w\|_{2,2} + (1 - \ln(L))\|w\|_\infty. \tag{25}$$

If $\|w\|_{2,2} \le 1$, we choose $L = 1$ and if $\|w\|_{2,2} > 1$, we take $L = \|w\|_{2,2}^{-2}$ so that (16) holds. This finishes the proof.

Before we end the subsection, we also note that a direct computation using the definition of Sobolev norms in frequency space (7) immediately yield

$$\|\mathbf{u}\|_{k+1,2} \lesssim \|w\|_{k,2} \tag{26}$$

for any $k \in \mathbb{N} \cup \{0\}$ where $w \in W^{k,2}(\mathbb{R}^2)$ is a given function in (15).

2.3 A Priori Estimate

In order to prove Theorem 2, we first need some preliminary estimates for (b, q). In the following, we define

$$\|(b, q)\| := \|b\|_{3,2} + \|q\|_{2,2}.$$

Lemma 1 *A strong solution of* (1)–(4) *satisfies the bound*

$$\frac{\mathrm{d}}{\mathrm{d}t}\|(b, q)\|^2 \lesssim \left(1 + \|\mathbf{u}\|_{1,\infty} + \|\nabla b\|_\infty + \|q\|_\infty\right)\left(1 + \|(b, q)\|^2\right). \tag{27}$$

Proof Since the space of smooth functions is dense in the space $W^{3,2}(\mathbb{R}^2) \times W^{2,2}(\mathbb{R}^2)$ of existence, in the following, we work with a smooth solution pair (b, q). To achieve our desired estimate, we apply ∂^β to (1) for $|\beta| \leq 3$ to obtain

$$\partial_t \partial^\beta b + \mathbf{u} \cdot \nabla \partial^\beta b = R_1 \tag{28}$$

where

$$R_1 := \mathbf{u} \cdot \partial^\beta \nabla b - \partial^\beta(\mathbf{u} \cdot \nabla b).$$

Now since $\operatorname{div}\mathbf{u} = 0$, if we multiply (28) by $\partial^\beta b$ and integrate over space, the second term on the left-hand side of (28) vanishes after integration by parts. On the other hand, we can use the commutator estimate (see for instant [4, Sect. 2.2]) to estimate the residual term R_1. Consequently, by multiplying (28) by $\partial^\beta b$, integrating over space, and summing over the multiindices β so that $|\beta| \leq 3$, we obtain

$$\frac{\mathrm{d}}{\mathrm{d}t}\|b\|_{3,2}^2 \lesssim \left(\|\nabla\mathbf{u}\|_\infty\|b\|_{3,2} + \|\nabla b\|_\infty\|\mathbf{u}\|_{3,2}\right)\|b\|_{3,2}$$

$$\lesssim (\|\nabla\mathbf{u}\|_\infty + \|\nabla b\|_\infty)(1 + \|(b,q)\|^2) \tag{29}$$

where we have used (26) for $w = q - f$ and $k = 2$.

Next, we find a bound for $\|q\|_{2,2}^2$. For this, we apply ∂^β to (2) for $|\beta| \leq 2$ and we obtain

$$\partial_t \partial^\beta q + \mathbf{u} \cdot \nabla \partial^\beta(q - b) + \mathbf{u}_h \cdot \nabla \partial^\beta b = R_2 + R_3 + R_4 \tag{30}$$

where

$$R_2 := \mathbf{u} \cdot \partial^\beta \nabla q - \partial^\beta(\mathbf{u} \cdot \nabla q),$$

$$R_3 := -\mathbf{u} \cdot \partial^\beta \nabla b + \partial^\beta(\mathbf{u} \cdot \nabla b),$$

$$R_4 := \mathbf{u}_h \cdot \partial^\beta \nabla b - \partial^\beta(\mathbf{u}_h \cdot \nabla b).$$

Now notice that for $\mathbb{U} := \nabla\mathbf{u}$, it follows from interpolation that

$$\|\nabla\mathbb{U}\|_4 \lesssim \|\mathbb{U}\|_\infty^{\frac{1}{2}}\|\nabla^2\mathbb{U}\|_2^{\frac{1}{2}}$$

and so,

$$\|\nabla^2\mathbf{u}\|_4 \lesssim \|\nabla\mathbf{u}\|_\infty^{\frac{1}{2}}\|\mathbf{u}\|_{3,2}^{\frac{1}{2}}.$$

Similarly

$$\|\nabla q\|_4 \lesssim \|q\|_\infty^{\frac{1}{2}}\|q\|_{2,2}^{\frac{1}{2}}.$$

Therefore,

$$\|\nabla q\|_4 \|\nabla^2 \mathbf{u}\|_4 \lesssim \|q\|_\infty \|\mathbf{u}\|_{3,2} + \|\nabla\mathbf{u}\|_\infty \|q\|_{2,2}.$$

By using this estimate, we deduce from (26) and commutator estimates that

$$\|R_2\|_2 \lesssim \|\nabla\mathbf{u}\|_\infty \|q\|_{2,2} + \|q\|_\infty (1 + \|q\|_{2,2}). \tag{31}$$

The commutators R_3 and R_4 are easy to estimate and are given by

$$\|R_3\|_2 \lesssim \|\nabla\mathbf{u}\|_\infty \|b\|_{3,2} + \|\nabla b\|_\infty (1 + \|q\|_{2,2}), \tag{32}$$

$$\|R_4\|_2 \lesssim \|b\|_{3,2} + \|\nabla b\|_\infty, \tag{33}$$

respectively, for a given $\mathbf{u}_h \in W^{3,2}(\mathbb{R}^2; \mathbb{R}^2)$. Next, by using $\mathrm{div}\,\mathbf{u} = 0$, we obtain

$$\left\langle (\mathbf{u}\cdot\nabla\partial^\beta q),\, \partial^\beta q \right\rangle = 0. \tag{34}$$

Additionally, the following estimates holds true

$$\left| \left\langle (\mathbf{u}\cdot\nabla\partial^\beta b),\, \partial^\beta q \right\rangle \right| \lesssim \|\mathbf{u}\|_\infty \|b\|_{3,2}^2 + \|\mathbf{u}\|_\infty \|q\|_{2,2}^2, \tag{35}$$

$$\left| \left\langle (\mathbf{u}_h\cdot\nabla\partial^\beta b),\, \partial^\beta q \right\rangle \right| \lesssim \|b\|_{3,2}^2 + \|q\|_{2,2}^2 \tag{36}$$

since $\mathbf{u}_h \in W^{3,2}(\mathbb{R}^2; \mathbb{R}^2)$. If we now collect the estimates above (keeping in mind that $f \in W^{2,2}(\mathbb{R}^2)$ and $\mathbf{u}_h \in W^{3,2}(\mathbb{R}^2; \mathbb{R}^2)$), we obtain by multiplying (2) by $\partial^\beta q$ and then summing over $|\beta| \leq 2$, the following

$$\frac{\mathrm{d}}{\mathrm{d}t}\|q\|_{2,2}^2 \lesssim \left(1 + \|\mathbf{u}\|_{1,\infty} + \|\nabla b\|_\infty + \|q\|_\infty\right)\left(1 + \|(b,q)\|^2\right). \tag{37}$$

Summing up (29) and (37) yields the desired result.

We now have all in hand to prove our main theorem, Theorem 2.

Proof of Theorem 2 In the following, we define the time-dependent function g as

$$g(t) := \mathrm{e} + \|(b,q)(t)\|, \quad \text{for} \quad t \in [0, T]. \tag{38}$$

Next, without loss of generality, we assume that $f = 0$ so that from Proposition 1, we obtain

$$\|\mathbf{u}(t)\|_{1,\infty} \lesssim 1 + (1 + \ln\|q(t)\|_{2,2})(\|\nabla b(t)\|_\infty + \|q(t)\|_\infty) \tag{39}$$

for $t \in [0, T]$. Using the monotonic properties of logarithms, it follows from the above that

$$\|\mathbf{u}(t)\|_{1,\infty} \lesssim 1 + \ln[g(t)](\|\nabla b(t)\|_\infty + \|q(t)\|_\infty). \tag{40}$$

Furthermore, since $1 \leq \ln(e+|x|)$ for any $x \in \mathbb{R}$, we can deduce from the inequality above that

$$\|\mathbf{u}(t)\|_{1,\infty} + \|\nabla b(t)\|_\infty + \|q(t)\|_\infty \lesssim 1 + \ln[g(t)](\|\nabla b(t)\|_\infty + \|q(t)\|_\infty). \tag{41}$$

On the other hand, it follows from Lemma 1 that

$$g(t) \leq g(0) \exp\left(c \int_0^t \left(1 + \|\mathbf{u}(s)\|_{1,\infty} + \|\nabla b(s)\|_\infty + \|q(s)\|_\infty\right) ds\right) \tag{42}$$

for any $t \in [0, T]$. Combining (41) and (42) yields

$$g(t) \leq g(0) \exp\left(c \int_0^t \left(1 + \ln[g(s)](\|\nabla b(s)\|_\infty + \|q(s)\|_\infty)\right) ds\right). \tag{43}$$

We can now take logarithm of both sides and apply Grönwall's lemma to the resulting inequality to obtain

$$\ln[g(t)] \leq \left(\ln[g(0)] + cT\right) \exp\left(c \int_0^t (\|\nabla b(s)\|_\infty + \|q(s)\|_\infty) ds\right). \tag{44}$$

At this, point, we can now utilize (8), take exponentials in (44) and obtain

$$\|(b, q)(t)\| \leq [g(0)]^{\exp(cK)} \exp[cT \exp(cK)] \tag{45}$$

for any $t \in [0, T]$. Since the right-hand side is finite, it follows that the solution (b, q) can be continued on some interval $[0, T')$ for some $T' > T$. This finishes the proof.

Acknowledgments This work has been supported by the European Research Council (ERC) Synergy grant STUOD-DLV-856408.

References

1. Arfken, G.B., Weber, H.J.: Mathematical methods for physicists, sixth edn. Elsevier/Academic Press (2005)
2. Beale, J.T., Kato, T., Majda, A.: Remarks on the breakdown of smooth solutions for the 3-D Euler equations. Comm. Math. Phys. **94**(1), 61–66 (1984)

3. Bhat, P., Curless, B., Cohen, M. and Zitnick, C.L.: Fourier analysis of the 2D screened Poisson equation for gradient domain problems. In European Conference on Computer Vision (pp. 114–128). Springer, Berlin, Heidelberg (2008)
4. Crisan, D., Holm, D.D., Luesink, E., Mensah, P.R. and Pan, W.: Theoretical and computational analysis of the thermal quasi-geostrophic model. arXiv preprint arXiv:2106.14850. (2021)
5. Gradshteyn, I.S., Ryzhik, I.M.: Table of integrals, series, and products, seventh edn. Elsevier/Academic Press, Amsterdam (2007). Translated from the Russian, Translation edited and with a preface by Alan Jeffrey and Daniel Zwillinger, With one CD-ROM (Windows, Macintosh and UNIX)

Modeling Under Location Uncertainty: A Convergent Large-Scale Representation of the Navier-Stokes Equations

Arnaud Debussche, Berenger Hug, and Etienne Mémin

Abstract We construct martingale solutions for the stochastic Navier-Stokes equations in the framework of the modelling under location uncertainty (LU). These solutions are pathwise and unique when the spatial dimension is 2D. We then prove that if the noise intensity goes to zero, these solutions converge, up to a subsequence in dimension 3, to a solution of the deterministic Navier-Stokes equation. This warrants that the LU Navier-Stokes equations can be interpreted as a large-scale model of the deterministic Navier-Stokes equation.

1 Introduction

For several years there has been a burst of activity to devise stochastic representations of fluid flow dynamics. These models are strongly motivated in particular by climate and weather forecasting issues and the need to provide accurate ensemble of large-scale flow realisations [2]. Yet, elaborating such stochastic dynamics on *ad hoc* grounds can be highly detrimental to the system of interest [4]. A minimal mathematical requirement for satisfactory large-scale flow dynamics representation is that a weak solution of the Large Eddy Simulation (LES) scheme converges toward a weak solution of the fine-scale deterministic Navier-Stokes equations in 3D and toward the unique solution for the 2D Navier-Stokes equations. The convergence of some classical LES models toward the true fine scale dynamics is well known in the deterministic case [3, 7]. However, the question of convergence of stochastic parametrization toward solutions of the deterministic equations at the limit of vanishing noise is not always clear.

A. Debussche
Univ Rennes, CNRS, IRMAR - UMR 6625, Rennes Cedex, France
e-mail: arnaud.debussche@ens-rennes.fr

B. Hug · E. Mémin (✉)
Inria/IRMAR Campus de Beaulieu, Rennes Cedex, France
e-mail: berenger.hug@ens-rennes.fr; etienne.memin@inria.fr

© The Author(s) 2023
B. Chapron et al. (eds.), *Stochastic Transport in Upper Ocean Dynamics*,
Mathematics of Planet Earth 10, https://doi.org/10.1007/978-3-031-18988-3_2

In this study we show that stochastic Navier-Stokes models defined within the modelling under location uncertainty principle (LU) [9] have martingale solutions in 3D and a unique strong solution—in the probabilistic sense—in 2D. Moreover, in 3D in the limit of vanishing noise there exists a subsequence converging in law toward a weak solution of the deterministic Navier-Stokes equations and in 2D the whole sequence converges toward the unique solution. As such these results enable to consider the LU representation as a valid large-scale stochastic representation of flow dynamics that is more amenable to ensemble forecasting and data assimilation than deterministic model due to an improved variability.

2 Modelling Under Location Uncertainty

The LU formulation relies mainly on the following time-scale separation assumption of the flow:

$$dX_t = u(X_t, t)\, dt + \sigma(X_t, t)\, dW_t, \tag{1}$$

where $X : \mathbb{R}^+ \times \Omega \to S$ is the Lagrangian displacement defined within the bounded domain $S \subset \mathbb{R}^d$ ($d = 2$ or 3) with smooth boundary, and $u : \mathbb{R}^+ \times S \times \Omega \to S$ denotes the large-scale velocity that is both spatially and temporally correlated, while $\sigma\, dW$ is a highly oscillating unresolved component (also called noise term) that is only correlated in space.

More precisely, we consider a cylindrical Wiener process W on $L^2(S, \mathbb{R}^d)$, the space of square integrable functions on S with values in $\mathbb{R}^d$,

$$W = \sum_{i \in \mathbb{N}} \hat{\beta}^i e_i,$$

where $(e_i)_{i \in \mathbb{N}}$ is a Hilbertian orthonormal basis of $L^2(S, \mathbb{R}^d)$ and $(\hat{\beta}_i)_{i \in \mathbb{N}}$ is a sequence of independent standard brownian motions on a stochastic basis $(\Omega, \mathcal{F}, (\mathcal{F}_t)_{t \in [0,T]}, \mathbb{P})$ ([11]). The above does not converge in $L^2(S, \mathbb{R}^d)$ but in any larger Hilbert space U such that the embedding of $L^2(S, \mathbb{R}^d)$ into U is Hilbert-Schmidt, for instance U can be the $L^2(S)$ based Sobolev space $H^{-\alpha}(S)$ for some $\alpha > d/2$.

The spatial structure of the noise is specified through a time dependent deterministic integral covariance operator σ_t defined from a bounded and symmetric kernel $\hat{\sigma}$:

$$\sigma_t f(x) := \int_S \hat{\sigma}(x, y, t)\, f(y)\, dy, \quad f \in L^2(S, \mathbb{R}^d).$$

For each (x, y, t), $\hat{\sigma}(x, y, t)$ is a $d \times d$ symmetric tensor. Since $\hat{\sigma}$ is bounded in x; y and t, $\sigma(x, t)$ maps $L^2(S, \mathbb{R}^d)$ into itself and is Hilbert-Schmidt. Then, the noise can be written as the Wiener process:

In this study we show that stochastic Navier-Stokes models defined within the modelling under location uncertainty principle (LU) [9] have martingale solutions in 3D and a unique strong solution—in the probabilistic sense—in 2D. Moreover, in 3D in the limit of vanishing noise there exists a subsequence converging in law toward a weak solution of the deterministic Navier-Stokes equations and in 2D the whole sequence converges toward the unique solution. As such these results enable to consider the LU representation as a valid large-scale stochastic representation of flow dynamics that is more amenable to ensemble forecasting and data assimilation than deterministic model due to an improved variability.

2 Modelling Under Location Uncertainty

The LU formulation relies mainly on the following time-scale separation assumption of the flow:

$$dX_t = u(X_t, t) dt + \sigma(X_t, t) dW_t, \tag{1}$$

where $X : \mathbb{R}^+ \times \Omega \to \mathcal{S}$ is the Lagrangian displacement defined within the bounded domain $\mathcal{S} \subset \mathbb{R}^d$ ($d = 2$ or 3) with smooth boundary, and $u : \mathbb{R}^+ \times \mathcal{S} \times \Omega \to \mathcal{S}$ denotes the large-scale velocity that is both spatially and temporally correlated, while σdW is a highly oscillating unresolved component (also called noise term) that is only correlated in space.

More precisely, we consider a cylindrical Wiener process W on $L^2(\mathcal{S}, \mathbb{R}^d)$, the space of square integrable functions on $\mathcal{S}$ with values in $\mathbb{R}^d$,

$$W = \sum_{i \in \mathbb{N}} \hat{\beta}^i e_i,$$

where $(e_i)_{i \in \mathbb{N}}$ is a Hilbertian orthonormal basis of $L^2(\mathcal{S}, \mathbb{R}^d)$ and $(\hat{\beta}_i)_{i \in \mathbb{N}}$ is a sequence of independent standard brownian motions on a stochastic basis $(\Omega, \mathcal{F}, (\mathcal{F}_t)_{t \in [0,T]}, \mathbb{P})$ ([11]). The above does not converge in $L^2(\mathcal{S}, \mathbb{R}^d)$ but in any larger Hilbert space U such that the embedding of $L^2(\mathcal{S}, \mathbb{R}^d)$ into U is Hilbert-Schmidt, for instance U can be the $L^2(\mathcal{S})$ based Sobolev space $H^{-\alpha}(\mathcal{S})$ for some $\alpha > d/2$.

The spatial structure of the noise is specified through a time dependent deterministic integral covariance operator σ_t defined from a bounded and symmetric kernel $\hat{\sigma}$:

$$\sigma_t f(x) := \int_{\mathcal{S}} \hat{\sigma}(x, y, t) f(y) \, dy, \ f \in L^2(\mathcal{S}, \mathbb{R}^d).$$

For each (x, y, t), $\hat{\sigma}(x, y, t)$ is a $d \times d$ symmetric tensor. Since $\hat{\sigma}$ is bounded in x; y and t, $\sigma(x, t)$ maps $L^2(\mathcal{S}, \mathbb{R}^d)$ into itself and is Hilbert-Schmidt. Then, the noise can be written as the Wiener process:

Modeling Under Location Uncertainty: A Convergent Large-Scale Representation of the Navier-Stokes Equations

Arnaud Debussche, Berenger Hug, and Etienne Mémin

Abstract We construct martingale solutions for the stochastic Navier-Stokes equations in the framework of the modelling under location uncertainty (LU). These solutions are pathwise and unique when the spatial dimension is 2D. We then prove that if the noise intensity goes to zero, these solutions converge, up to a subsequence in dimension 3, to a solution of the deterministic Navier-Stokes equation. This warrants that the LU Navier-Stokes equations can be interpreted as a large-scale model of the deterministic Navier-Stokes equation.

1 Introduction

For several years there has been a burst of activity to devise stochastic representations of fluid flow dynamics. These models are strongly motivated in particular by climate and weather forecasting issues and the need to provide accurate ensemble of large-scale flow realisations [2]. Yet, elaborating such stochastic dynamics on *ad hoc* grounds can be highly detrimental to the system of interest [4]. A minimal mathematical requirement for satisfactory large-scale flow dynamics representation is that a weak solution of the Large Eddy Simulation (LES) scheme converges toward a weak solution of the fine-scale deterministic Navier-Stokes equations in 3D and toward the unique solution for the 2D Navier-Stokes equations. The convergence of some classical LES models toward the true fine scale dynamics is well known in the deterministic case [3, 7]. However, the question of convergence of stochastic parametrization toward solutions of the deterministic equations at the limit of vanishing noise is not always clear.

A. Debussche
Univ Rennes, CNRS, IRMAR - UMR 6625, Rennes Cedex, France
e-mail: arnaud.debussche@ens-rennes.fr

B. Hug · E. Mémin (✉)
Inria/IRMAR Campus de Beaulieu, Rennes Cedex, France
e-mail: berenger.hug@ens-rennes.fr; etienne.memin@inria.fr

© The Author(s) 2023
B. Chapron et al. (eds.), *Stochastic Transport in Upper Ocean Dynamics*,
Mathematics of Planet Earth 10, https://doi.org/10.1007/978-3-031-18988-3_2

$$\sigma_t W_t = \sum_{i \in \mathbb{N}} \hat{\beta}_t^i \sigma_t e_i,$$

where the series converges in $L^2(\mathcal{S}, \mathbb{R}^d)$ almost surely and in $L^p(\Omega)$ for all $p \in \mathbb{N}$ and Eq. (1) should be understood in the Itô sense. We may further write the dependance of the Wiener process in terms of the other variables:

$$\sigma_t W_t(x, \omega) = \sum_{i \in \mathbb{N}} \hat{\beta}_t^i(\omega) \sigma_t e_i(x),$$

We consider a divergence free noise:

$$\nabla_x \cdot \hat{\sigma}(x, y, t) = 0, \ x, y \in \mathcal{S}, \ t \geq 0.$$

Also, for each $t \in \mathbb{R}^+$, there exists $(\phi_n(t))_n$ a complete orthogonal system composed by eigenfunctions of the covariance operator at each time $t \in \mathbb{R}$ and another sequence of independent standard brownian motions, on the same stochastic basis $(\Omega, \mathcal{F}, (\mathcal{F}_t)_{t \in [0,T]}, \mathbb{P})$, such that we have the representation:

$$\sigma_t W_t = \sum_{k=0}^{\infty} \phi_k(t) \, \beta_t^k.$$

This Gaussian random field is associated to the two-times, two-points covariance tensor given by

$$Q(x, y, t, t') = \mathbb{E}\left(\sigma_t dW_t(x) \, [\sigma_{t'} \, dW_{t'}]^T(y)\right) = \int_{\mathcal{S}} \hat{\sigma}(x, z, t) \, \hat{\sigma}(y, z, t') dy \, \delta(t - t'),$$

with the diagonal part (i.e one time auto-correlation), referred to in the following as the variance tensor, and denoted by

$$a(x, t) = \int_{\mathcal{S}} \hat{\sigma}(x, y, t) \, \hat{\sigma}(x, y, t) dy = \sum_{k=0}^{\infty} \phi_k(x, t) \, \phi_k^T(x, t). \tag{2}$$

In a way similar to the classical derivation of Navier-Stokes equations, the LU setting is based on a stochastic representation of the Reynolds transport theorem (SRTT) [9], describing the rate of change of a random scalar q within a volume $V(t)$ transported by the stochastic flow (1). For incompressible unresolved flows, (i.e. $\nabla \cdot \sigma = 0$), the SRTT reads

$$d\left(\int_{V(t)} q(x, t) \, dx\right) = \int_{V(t)} \left(\mathbb{D}_t q + q \nabla \cdot (u - u_s) dt\right) dx, \tag{3a}$$

$$\mathbb{D}_t q = d_t q + (u - u_s) \cdot \nabla q \, dt + \sigma dW_t \cdot \nabla q - \frac{1}{2} \nabla \cdot (a \nabla q) \, dt, \tag{3b}$$

where $d_t q(x, t) = q(x, t + dt) - q(x, t)$ stands for the forward time-increment of q at a fixed point x, $\mathbb{D}_t$ is introduced as the stochastic transport operator in [9, 12] and plays the role of the material derivative. Recall that u is the large-scale velocity used in (1) and a is defined in (2). Note also that we omit to mention the dependance of σ on time.

This operator is derived from the Itô-Wentzell formula [8] to express the differentiation of a stochastic process transported by the flow [9]. The drift $u_s = \frac{1}{2}\nabla \cdot a$, coined as the Itô-Stokes drift (ISD) in [1], represents through the divergence of the variance tensor, the effects of the small-scale inhomogeneity on the large-scale flow component. This term can be understood as a generalization of the Stokes drift associated to the waves orbital motion. In addition to this modified advection, the stochastic transport operator involves an inhomogeneous diffusion driven by the variance tensor, which can be interpreted as a subgrid diffusion term attached to the mixing operated by the small scales. It can be noticed that this term would only be implicitly represented in Stratonovich integral form. However, the ISD would remain [1]. The remaining term corresponds to the advection by the random term. It can be observed by a direct application of Itô on the norm of the scalar that the positive energy brought by this (backscattering) term is exactly compensated by the energy loss by the diffusion [12]. Due to that, for a transported quantity, its energy is conserved pathwise, or in other words: for any realization of the flow.

The above SRTT (3a) and Newton's second principle (in a distributional sense) allow us to derive the following stochastic equations of motions (see Sect. 5 of [9] or Sect. 2.2–2.3 of [10]), which for any noise scaling $\varepsilon > 0$ parameter and for all points of $\mathcal{S}$ reads, using σ, u_s, a introduced above:

$$d_t u + (u - \varepsilon^2 u_s) \cdot \nabla u \, dt + \varepsilon \sigma \, dW_t \cdot \nabla u - \frac{1}{2} \varepsilon^2 \nabla \cdot (a \nabla u) \, dt$$

$$= -\frac{1}{\rho} \nabla(p \, dt + dp_t^\sigma) + \frac{1}{R_e} \Delta(u \, dt + \varepsilon \sigma \, dW_t), \tag{4}$$

with the incompressibility conditions

$$\nabla \cdot (u - \varepsilon^2 u_s) = 0 \quad , \quad \nabla \cdot \sigma = 0 , \tag{5}$$

and associated with Dirichlet boundary condition $u(t, x) = 0$ and $\widehat{\sigma}(x, y, t) = 0$ for all $x \in \partial \mathcal{S}$ and $t > 0$. The initial condition is denoted by $u(0, x) = u_0(x)$ for all $x \in \mathcal{S}$. As usual, $u(t, x) = (u_1(t, x), \ldots, u_d(t, x))$ and $p(t, x)$ stands for the velocity and the pressure of the fluid, respectively. The term dp_t^σ corresponds to the Brownian (martingale) part of the pressure. The Ito-Stokes drift u_s is defined as $u_s := \frac{1}{2}\nabla \cdot a$ and ρ stands for the fluid density. The dimensioning constant $R_e = UL/\nu$ denotes the Reynolds number, sets from the ratio of the product of characteristic length and velocity scales, UL, with the kinematics viscosity ν. As for the noise scaling parameter, ϵ, it encodes a scale of the unresolved energy and should converge to zero when all the flow components are resolved. Meaning thus there is no noise and the system corresponds trivially to the deterministic Navier-Stokes system.

Although the system corresponds to the Navier-Stokes for zero noise, the convergence toward weak (strong) solutions of the 3D (2D) deterministic Navier-Stokes, respectively, at the limit of vanishing noise needs to be assessed. This is the results we aim to prove in this paper.

First of all, in order to work with a pressure-free system through a divergence-free Leray projection, we proceed to the change of variable $v := u - \varepsilon^2 u_s$ in (4) to rewrite the system with a classical incompressibility condition on v:

$$d_t v + v \cdot \nabla v \, dt - \frac{1}{R_e} \Delta v \, dt + \varepsilon^2 (v \cdot \nabla) u_s \, dt - \frac{\varepsilon^2}{2} \nabla \cdot (a \nabla v) \, dt$$

$$- \frac{\varepsilon^4}{2} \nabla \cdot (a \nabla u_s) \, dt - \frac{\varepsilon^2}{R_e} \Delta u_s \, dt + \varepsilon^2 \partial_t u_s dt = -\frac{1}{\rho} \nabla(p \, dt + dp_t^\sigma) -$$

$$(\varepsilon \sigma dW_t \cdot \nabla) v - (\varepsilon^3 \sigma dW_t \cdot \nabla) u_s + \frac{\varepsilon}{R_e} \Delta(\sigma \, dW_t), \qquad (6)$$

with the incompressibility conditions

$$\nabla \cdot v = 0 \qquad \nabla \cdot \sigma = 0, \qquad (7)$$

for all points in $\mathcal{S}$ together with Dirichlet boundary conditions $v(t, x) = 0$, $\widehat{\sigma}(x, y, t) = 0$ for all $x \in \partial \mathcal{S}$ and $t > 0$ and the initial condition $v(0, x) = v_0(x) := u_0(x) - \varepsilon^2 u_s(0, x)$ for all $x \in \mathcal{S}$. In the following section we specify the spaces on which this system is defined, rewrite it in an equivalent abstract form and state our main result.

3 Notations and Main Result

Let $\mathcal{V}$ be the space of infinitely differentiable d-dimensional vector fields u on $\mathcal{S}$, with compact support strictly contained in $\mathcal{S}$, and satisfying $\nabla \cdot u = 0$. We denote by H the closure of $\mathcal{V}$ in $L^2(\mathcal{S}, \mathbb{R}^d)$ and V the closure of $\mathcal{V}$ in the Sobolev space $H^1(\mathcal{S}, \mathbb{R}^d)$. The space H is endowed with the $L^2(\mathcal{S}, \mathbb{R}^d)$ inner product. This inner product and its induced norm are noted:

$$(u, v)_H := (u, v)_{L^2(\mathcal{S})} \quad \text{and} \quad |u|_H := \|u\|_{L^2(\mathcal{S})}.$$

As for space V, thanks to Poincaré inequality, it is endowed with the $H_0^1(\mathcal{S}, \mathbb{R}^d)$ inner product and its associated norm, denoted respectively as

$$((u, v))_V := (\nabla u, \nabla v)_{L^2(\mathcal{S})} \quad \text{and} \quad \|u\|_V := \|\nabla u\|_{L^2(\mathcal{S})}.$$

We may define then the Gelfand triple $V \subset H \subset V'$ where V' is the dual space of V relative to H. We denote by $\langle \cdot, \cdot \rangle_{V' \times V}$ the duality pairing between V' and V. The space of Hilbert-Schmidt operators from H to H is denoted by $\mathcal{L}_2(H)$ and $\| \cdot \|_{\mathcal{L}_2}$ is its norm.

System (4) may be rewritten in an equivalent simplified pressure-free formulation by using the Leray projection $P : L^2(\mathcal{S}, \mathbb{R}^d) \to H$ of $L^2(\mathcal{S}, \mathbb{R}^d)$ onto the space H of divergence-free vectorial functions. Applying Leray's projector to (6), we obtain

$$d_t v - \frac{1}{R_e} P(\Delta v dt) + P(v \cdot \nabla v \, dt)$$

$$+ P\left(\varepsilon^2 (v \cdot \nabla) u_s \, dt - \frac{\varepsilon^2}{2} \nabla \cdot (a \nabla v) dt - \frac{\varepsilon^4}{2} \nabla \cdot (a \nabla u_s) dt - \frac{\varepsilon^2}{R_e} \Delta u_s dt + \varepsilon^2 \partial_t u_s dt \right)$$

$$= P\left(\frac{\varepsilon}{R_e} \Delta(\sigma \, dW_t) - (\varepsilon \sigma dW_t \cdot \nabla) v - (\varepsilon^3 \sigma dW_t \cdot \nabla) u_s \right). \qquad (8)$$

This system can finally be rewritten in the following simplified abstract form

$$\begin{cases} d_t v(t) + A v(t) \, dt + B v(t) \, dt + F_\varepsilon v(t) \, dt = G_\varepsilon v(t) \, dW_t, \\ v(0) = v_0. \end{cases} \qquad (9)$$

The deterministic terms A, B, F_ε and the stochastic term G_ε are described below.

Several kinds of solutions can be defined for stochastic partial differential equations. As for deterministic PDEs, these can be strong, weak or mild (semi-group) solutions. When the solutions are constructed for a fixed Wiener process W on a given stochastic basis $(\Omega, \mathcal{F}, (\mathcal{F}_t)_{t \in [0,T]}, \mathbb{P})$, they are strong in the probabilistic sense. As usual in 3D, due to the lack of uniqueness, we work with weaker solutions, called martingale solutions, that consists in looking for solutions defined as a triplet composed of a stochastic basis, a Wiener process and an adapted process.

More precisely, we say that there is a martingale solution of system (9) if there exists a stochastic basis $(\Omega, \mathcal{F}, (\mathcal{F}_t)_{t \in [0,T]}, \mathbb{P})$, a cylindrical Wiener process W on $L^2(\mathcal{S}; \mathbb{R}^d)$ and a progressively measurable process $v : [0, T] \times \Omega \to H$, with

$$v \in L^2\left(\Omega \times [0, T]; V \right) \cap L^2\left(\Omega, C^0([0, T]; H) \right),$$

such that $\mathbb{P} - a.e,\ v$ satisfies for all time $t \in [0, T]$

$$v(t) + \int_0^t A v(s) \, ds + \int_0^t B v(s) \, ds + \int_0^t F v(s) \, ds = v_0 + \int_0^t G(v(s)) \, dW_s, \qquad (10)$$

where the equality must be understood in the weak sense. We will show, for all $\varepsilon > 0$, the existence in 3D of a martingale solution for the LU representation of the Navier-Stokes equations for noises associated with a smooth enough diffusion tensor kernel $\hat{\sigma}$ in space and time. In 2D, this solution is unique and strong in the probabilistic sense. This result is summarized in the following theorem.

Theorem 1 *Let $d = 2$ or 3 and assume that the noise is smooth enough in the sense that its variance tensor and Ito-Stokes drift are such that*

$$\sup_{t \in [0,T]} \sum_{k=0}^{\infty} \|\phi_k(t)\|_{H^3(\mathcal{S})}^2 < \infty, \tag{11}$$

$$u_s \in L^\infty(0, T; H^3(\mathcal{S}, \mathbb{R}^d)); \ \partial_t u_s \in L^\infty(0, T; H) \ and \ a\nabla u_s \in L^\infty(0, T; V). \tag{12}$$

Then, for all $\varepsilon > 0$, Eq. (10) admits a martingale solution. Moreover, for $d = 2$, any solution of (10) is strong in the probabilistic sense and unique.

Morever, when $\varepsilon \to 0$, for $d = 3$, there exists a subsequence of $(u_\varepsilon)_{\varepsilon>0}$ which converges in law to a solution of the deterministic Navier-Stokes equation. For $d = 2$, the whole sequence converges to the unique solution of the Navier-Stokes equation.

The condition of Theorem 1 simplifies when the covariance operator does not depend on time or if the ISD is divergence free. In both cases the condition on the temporal derivative of the ISD are not necessary. We note also, that for a spatially homogeneous noise, the variance tensor is constant and the ISD cancels. However this may happen only on a periodic domain or on the full space. The assumptions on the noise are anyway non optimal but it is not the purpose of this paper to consider non spatially smooth noise since in practice it is smooth.

Note that condition (11) is satisfied for instance if we choose σ independent on t and equal to A^{-r} with r large enough where A is the Stokes operator defined below. Indeed, in this case $\phi_k = \lambda_k^{-r} e_k$ where $(e_k)_k$ is an orthonormal complete system of eigenvectors of A associated to the eigenvalues $(\lambda_k)_k$ and $\|\phi_k(t)\|_{H^3(\mathcal{S})}^2 = \lambda_k^{3-2r}$. The behavior of the eigenvalues: $\lambda_k \sim k^{2/d}$ allows to conclude that (11) follows. Since $u_s = \frac{1}{2}\nabla \cdot a$ and a is defined by (2), (12) holds also for r large enough since $\|u_s\|_{H^3(\mathcal{S})} \leq \sum_{k=0}^{\infty} \|\phi_k(t)\|_{H^4(\mathcal{S})}^2$. Finally, since A^{-r} is self-adjoint and Hilbert-Schmidt for $r > d/4$, it is associated to a symmetric kernel $\hat{\sigma}$ which is bounded for r large enough.

These convergence results open new interesting possibilities for the study of turbulence or for the proposition of new large-scale representations of fluid dynamics. From the theoretical point of view, it might be interesting to explore multiscale versions of the LU representation based on spatial filtering together with nested noise models. This would generalize classical large eddy models in which the noise would depend on the spatial filtering applied. The coarser the filtering the larger the noise. Energy transfer between scales would then be very interesting to study in this probabilistic setting. Stochastic Karman-Howarth-Monin equations for energy exchanges across scales could be obtained by this way. From a practical point of view, these convergence results justify the setting of such stochastic models to represent large-scale solutions of the Navier-Stokes equations.

4   Proofs of the Main Result

We introduce the Stokes operator: $Av := -\frac{1}{R_e} P(\Delta v)$ on the domain $\mathcal{D}(A) := V \cap H^2(\mathcal{S}, \mathbb{R}^d)$. Let b be the trilinear form and B the bilinear operator defined for all u, v and $w \in V$ by

$$b(u, v, w) = \int_{\mathcal{S}} w(x) \, [u(x) \cdot \nabla] \, v(x) \, dx = (B(u, v), w)_H.$$

Recall that for all u, v and $w \in V$: $b(u, v, w) = -b(u, w, v)$. As usual, we set $B(u) = B(u, u)$. We then define F by:

$$F(v) = \varepsilon^2 B(v, u_s) - \frac{\varepsilon^2}{2} P\nabla \cdot (a\nabla v) - \frac{\varepsilon^4}{2} P\nabla \cdot (a\nabla u_s) - \varepsilon^2 A u_s \qquad (13)$$

$$+ \varepsilon^2 \partial_t u_s, \quad v \in V.$$

It can be seen that $F(v) \in V'$. We next write the noise term as

$$G(v) \, dW_t = \sum_{k=0}^{\infty} \Big(-\varepsilon \, A\phi_k - \varepsilon B(\phi_k, v) - \varepsilon^3 B(\phi_k, u_s) \Big) \, d\beta_{t,k},$$

where, as for σ, we omit to write dependance of ϕ_k on t. With these notations, (8) may indeed be rewritten as (9).

Let $(e_i)_{i \geq 0}$ be the Hilbertian basis of H consisting of eigenvectors of A. We use the finite dimensional orthogonal projector P_n, $n \in \mathbb{N}$, onto $Span(e_0, \dots, e_n)$ and the projected operators:

$$B^n := P_n B \qquad F^n = P_n F \qquad G^n = P_n G \, .$$

The Galerkin approximation of (9) is given by:

$$\begin{cases} d_t v_n(t) + A v_n(t) \, dt + B^n[v_n(t)] \, dt + F^n[v_n(t)] \, dt = G^n[v_n(t)] \, dW_t, \\ v_n(0) = P_n(v_0). \end{cases}$$
$$(14)$$

This is a finite dimensional system of a stochastic differential equation with smooth coefficients. It has a unique local solution, by the estimate (17) below it is global.

Apply Itô formula to $F(x) = |x|_H^p$ for $p \geq 2$:

$$d_t |v_n(t)|_H^p = p |v_n(t)|_H^{p-2} \big(v_n(t), \, G^n(v_n(t)) dW_t \big)_H$$

$$- p |v_n(t)|_H^{p-2} \big(v_n(t), \, A v_n(t) + B^n v_n(t) + F^n v_n(t) \big)_H \, dt$$

$$+ \frac{p(p-2)}{2} \big(G^n v_n(t), v_n(t) \big)_H^2 \, |v_n(t)|_H^{p-4} dt + \frac{p}{2} \|G^n v_n(t)\|_{\mathcal{L}_2(H)}^2 |v_n(t)|_H^{p-2} dt.$$
$$(15)$$

We have $(v_n(t), Av_n(t))_H = \dfrac{1}{R_e}\|v_n(t)\|_V^2$, $(v_n(t), B^n v_n(t))_H = 0$ and

$$\left(v_n(t), F^n v_n(t)\right)_H = \varepsilon^2 \left([v_n(t)\cdot\nabla]u_s, v_n(t)\right)_H - \frac{\varepsilon^2}{2}\left(v_n(t), \nabla\cdot(a\nabla v_n(t))\right)_H$$

$$- \frac{\varepsilon^4}{2}\left(v_n(t), \nabla\cdot(a\nabla u_s)\right)_H + \varepsilon^2\left(Au_s, v_n(t)\right)_H + \varepsilon^2\left(\partial_t u_s, v_n(t)\right)_H$$

$$:= F_1^n + F_2^n + F_3^n + F_4^n + F_5^n.$$

Under the assumption (12) in Theorem 1, we have the estimate:

$$|F_1^n + F_3^n + F_4^n + F_5^n| \le C\,(\varepsilon^2 + \varepsilon^4)\,|v_n(t)|_H^2 + C(\varepsilon^2 + \varepsilon^4)$$

with $C > 0$ a finite constant. And by the definition of a, we have

$$F_2^n = \frac{\varepsilon^2}{2}\sum_{k=0}^{\infty}|(\phi_k\cdot\nabla)v_n(t)|_{L^2(S)}^2.$$

Furthermore, using (11),

$$\frac{1}{2}\|G^n v_n(t)\|_{\mathcal{L}_2(l^2(H))}^2 \le \frac{\varepsilon^2}{2}\sum_{k=0}^{\infty}|(\phi_k\cdot\nabla)v_n(t)|_{L^2(S)}^2 + C\varepsilon^2 + 2\varepsilon^2|v_n(t)|_H^2$$

and the first term corresponds exactly to F_2^n. Finally, using again (11),

$$(G^n v_n(t), v_n(t))_H^2 \le 2\,C\,(\varepsilon^2 + \varepsilon^6)\,|v_n(t)|_H^2.$$

Hence

$$d_t|v_n(t)|_H^p + \frac{p}{R_e}|v_n(t)|_H^{p-2}\|v_n(t)\|_V^2 \le p|v_n(t)|_H^{p-2}\left(v_n(t), G^n(v_n(t))dW_t\right)_H$$

$$+ C\,(\varepsilon^2 + \varepsilon^4)\,|v_n(t)|_H^p + C\,[(\varepsilon^2 + \varepsilon^6)^\alpha + (\varepsilon^2 + \varepsilon^4)] \qquad (16)$$

with $C > 0$ depending on p (and not on ε and n). We then use classical arguments based in particular on Burkholder-Davis-Gundy inequality to deduce:

$$\frac{1}{2}\mathbb{E}\left[\sup_{0\le t\le T}|v_n(t)|_H^p + \int_0^T |v_n(t)|_H^{p-2}\|v_n(t)\|_V^2\right] \le \mathbb{E}\left[|v_0|_H^p\right] + C\,\varepsilon^2. \qquad (17)$$

Arguing as in [6], we prove that the laws $(\mathcal{L}(v_n))_n$ are tight in $L^2([0, T]; H)$ and in $C^0([0, T]; \mathcal{D}(A^{-3/2}))$.

By the Skorohod's embedding theorem, there exists a stochastic basis $(\overline{\Omega}, \overline{\mathcal{F}}, (\overline{\mathcal{F}}_t)_t, \overline{\mathbb{P}})$ with $L^2([0, T]; H) \cap C^0([0, T]; \mathcal{D}(A^{-3/2}))$-valued random variables $\overline{v}_n$ for $n \geq 1$ and $\overline{v}$ such that $\overline{v}_n$ has the same law as v_n on $L^2([0, T]; H) \cap C^0([0, T]; \mathcal{D}(A^{-3/2}))$ and $C^0([0, T], U_0)$ cylindrical Wiener processes $\overline{W}^n$ for $n \geq 1$ together with $\overline{W}$ such that (by thinning the sequences)

$$\overline{v}_n \to \overline{v} \text{ in } L^2([0, T]; H) \cap C^0([0, T]; \mathcal{D}(A^{-3/2})) \qquad \overline{\mathbb{P}} \text{ a.s} \tag{18}$$

$$\overline{W}^n \to \overline{W} \text{ in } C^0([0, T], U_0) \qquad \overline{\mathbb{P}} \text{ a.s .} \tag{19}$$

For all integers n, $\overline{v}_n$ verifies

$$\overline{v}_n(t) - P_n(v_0) + \int_0^t \left[A\overline{v}_n(r) + B^n \overline{v}_n(r) + F^n \overline{v}_n(r) \right] dr = \int_0^t G^n(\overline{v}_n(r)) d\overline{W}_r^n. \tag{20}$$

We may let $n \to \infty$ in this equation and prove that $\overline{v}$ verifies for almost surely $(t, \omega) \in [0, T] \times \overline{\Omega}$

$$\overline{v}(t) - v_0 + \int_0^t \left(A\overline{v}(r) + B\overline{v}(r) + F\overline{v}(r) \right) dr = \int_0^t G(\overline{v}(r)) d\overline{W}_r \tag{21}$$

in the weak sense. For instance, let w be a smooth test function, then:

$$\int_0^t (B^n(\overline{v}_n(r)), w)_H dr = \int_0^t b((\overline{v}_n(r), (\overline{v}_n(r), w) dr = -\int_0^t b((\overline{v}_n(r), w, (\overline{v}_n(r)) dr$$

and by the almost sure strong convergence in $L^2(0, T, H)$ this converges to $-\int_0^t b((\overline{v}(r), w, (\overline{v}(r)) dr$ when $n \to \infty$.

It can be shown that (17) holds for $\overline{v}_n$ and letting $n \to \infty$ we obtain a bound on $\overline{v}$. In particular, $\overline{v} \in L^2(\overline{\Omega}; L^2([0, T], V)) \cap L^2(\overline{\Omega}; L^\infty([0, T], H))$. We then use the mild form of this equation to prove that $\overline{v} \in C^0([0, T], H)$ almost surely.

For $d = 2$, we consider v_1 and v_2 two solutions of (9) on the same probability space $(\Omega, \mathcal{F}, (\mathcal{F}_t)_t, \mathbb{P})$ and, using Ito formula and classical estimates, prove that

$$\mathbb{E} \left[\sup_{0 \leq r \leq T} e(r) |(v_1 - v_2)(r)|_H^2 \right] = 0,$$

where $e(t) := \exp\left(-\alpha \int_0^t \|v_2(r)\|_V^2 \, dr \right)$ for a well chosen α. As $\mathbb{E} \int_0^T \|v_2(r)\|_V^2 \, dr < \infty$, we deduce $\mathbb{P}$ a.s, $v_1 = v_2$ for all $t \in [0, T]$. We have proved that pathwise uniqueness holds for $d = 2$. Then, using an argument due to Gyongy and Krylov (see for instance [5], Sect. 5), we conclude that the whole sequence $(v_n)_n$ converges to a unique solution of (21).

Let $v_0 \in H$. For all $\varepsilon > 0$, we have proved that the abstract problem (8) admits martingale solutions $(v_\varepsilon)_{\varepsilon > 0}$. We then study if $(v_\varepsilon)_{\varepsilon > 0}$ converges when $[\varepsilon \to 0^+]$ to a solution v of the following deterministic Navier-Stokes equation

$$\begin{cases} d_t v(t) + A v(t)\, dt + B v(t)\, dt = 0 \\ v(0) = v_0 \, . \end{cases} \tag{22}$$

When $d = 2$, the solution v_ε is strong and unique. The deterministic Eq. (22) admits also a unique weak solution v. By classical estimate, we prove:

$$\mathbb{E}_\varepsilon \left[\sup_{0 \le t \le T} e(t) \, |v_\varepsilon(t) - v(t)|_H^2 \right] \xrightarrow[\varepsilon \to 0^+]{} 0,$$

where $e(t) := \exp\left(-\alpha \int_0^t \|v(r)\|_V^2 \, dr \right)$ for some $\alpha > 0$.

When $d = 3$, inequality (17) shows that $\big(\mathcal{L}(v_{\varepsilon_n})\big)_n$ are tight in $L^2([0, T]; H) \cap C^0([0, T]; \mathcal{D}(A^{-3/2}))$. Using Skorohod embedding theorem, we show that a subsequence converges to the law a weak solution of (22).

References

1. W. Bauer, P. Chandramouli, B. Chapron, L. Li, and E. Mémin. Deciphering the role of small-scale inhomogeneity on geophysical flow structuration: A stochastic approach. *Journal of Physical Oceanography*, 50(4):983 – 1003, 01 Apr. 2020.

2. J. Berner and Coauthors. Stochastic parameterization: Toward a new view of weather and climate models. *Bull. Amer. Meteor. Soc.*, 98:565–588, 2017.

3. L. Berselli, T. Iliescu, and W. Layton. *Mathematics of Large Eddy Simulation of Turbulent Flows, First Edition*. Springer-Verlag, 2010.

4. B. Chapron, P. Dérian, E. Mémin, and V. Resseguier. Large-scale flows under location uncertainty: a consistent stochastic framework. *QJRMS*, 144(710):251–260, 2018.

5. A. Debussche, N. Glatt-Holtz, and R. Temam. Local martingale and pathwise solutions for an abstract fluids model. *Physica D: Nonlinear Phenomena*, 240(14):1123–1144, 2011.

6. F. Flandoli and D. Gatarek. Martingale and stationary solutions for stochastic Navier-Stokes equations. *Probability Theory and Related Fields*, 102(3):367–391, 1995.

7. J. L. Guermond, J. T. Oden, and S. Prudhomme. Mathematical perspectives on large eddy simulation models for turbulent flows. *Journal of Mathematical Fluid Mechanics*, 6(2):194–248, 2004.

8. H. Kunita. *Stochastic flows and stochastic differential equations*. Cambridge University Press, 1990.

9. E. Mémin. Fluid flow dynamics under location uncertainty. *Geophys. & Astro. Fluid Dyn.*, 108(2):119–146, 2014.

10. R. Mikulevicius and B. Rozovskii. Stochastic Navier-Stokes equations for turbulent flows. *SIAM J. Math. Anal.*, 35(4):1250–1310, 2004.

11. G. D. Prato and J. Zabczyk. *Stochastic equations in infinite dimensions*. Cambridge University Press, 1992.

12. V. Resseguier, E. Mémin, and B. Chapron. Geophysical flows under location uncertainty, Part I Random transport and general models. *Geophys. & Astro. Fluid Dyn.*, 111(3):149–176, 2017.

A Stochastic Benjamin-Bona-Mahony Type Equation

Evgueni Dinvay

Abstract Considered herein is a particular nonlinear dispersive stochastic equation. It was introduced recently in Dinvay and Mémin (Proc. R. Soc. A. 478:20220050, 2022), as a model describing surface water waves under location uncertainty. The corresponding noise term is introduced through a Hamiltonian formulation, which guarantees the energy conservation of the flow. Here the initial-value problem is studied.

Keywords Water waves · BBM equation · multiplicative noise

2010 Mathematics Subject Classification 35Q53, 35Q60, 60H15

1 Introduction

Consideration is given to the following Stratonovich one-dimensional BBM-type equation

$$du = -\partial_x K \left(u + K u^2 \right) dt + \sum_j \gamma_j \partial_x \left(u + K u^2 \right) \circ dW_j \tag{1}$$

introduced in [4], as a model describing surface waves of a fluid layer. It is supplemented with the initial condition $u(0) = u_0$. Equation (1) has a Hamiltonian structure with the energy

$$\mathcal{H}(u) = \int_{\mathbb{R}} \left(\frac{1}{2} (K^{-1/2} u)^2 + \frac{1}{3} u^3 \right) dx. \tag{2}$$

E. Dinvay (✉)
Inria Rennes - Bretagne Atlantique, Campus universitaire de Beaulieu Avenue du Général Leclerc, Rennes Cedex, France
e-mail: Evgueni.Dinvay@inria.fr

© The Author(s) 2023
B. Chapron et al. (eds.), *Stochastic Transport in Upper Ocean Dynamics*,
Mathematics of Planet Earth 10, https://doi.org/10.1007/978-3-031-18988-3_3

The Fourier multiplier operator K, defined in the space of tempered distributions $\mathcal{S}'(\mathbb{R})$, has an even symbol of the form

$$K(\xi) \simeq (1 + \xi^2)^{-\sigma_0} \tag{3}$$

with $\sigma_0 > 1/2$. Expression (3) means that the symbol $K(\xi)$ is bounded from below and above by RHS(3) multiplied by some positive constants. In other words the operator K essentially behaves as the Bessel potential of order $2\sigma_0$, see [6]. The space variable is $x \in \mathbb{R}$ and the time variable is $t \geqslant 0$. The unknown u is a real valued function of these variables and of the probability variable $\omega \in \Omega$, representing the free surface elevation in the fluid layer. The scalar sequence $\{\gamma_j\}$ satisfies the restriction $\sum_j \gamma_j^2 < \infty$, and $\{W_j\}$ is a sequence of independent scalar Brownian motions on a filtered probability space $(\Omega, \mathcal{F}, \{\mathcal{F}_t\}, \mathbb{P})$.

Model (1) was introduced in [4], where an attempt to extend an elegant Hamiltonian formulation of [1] to the stochastic setting was made. We will just briefly comment on the methodology of [4]. The white noise is firstly introduced via the stochastic transport theory presented in [8], which is based on splitting of fluid particle motion into smooth and random movements. Then it is restricted to a particular Stratonovich form in order to respect the energy conservation. In particular, it provides us with a model having multiplicative noise of Hamiltonian structure. Finally, a long wave approximation results in simplified models as (1), for example.

One may notice that after discarding the nonlinear terms in Eq. (1), the details can be seen in [4], the corresponding linearised initial-value problem can be solved exactly with the help of the fundamental multiplier operator

$$\mathcal{S}(t, t_0) = \exp\left[-\partial_x K(t - t_0) + \sum_j \gamma_j \partial_x (W_j(t) - W_j(t_0)) \right], \tag{4}$$

where $t_0, t \in \mathbb{R}$. Note that it can be factorised as $\mathcal{S}(t, t_0) = S(t - t_0)S_W(t, t_0)$, where $S(t) = \exp(-\partial_x K t)$ is a unitary semi-group and S_W containing all the randomness coming from the Wiener process is unitary as well. They obviously commute as bounded differential operators. We recall that $S(t)$ is defined via the Fourier transform $\mathfrak{F}(S(t)\psi) = \exp(-i\xi K(\xi)t)\widehat{\psi}(\xi)$ for any $\psi \in \mathcal{S}'(\mathbb{R})$ and $\widehat{\psi} = \mathfrak{F}\psi$. Similarly, $S_W(t, t_0)$ is defined by the line

$$S_W(t, t_0)\psi = \mathfrak{F}^{-1}\left(\xi \mapsto \exp\left(i\xi \sum_j \gamma_j(W_j(t) - W_j(t_0)) \right) \widehat{\psi}(\xi) \right).$$

It allows us to represent (1) in the Duhamel form

$$u(t) = \mathcal{S}(t, 0)\left(u_0 + \int_0^t \mathcal{S}(0, s)f(u(s))ds + \sum_j \gamma_j \int_0^t \mathcal{S}(0, s)g(u(s))dW_j(s) \right), \tag{5}$$

where

$$f(u) = -\partial_x K^2 u^2 + \sum_j \gamma_j^2 \partial_x K(u \partial_x K u^2)$$

and

$$g(u) = \partial_x K u^2.$$

Existence and uniqueness of solution to Eq. (5) is under consideration. It is worth to point out that both S_W and the stochastic integral in (5) are well defined. Indeed, appealing to Doobs' inequalities for the submartingale $\left| \sum_{j=n}^{n+m} \gamma_j W_j \right|$ and the Itô-Nisio theorem one can show that $\sum_j \gamma_j W_j$ converges uniformly in time almost surely, in probability and in L^2 sense. If the integrand of the stochastic integral in (5) is in some Sobolev space $H^\sigma(\mathbb{R})$ for each s and a.e. ω, then we can understand this sum of integrals as an integration with respect to a Q-Wiener process associated with a Hilbert space H and a non-negative symmetric trace class operator Q having eigenvalues γ_j^2 and eigenfunctions e_j forming an orthonormal basis in H. Then the corresponding integrand is the unbounded linear operator between H and $H^\sigma(\mathbb{R})$ that maps all e_j to the same element of $H^\sigma(\mathbb{R})$, namely, to $S(0, s)g(u(s))$. In particular, it explains why we need the summability condition $\sum_j \gamma_j^2 < \infty$.

Before we formulate the main result it is left to introduce a notation as follows. By $C(0, T; H^\sigma(\mathbb{R}))$ we will notate the space of continuous functions on $[0, T]$ having values in $H^\sigma(\mathbb{R})$ with the usual supremum norm.

Theorem 1 *Let $\sigma_0 > 1/2$ and $\sigma \geqslant \max\{\sigma_0, 1\}$. Then for any $\mathcal{F}_0$-measurable $u_0 \in L^2(\Omega; H^\sigma(\mathbb{R})) \cap L^\infty(\Omega; H^{\sigma_0}(\mathbb{R}))$ with sufficiently small $L^\infty H^{\sigma_0}$-norm and any $T_0 > 0$ Eq. (5) has a unique adapted solution $u \in L^2(\Omega; C(0, T_0; H^\sigma(\mathbb{R}))) \cap L^\infty(\Omega; C(0, T_0; H^{\sigma_0}(\mathbb{R})))$. Moreover, $\mathcal{H}(u(t)) = \mathcal{H}(u_0)$ for each $t \in [0, T_0]$ almost surely on Ω.*

The conservation of energy (2) plays a crucial role in the proof. So it will be a bit more convenient to regard the energy norm defined by

$$\|u\|_{\mathcal{H}}^2 = \frac{1}{2} \int_{\mathbb{R}} \left(K^{-1/2} u \right)^2 dx$$

instead of the spatial H^{σ_0}-norm. They are obviously equivalent.

The proof is essentially based on the contraction mapping principle. We do not exploit much smoothing properties of the group $S(t, t_0)$, as for example is done in [2] for analysis of a stochastic nonlinear Schrödinger equation. It is enough to know that the absolute value of its symbol equals one, and that $S(t)$ is a unitary semigroup. However, in order to appeal to the fixed point theorem we have to truncate both deterministic f and random g nonlinearities. There are a couple of technical difficulties related to implementation of the energy conservation in our case. Firstly, for the truncated equation we can claim $\mathcal{H}$-conservation only until a particular stopping time. Secondly, one can control $\|u\|_{\mathcal{H}}$ with $\mathcal{H}(u)$ only provided $\|u\|_{\mathcal{H}}$ is small. These additional difficulties make us repeat the arguments of the last section in the paper iteratively in order to construct solution on the whole time interval $[0, T_0]$.

As a final remark we point out that the noise in Eq. (1) can be gathered in one dimensional $\partial_x \left(u + Ku^2\right) \circ dB$ with the scalar Brownian motion $B = \sum_j \gamma_j W_j$. However, this does not affect the proof below anyhow, so we continue to stick to the original formulation (1). In future works we are planning to extend it to γ_j being either Fourier multipliers or space-dependent coefficients.

2   Truncation

The Sobolev space $H^\sigma(\mathbb{R})$ consists of tempered distributions u having the finite square norm $\|u\|_{H^\sigma}^2 = \int |\widehat{u}(\xi)|^2 \left(1 + \xi^2\right)^\sigma d\xi < \infty$. Let $\theta \in C_0^\infty(\mathbb{R})$ with $\operatorname{supp}\theta \in [-2, 2]$ being such that $\theta(x) = 1$ for $x \in [-1, 1]$ and $0 \leqslant \theta(x) \leqslant 1$ for $x \in \mathbb{R}$. For $R > 0$ we introduce the cut off $\theta_R(x) = \theta(x/R)$ and

$$f_R(u) = \theta_R(\|u\|_{H^\sigma})f(u), \quad g_R(u) = \theta_R(\|u\|_{H^\sigma})g(u)$$

that we substitute in (5) instead of $f(u)$, $g(u)$, respectively. The new R-regularisation of (5) reads as

$$u(t) = \mathcal{S}(t, t_0)\left(u(t_0) + \int_{t_0}^t \mathcal{S}(t_0, s)f_R(u(s))ds + \sum_j \gamma_j \int_{t_0}^t \mathcal{S}(t_0, s)g_R(u(s))dW_j(s)\right).$$

$$(6)$$

In this section without loss of generality we can set $t_0 = 0$ and $u(t_0) = u_0$. We will vary time moments t_0 below in the next section. Equation (6) can be solved with a help of the contraction mapping principle in $L^2(\Omega; C(0, T; H^\sigma(\mathbb{R})))$.

Proposition 1 *Let $\sigma > 1/2$, $u_0 \in L^2(\Omega; H^\sigma(\mathbb{R}))$ be $\mathcal{F}_0$-measurable and $T_0 > 0$. Then (6) has a unique adapted solution $u \in L^2(\Omega; C(0, T_0; H^\sigma(\mathbb{R})))$. Moreover, it depends continuously on the initial data u_0.*

Proof We set $\mathcal{T}u(t) = $ RHS(6). We will show that $\mathcal{T}$ is a contraction mapping in $X_T = L^2(\Omega; C(0, T; H^\sigma(\mathbb{R})))$, provided $T > 0$ is sufficiently small, depending only on R. Let u_1, u_2 be two adapted processes in X_T. Firstly, one can notice that

$$\|f_R(u_1) - f_R(u_2)\|_{H^\sigma} \leqslant C \left(1 + R\right)^2 \|u_1 - u_2\|_{H^\sigma},$$

$$\|g_R(u_1) - g_R(u_2)\|_{H^\sigma} \leqslant C R \|u_1 - u_2\|_{H^\sigma}.$$

Indeed, $H^\sigma(\mathbb{R})$ poses an algebraic property for $\sigma > 1/2$ and $\partial_x K$ is bounded in $H^\sigma(\mathbb{R})$. Then assuming $\|u_1\|_{H^\sigma} \geqslant \|u_2\|_{H^\sigma}$ without loss of generality one deduces

$$\|g_R(u_1) - g_R(u_2)\|_{H^\sigma} \leqslant C \left\| \theta_R(\|u_1\|_{H^\sigma}) u_1^2 - \theta_R(\|u_2\|_{H^\sigma}) u_2^2 \right\|_{H^\sigma}$$

$$\leqslant C\theta_R(\|u_1\|_{H^\sigma}) \left\| u_1^2 - u_2^2 \right\|_{H^\sigma}$$

$$+ |\theta_R(\|u_1\|_{H^\sigma}) - \theta_R(\|u_2\|_{H^\sigma})| \left\| u_2^2 \right\|_{H^\sigma}$$

$$\leqslant CR \|u_1 - u_2\|_{H^\sigma},$$

where we have used the estimate $|\theta_R(\|u_1\|_{H^\sigma}) - \theta_R(\|u_2\|_{H^\sigma})| \leqslant \|\theta'\|_{L^\infty} R^{-1}$ $\|u_1 - u_2\|_{H^\sigma}$ following obviously from the mean value theorem. The difference between $f_R(u_1)$ and $f_R(u_2)$ can be obtained in the same way. Thus

$$\|\mathcal{T}u_1(t) - \mathcal{T}u_2(t)\|_{H^\sigma} \leqslant \left\| \int_0^t \mathcal{S}(0,s)(f_R(u_1(s)) - f_R(u_2(s)))ds \right\|_{H^\sigma}$$

$$+ \left\| \sum_j \gamma_j \int_0^t \mathcal{S}(0,s)(g_R(u_1(s)) - g_R(u_2(s)))dW_j(s) \right\|_{H^\sigma} = I + II.$$

The first integral is estimated straightforwardly as

$$I \leqslant \int_0^T \|f_R(u_1(s)) - f_R(u_2(s))\|_{H^\sigma} ds \leqslant C(1+R)^2 T \|u_1 - u_2\|_{C(0,T;H^\sigma)}.$$

The second one is estimated with the use of the Burkholder inequality [5] as

$$\mathbb{E} \sup_{0 \leqslant t \leqslant T} II^2 \leqslant C\mathbb{E} \int_0^T \|g_R(u_1(s)) - g_R(u_2(s))\|_{H^\sigma}^2 ds \leqslant CR^2 T\mathbb{E} \|u_1 - u_2\|_{C(0,T;H^\sigma)}^2.$$

It is clear that time-continuity of $\mathcal{T}u_1, \mathcal{T}u_2$ follows from the factorisation $\mathcal{S} = \mathcal{S}\mathcal{S}_W$ and the estimate $\|\mathcal{S}_W g_R(u)\|_{H^\sigma} \leqslant CR^2$, so we have a stochastic convolution as in [5, Lemma 3.3]. Thus

$$\|\mathcal{T}u_1 - \mathcal{T}u_2\|_{X_T} \leqslant C \left((1+R)^2 T + R\sqrt{T}\right) \|u_1 - u_2\|_{X_T},$$

and so there exists a small T depending only on R such that $\mathcal{T}$ has a unique fixed point in X_T. Moreover, this estimate also gives us continuous dependence of solution in X_T on the initial data $u_0 \in L^2(\Omega; H^\sigma(\mathbb{R}))$, obviously. Clearly, the solution can be extended to the whole interval $[0, T_0]$. $\qquad\square$

The regularisation affects the energy conservation. Indeed, in the Itô differential form Eq. (6) reads

$$du = \left(-\partial_x K u + \frac{1}{2} \sum_j \gamma_j^2 \partial_x^2 u + f_R(u) + \sum_j \gamma_j^2 \partial_x g_R(u) \right) dt \qquad (7)$$

$$+ \sum_j \gamma_j \left(\partial_x u + g_R(u) \right) dW_j,$$

and so applying the Itô formula to the energy functional $\mathcal{H}(u(t))$ defined by (2) with the use of (7), one can easily obtain

$$d\mathcal{H}(u) = \left((\theta_R - 1) \int u^2 \partial_x K u \, dx + \theta_R (\theta_R - 1) \sum_j \gamma_j^2 \int \left(\frac{1}{2} g(u) K^{-1} g(u) + u g^2(u) \right) dx \right) dt. \qquad (8)$$

Indeed, assuming $\sigma \geq \sigma_0 + 2$ at first, we notice that the solution u given by Proposition 1 solves Eq. (7). Let us introduce the following notations

$$\Psi(t)dt + \Phi(t)dW = \Psi(t)dt + \sum_j \gamma_j \Phi(t) e_j dW_j = \text{RHS}(7).$$

Then Itô's formula reads

$$\mathcal{H}(u(t)) = \mathcal{H}(u_0) + \int_0^t \partial_u \mathcal{H}(u(s))\Psi(s)ds + \int_0^t \partial_u \mathcal{H}(u(s))\Phi(s)dW(s)$$

$$+ \frac{1}{2} \int_0^t \text{tr}\, \partial_u^2 \mathcal{H}(u(s))(\Phi(s), \Phi(s))ds,$$

where the Fréchet derivatives are defined by

$$\partial_u \mathcal{H}(u)\phi = \int_{\mathbb{R}} \left(K^{-1/2} u K^{-1/2} \phi + u^2 \phi \right) dx,$$

$$\partial_u^2 \mathcal{H}(u)(\phi, \psi) = \int_{\mathbb{R}} \left(K^{-1/2} \phi K^{-1/2} \psi + 2u\phi\psi \right) dx$$

at every $\phi, \psi \in H^{\sigma_0}(\mathbb{R})$. Substituting these expressions together with the definitions of Φ and Ψ into the Itô's formula one obtains (8). Let us, for example, calculate the stochastic integral

$$\int_0^t \partial_u \mathcal{H}(u(s))\Phi(s)dW(s) = \sum_j \gamma_j \int_0^t \int_{\mathbb{R}} \left(K^{-1/2} u K^{-1/2} + u^2 \right)$$

$$\left(\partial_x u + \theta_R(\|u\|_{H^\sigma})\partial_x K u^2 \right) dx dW_j$$

that equals zero as one can see integrating by parts in the space integral. Similarly, one calculates the other two integrals in the Itô formula. Thus we have proved (8) for $\sigma \geqslant \sigma_0 + 2$. In order to lower the bound for σ, one would like to argue here by approximation of initial value u_0 via smooth functions and appeal to the continuous dependence on u_0, however, there is a problem here, since θ_R in (8) contains the dependence on σ. So even for a smooth initial data the corresponding solution lies a priori only in H^σ. This difficulty is overcome in the next statement, where we argue similar to [3].

Proposition 2 *Let $\sigma_0 > 1/2$ and $\sigma \geqslant \max\{\sigma_0, 1\}$. Then (8) holds almost surely for u satisfying Eq. (6) given by Proposition 1.*

Proof The main idea is to cut off high frequencies of the differential operator ∂_x in (7) as follows. Let P_λ be a Fourier multiplier with the symbol θ_λ, $\lambda > 0$. It is defined by the expression $\mathfrak{F}(P_\lambda \psi) = \theta_\lambda \widehat{\psi}$. Now we consider instead of (7) the following regularisation

$$
du = \left(-\partial_x K u + \frac{1}{2} \sum_j \gamma_j^2 \partial_x^2 P_\lambda^2 u + f_R(u) + \sum_j \gamma_j^2 \partial_x P_\lambda g_R(u) \right) dt \qquad (9)
$$
$$
+ \sum_j \gamma_j \left(\partial_x P_\lambda u + g_R(u) \right) dW_j
$$

that has a strong solution. Indeed, it contains only bounded operators and the corresponding mild equation has exactly the same form as Eq. (6) with $\mathcal{S}^\lambda = S S_W^\lambda$ now instead of $\mathcal{S}$, where

$$
S_W^\lambda = \exp \left[\sum_j \gamma_j \partial_x P_\lambda (W_j(t) - W_j(t_0)) \right].
$$

So we can actually apply Proposition 1 to obtain $u = u_\lambda$ solving (9). Let $u = u_\infty$ stay for the solution of the original Eq. (6). Firstly, we will check that $u_\lambda \to u_\infty$ in $L^2(\Omega; L^2(0, T_0; H^\sigma(\mathbb{R})))$ for any $\sigma > 1/2$ as $\lambda \to \infty$.

Let $0 \leqslant t \leqslant T \leqslant T_0$, where a positive small enough time moment T is to be chosen below. Then

$$
\begin{aligned}
\|u_\lambda(t) - u_\infty(t)\|_{H^\sigma} &= \left\| \mathcal{T}^\lambda u_\lambda(t) - \mathcal{T}^\infty u_\infty(t) \right\|_{H^\sigma} \\
&\leqslant \left\| \left(\mathcal{S}^\lambda(t, 0) - \mathcal{S}^\infty(t, 0) \right) u_0 \right\|_{H^\sigma} \\
&\quad + \left\| \int_0^t \left(\mathcal{S}^\lambda(t, s) - \mathcal{S}^\infty(t, s) \right) f_R(u_\infty(s)) ds \right\|_{H^\sigma} \\
&\quad + \left\| \int_0^t \mathcal{S}^\lambda(t, s) (f_R(u_\lambda(s)) - f_R(u_\infty(s))) ds \right\|_{H^\sigma}
\end{aligned}
$$

$$+ \left\| (\mathcal{S}^\lambda(t,0) - \mathcal{S}^\infty(t,0)) \sum_j \gamma_j \int_0^t \mathcal{S}^\infty(0,s) g_R(u_\infty(s)) dW_j(s) \right\|_{H^\sigma}$$

$$+ \left\| \sum_j \gamma_j \int_0^t (\mathcal{S}^\lambda(0,s) - \mathcal{S}^\infty(0,s)) g_R(u_\infty(s)) dW_j(s) \right\|_{H^\sigma}$$

$$+ \left\| \sum_j \gamma_j \int_0^t \mathcal{S}^\lambda(0,s)(g_R(u_\lambda(s)) - g_R(u_\infty(s))) dW_j(s) \right\|_{H^\sigma}$$

$$= I_1 + \ldots + I_6.$$

The terms I_3 and I_6 are estimated exactly as the analogous integrals I and II in the proof of Proposition 1, namely,

$$I_3 \leqslant C(1 + R)^2 \sqrt{T} \|u_\lambda - u_\infty\|_{L^2(0,T;H^\sigma)}$$

and

$$\mathbb{E} \sup_{0 \leqslant t \leqslant T} I_6^2 \leqslant C \mathbb{E} \int_0^T \|g_R(u_\lambda(s)) - g_R(u_\infty(s))\|_{H^\sigma}^2 \, ds$$

$$\leqslant C R^2 \mathbb{E} \|u_\lambda - u_\infty\|_{L^2(0,T;H^\sigma)}^2.$$

Thus

$$\mathbb{E} \int_0^T \left(I_3^2 + I_6^2 \right) dt \leqslant C \left((1 + R)^4 T^2 + R^2 T \right) \mathbb{E} \|u_\lambda - u_\infty\|_{L^2(0,T;H^\sigma)}^2,$$

and so there exists a small $T > 0$ depending only on R such that

$$\mathbb{E} \|u_\lambda - u_\infty\|_{L^2(0,T;H^\sigma)}^2 \leqslant C \mathbb{E} \int_0^T \left(I_1^2 + I_2^2 + I_4^2 + I_5^2 \right) dt.$$

One needs to show that the right hand side of this expression tends to zero when $\lambda \to \infty$. All these four integrals are treated similarly. Indeed, let us regard more closely the first one

$$I_1^2 = \int \left| \exp \left(i\xi\theta_\lambda(\xi) \sum_j \gamma_j W_j(t) \right) - \exp \left(i\xi \sum_j \gamma_j W_j(t) \right) \right|^2 |\widehat{u_0}(\xi)|^2 \left(1 + \xi^2 \right)^\sigma d\xi$$

that obviously tends to zero as $\lambda \to \infty$ for a.e. ω and any t. Hence $\mathbb{E} \int_0^T I_1^2 dt \to 0$ by the dominated convergence theorem, sine $I_1 \leqslant 2 \|u_0\|_{H^\sigma}$. The integral of I_4^2

is estimated exactly in the same manner with the stochastic integral of $\mathcal{S}^\infty g_R(u_\infty)$ standing in place of u_0. The second integral

$$\mathbb{E}\int_0^T I_2^2 dt \leqslant T\mathbb{E}\int_0^T\int_0^T \left\|\left(\mathcal{S}^\lambda(t,s)-\mathcal{S}^\infty(t,s)\right)f_R(u_\infty(s))\right\|_{H^\sigma}^2 ds dt \to 0$$

by the dominated convergence theorem, since $\|\ldots\|_{H^\sigma}^2 \leqslant CR^2(1+R)^4$. Finally, the last integral

$$\mathbb{E}\int_0^T I_5^2 dt \leqslant T\mathbb{E}\sup_{t\in[0,T]} I_5^2 \leqslant CT\mathbb{E}\int_0^T \left\|\left(\mathcal{S}^\lambda(0,s)-\mathcal{S}^\infty(0,s)\right)g_R(u_\infty(s))\right\|_{H^\sigma}^2 ds \to 0$$

by the Burkholder inequality and the dominated convergence theorem, since $\|\ldots\|_{H^\sigma}^2 \leqslant CR^4$.

Repeating this argument iteratively on subintervals of $[0,T_0]$ of the size T one obtains that $u_\lambda \to u_\infty$ in $L^2(\Omega \times [0,T_0]; H^\sigma(\mathbb{R}))$.

Let us calculate each term in the Itô formula for $u = u_\lambda$. As we shall see the corresponding stochastic integral is not zero, and moreover, it is difficult to pass to the limit $\lambda \to \infty$ treating the stochastic part. So instead of $\mathcal{H}$ we consider at first a sequence $\mathcal{H}_n$, $n \in \mathbb{N}$, with the cubic term being cut off in the following way

$$\mathcal{H}_n(u) = \|u\|_{\mathcal{H}}^2 + \frac{1}{3}\theta_n\left(\|u\|_{\mathcal{H}}^2\right)\int u^3 dx$$

that clearly tends to $\mathcal{H}(u)$ almost surely at any fixed time moment. The corresponding Fréchet derivatives are defined by

$$\partial_u \mathcal{H}_n(u)\phi = \int_{\mathbb{R}}\left[\left(1+\frac{1}{3}\theta_n'\left(\|u\|_{\mathcal{H}}^2\right)\int u^3 dy\right)K^{-1/2}uK^{-1/2}\phi + \theta_n\left(\|u\|_{\mathcal{H}}^2\right)u^2\phi\right]dx,$$

$$\partial_u^2 \mathcal{H}_n(u)(\phi,\psi) = \int_{\mathbb{R}}\left[\left(1+\frac{1}{3}\theta_n'\left(\|u\|_{\mathcal{H}}^2\right)\int u^3 dx\right)K^{-1/2}\phi K^{-1/2}\psi + 2\theta_n\left(\|u\|_{\mathcal{H}}^2\right)u\phi\psi\right]dx$$

$$+\theta_n'\left(\|u\|_{\mathcal{H}}^2\right)\int u^2\phi dx\int K^{-1/2}uK^{-1/2}\psi dy$$

$$+\frac{1}{3}\theta_n''\left(\|u\|_{\mathcal{H}}^2\right)\int u^3 dx\int K^{-1/2}uK^{-1/2}\phi dy$$

$$\int K^{-1/2}uK^{-1/2}\psi dz$$

at every $\phi, \psi \in H^{\sigma_0}(\mathbb{R})$. Substituting it to the stochastic integral one obtains the following expression that can be simplified by integration by parts

$$\int_0^t \partial_u \mathcal{H}_n(u(s))\Phi(s)dW(s)$$

$$= \sum_j \gamma_j \int_0^t \int_{\mathbb{R}} \left[\left(1 + \frac{1}{3}\theta_n' \left(\|u\|_{\mathcal{H}}^2 \right) \int u^3 dy \right) K^{-1/2} u K^{-1/2} + \theta_n \left(\|u\|_{\mathcal{H}}^2 \right) u^2 \right]$$

$$\left(\partial_x P_\lambda u + \theta_R(\|u\|_{H^\circ})\partial_x K u^2 \right) dx dW_j = \sum_j \gamma_j \int_0^t \theta_n \left(\|u\|_{\mathcal{H}}^2 \right) \int_{\mathbb{R}} u^2 \partial_x P_\lambda u \, dx dW_j,$$

where $u = u_\lambda$. We will show that this integral tends to zero as $\lambda \to \infty$. That is exactly the place where we need the cut off θ_n. Applying some algebraic manipulations to the space integral and the Burkholder inequality to the stochastic integral, one deduces the estimate

$$\mathbb{E} \sup_{0 \leqslant t \leqslant T_0} \left| \int_0^t \partial_u \mathcal{H}_n(u(s))\Phi(s)dW(s) \right|^2$$

$$\leqslant C\mathbb{E} \int_0^{T_0} \theta_n^2 \left(\|u_\lambda(t)\|_{\mathcal{H}}^2 \right) \left(\int_{\mathbb{R}} u_\lambda^2(t)\partial_x(P_\lambda - 1)u_\lambda(t)dx \right)^2 dt$$

$$\leqslant C\mathbb{E} \int_0^{T_0} \theta_n^2 \left(\|u_\lambda(t)\|_{\mathcal{H}}^2 \right) \|u_\lambda(t)\|_{\mathcal{H}}^4$$

$$\left(\|(P_\lambda - 1)u_\infty(t)\|_{H^{1/2}}^2 + \|(P_\lambda - 1)(u_\lambda(t) - u_\infty(t))\|_{H^{1/2}}^2 \right) dt$$

$$\leqslant Cn^4\mathbb{E} \int_0^{T_0} \left(\|(P_\lambda - 1)u_\infty(t)\|_{H^{1/2}}^2 + \|(u_\lambda(t) - u_\infty(t))\|_{H^{1/2}}^2 \right) dt \to 0$$

as $\lambda \to 0$ for each fixed $n \in \mathbb{N}$. Note that the use of the functional $\mathcal{H}_n$ instead of $\mathcal{H}$ is important here. Similarly, we calculate the rest two terms in the Itô formula

$$\partial_u \mathcal{H}_n(u)\Phi + \frac{1}{2} \operatorname{tr} \partial_u^2 \mathcal{H}(u)(\Phi, \Phi)$$

$$= (\theta_R - \theta_n) \int u^2 \partial_x Ku \, dx + \theta_n \theta_R (\theta_R - 1) \sum_j \gamma_j^2 \int ug^2(u)dx$$

$$+ \frac{\theta_R(\theta_R - 1)}{2} \sum_j \gamma_j^2 \int g(u)K^{-1}g(u)dx$$

$$+ \frac{\theta_n}{2} \sum_j \gamma_j^2 \int \left(u^2 \partial_x^2 P_\lambda^2 u + 2u(\partial_x P_\lambda u)^2 \right) dx$$

$$+ \theta_n \theta_R \sum_j \gamma_j^2 \left(2 \int u(\partial_x P_\lambda u)g(u)dx - \int g(u)P_\lambda K^{-1}g(u)dx \right)$$

$$+ \frac{1}{3}\theta_R\theta_n' \int u^3 dy \left(\frac{\theta_R - 1}{2} \sum_j \gamma_j^2 \int g(u)K^{-1}g(u)dx - \int ug(u)dx \right)$$

$$= J_1 + \ldots + J_6,$$

where as above $u = u_\lambda$. One can prove that for a.e. $\omega \in \Omega$ and $t \in [0, T_0]$ the first three terms $J_1 + J_2 + J_3$ tend to the integrand of the right hand side of Expression (8) in the subsequent limits, firstly, as $\lambda \to \infty$ and then as $n \to \infty$. Both J_4 and J_5 tend to zero as $\lambda \to \infty$. Meanwhile the last term J_6 stays bounded by C/n, and so $\lim_{n\to\infty} \lim_{\lambda\to\infty} J_6 = 0$. Let us show, for example, that $J_4 \to 0$ which is the most troublesome term in the sum, since here is the only place in the paper where we make use of the fact $\sigma \geqslant 1$. The rest are treated similarly without this additional restriction. Indeed,

$$J_4 \leqslant C \left| \int (u\partial_x P_\lambda u - P_\lambda(u\partial_x u)) (P_\lambda - 1)\partial_x u dx \right|$$

$$\leqslant C \|u_\lambda\|_{H^1}^2 \left(\|(P_\lambda - 1)u_\infty\|_{H^1} + \|u_\lambda - u_\infty\|_{H^1} \right)$$

that obviously tends to zero as $\lambda \to \infty$. This concludes the proof. $\qquad\square$

At this stage one cannot claim the energy conservation yet, so we will prove a weaker result that will be sharpened later. Note that there exists $C_\mathcal{H} > 0$ such that

$$\|u\|_{\mathcal{H}}^2 (1 - C_\mathcal{H} \|u\|_{\mathcal{H}}) \leqslant \mathcal{H}(u) \leqslant \|u\|_{\mathcal{H}}^2 (1 + C_\mathcal{H} \|u\|_{\mathcal{H}}), \qquad (10)$$

following from the well-known embedding $H^{\sigma_0}(\mathbb{R}) \hookrightarrow L^\infty(\mathbb{R})$, recall that $\sigma_0 > 1/2$.

Lemma 1 *There exists a constant $T_1 > 0$ independent of ω such that if u solving Eq. (6) has $\|u\|_\mathcal{H} \leqslant \frac{1}{2C_\mathcal{H}}$ on some interval $[0, \tau]$ then $\mathcal{H}(u) \leqslant 2\mathcal{H}(u(0))$ on $[0, T_1 \wedge \tau]$.*

Proof At first one can notice that as long as $\|u\|_\mathcal{H}$ stays bounded by $(2C_\mathcal{H})^{-1}$, we have

$$\frac{1}{2} \|u\|_\mathcal{H}^2 \leqslant \mathcal{H}(u) \leqslant \frac{3}{2} \|u\|_\mathcal{H}^2.$$

Moreover, one can as well easily deduce from (8) the following bound

$$\mathcal{H}(u(t)) \leqslant \mathcal{H}(u(0)) + C \int_0^t \mathcal{H}(u(s))ds,$$

and so the proof is concluded by Grönwall's lemma. $\qquad\square$

3   Proof of the Main Result

We construct a solution u of (5) iteratively on the intervals $[0, T_1], [T_1, 2T_1]$ and so on. Here the interval size T_1 is defined by Lemma 1. Staying under the assumptions of Theorem 1, we denote by u_m solutions of Eq. (6) with $R = m \in \mathbb{N}$ given by Proposition 1, where we subsequently set $t_0 = 0, T_1, 2T_1, \ldots$. We define the stopping times

$$\tau_m = \tau_m^{t_0} = \inf\{t \in [t_0, T_0] : \|u_m(t)\|_{H^\sigma} > m\} \tag{11}$$

with the agreement $\inf \emptyset = T_0$. Starting with $t_0 = 0$ we firstly show the following result.

Lemma 2 *For a.e. $\omega \in \Omega$, any $m \in \mathbb{N}$ and each $t \in [0, \tau]$ with $\tau(\omega) = \min\{\tau_m(\omega), \tau_{m+1}(\omega)\}$, it holds true that $u_m(t) = u_{m+1}(t)$.*

Proof We define

$$\widetilde{u}_i(t) = \begin{cases} u_i(t) & \text{if } t \in [0, \tau] \\ S(t, \tau)u_i(\tau) & \text{if } t \in [\tau, T_0] \end{cases}, \quad i = m, m+1.$$

At first we will show that $\widetilde{u}_m$ and $\widetilde{u}_{m+1}$ coincide in X_T provided T is sufficiently small. Then we will finish the proof by an iteration procedure. The difference of these functions has the form

$$\widetilde{u}_{m+1}(t) - \widetilde{u}_m(t) = S(t, 0) \int_0^{t \wedge \tau} S(0, s) \left(f(\widetilde{u}_{m+1}(s)) - f(\widetilde{u}_m(s)) \right) ds$$

$$+ S(t, 0) \sum_j \gamma_j \int_0^{t \wedge \tau} S(0, s) \left(g(\widetilde{u}_{m+1}(s)) - g(\widetilde{u}_m(s)) \right) dW_j(s),$$

where the stochastic integral is estimated via

$$\mathbb{E} \sup_{0 \leqslant t \leqslant T} \left\| S_W(t, 0) \sum_j \gamma_j \int_0^t S(t-s) \chi_{\{s \leqslant \tau\}}(s) S_W(0, s) \left(g(\widetilde{u}_{m+1}(s)) - g(\widetilde{u}_m(s)) \right) dW_j(s) \right\|_{H^\sigma}^2$$

$$\leqslant C\mathbb{E} \int_0^T \chi_{\{s \leqslant \tau\}}(s) \left\| S_W(0, s) \left(g(\widetilde{u}_{m+1}(s)) - g(\widetilde{u}_m(s)) \right) \right\|_{H^\sigma}^2 ds$$

$$\leqslant C\mathbb{E} \int_0^T \chi_{\{s \leqslant \tau\}}(s) \left(\|\widetilde{u}_{m+1}(s)\| + \|\widetilde{u}_m(s)\|_{H^\sigma} \right)^2 \|\widetilde{u}_{m+1}(s) - \widetilde{u}_m(s)\|_{H^\sigma}^2 ds$$

$$\leqslant C(2m+1)^2 T \mathbb{E} \sup_{[0,T]} \|\widetilde{u}_{m+1} - \widetilde{u}_m\|_{H^\sigma}^2$$

with the help of the Burkholder inequality for convolution with the unitary group S, see [5, Lemma 3.3]. The first integral is estimated more straightforwardly, notice a similar argument employed to I in the proof of Proposition 1, and so one obtains

$$\|\widetilde{u}_{m+1} - \widetilde{u}_m\|_{X_T} \leqslant C(m)\sqrt{T}\,\|\widetilde{u}_{m+1} - \widetilde{u}_m\|_{X_T}.$$

Hence $\widetilde{u}_{m+1} = \widetilde{u}_m$ on $[0, T]$ for a.e. $\omega \in \Omega$ provided T is chosen sufficiently small depending only on m. Thus we can iterate this procedure to show that $\widetilde{u}_{m+1} = \widetilde{u}_m$ on the whole interval $[0, T_0]$, which concludes the proof of the lemma. $\qquad\square$

Our goal is to bound $\|u_m\|_{L^2 C(0,T_1;H^\sigma)}$ by a constant independent of $m \in \mathbb{N}$, and so we will need to estimate $\|f(u_m)\|_{H^\sigma}$, $\|g(u_m)\|_{H^\sigma}$, in particular. This can be easily done with the help of

$$\|\phi\psi\|_{H^\sigma} \leqslant C(\sigma, \sigma_0)\left(\|\phi\|_{H^\sigma}\|\psi\|_{H^{\sigma_0}} + \|\phi\|_{H^{\sigma_0}}\|\psi\|_{H^\sigma}\right)$$

being true for any $\sigma \geqslant 0$ and $\sigma_0 > 1/2$, see for example [7, Estimate (3.12)].

For a.e. $\omega \in \Omega$ and any $m \in \mathbb{N}$, $t \in [0, T_0]$ we have

$$\|u_m(t)\|_{H^\sigma} \leqslant \|u_0\|_{H^\sigma} + \int_0^t \|f(u_m(s))\|_{H^\sigma}\,ds + \left\|\sum_j \gamma_j \int_0^t S(0, s)g_m(u_m(s))dW_j(s)\right\|_{H^\sigma},$$

where $\|f(u_m(s))\|_{H^\sigma} \leqslant C\left(\|u_m(s)\|_{H^{\sigma_0}} + \|u_m(s)\|_{H^{\sigma_0}}^2\right)\|u_m(s)\|_{H^\sigma}$. Now taking into account that $\|S(0, s)g_m(u_m(s))\|_{H^\sigma} \leqslant C\|u_m(s)\|_{H^{\sigma_0}}\|u_m(s)\|_{H^\sigma}$, the stochastic integral can be estimated by the Burkholder inequality, and so we obtain for any $0 < T \leqslant T_0$ the following inequality

$$\mathbb{E}\sup_{t\in[0,T]}\|u_m(t)\|_{H^\sigma}^2 \leqslant 3\mathbb{E}\|u_0\|_{H^\sigma}^2 + C\mathbb{E}\int_0^T\left(\|u_m(t)\|_{H^{\sigma_0}}^2 + \|u_m(t)\|_{H^{\sigma_0}}^4\right)\|u_m(t)\|_{H^\sigma}^2\,dt, \tag{12}$$

where C depends only on σ_0, σ, T_0, $\sum_j \gamma_j^2$. This inequality we will use iteratively on the intervals $[0, T_0 \wedge kT_1]$, $k \in \mathbb{N}$, with T_1 found in Lemma 1. Let $\|u_0\|_{\mathcal{H}} \leqslant (5C_{\mathcal{H}})^{-1}$ a.e. on Ω. Consider the following stopping time

$$T_2^m = \inf\left\{t \in [0, T_0] : \|u_m(t)\|_{\mathcal{H}} > (2C_{\mathcal{H}})^{-1}\right\}.$$

Then a.e. $T_1 \leqslant T_2^m$. Indeed, assuming the contrary $T_1 > T_2^m$ one can deduce from (10) and Lemma 1 that

$$\left\|u_m(T_2^m)\right\|_{\mathcal{H}} \leqslant \sqrt{2\mathcal{H}(u_m(T_2^m))} \leqslant 2\sqrt{\mathcal{H}(u_0)} \leqslant 2\sqrt{1 + C_{\mathcal{H}}\|u_0\|_{\mathcal{H}}}\,\|u_0\|_{\mathcal{H}}$$

$$\leqslant \sqrt{\frac{24}{125}C_{\mathcal{H}}^{-1}} < (2C_{\mathcal{H}})^{-1},$$

which contradicts to the definition of the stopping time T_2^m due to continuity of $\|u_m\|_{\mathcal{H}}$. As a result $\|u_m\|_{\mathcal{H}}$ stays bounded by $(2C_{\mathcal{H}})^{-1}$ on the interval $[0, T_1]$ for a.e. ω, and this simplifies (12) in the following way

$$\mathbb{E} \sup_{t \in [0,T]} \|u_m(t)\|_{H^\sigma}^2 \leq 3\mathbb{E}\|u_0\|_{H^\sigma}^2 + C \int_0^T \mathbb{E} \sup_{s \in [0,t]} \|u_m(s)\|_{H^\sigma}^2 \, dt$$

holding true for any $0 < T \leq T_1$. Hence by Grönwall's lemma we obtain

$$\|u_m\|_{L^2C(0,T_1;H^\sigma)}^2 \leq 3\|u_0\|_{L^2 H^\sigma}^2 e^{CT_1} = M,$$

where M does not depend on $m \in \mathbb{N}$. Hence

$$\mathbb{P}(\tau_m \geq T_1) = \mathbb{P}\left(\|u_m\|_{C(0,T_1;H^\sigma)} \leq m\right) \geq 1 - \frac{1}{m^2}\mathbb{E}\|u_m\|_{C(0,T_1;H^\sigma)}^2 \geq 1 - \frac{M}{m^2},$$

and so $[0, T_1] \subset \cup_{m \in \mathbb{N}}[0, \tau_m(\omega)]$ for a.e. $\omega \in \Omega$. Thus we can define u on $[0, T_1]$ by assigning $u = u_m$ on $[0, \tau_m]$. This is obviously a solution of (5) on $[0, T_1]$ satisfying $d\mathcal{H}(u) = 0$ and $\|u\|_{\mathcal{H}} < (2C_{\mathcal{H}})^{-1}$ for a.e. $\omega \in \Omega$.

Now one can repeat the argument on $[T_1, 2T_1]$ by constructing new solutions u_m of Eq. (6) with the initial data $u(T_1)$ given at the time moment $t_0 = T_1$. The stopping times τ_m are defined by (11) with $t_0 = T_1$. The fact that $\|u_m\|_{\mathcal{H}}$ does not exceed the level $(2C_{\mathcal{H}})^{-1}$, is guaranteed by the energy conservation, namely by $\mathcal{H}(u(T_1)) = \mathcal{H}(u_0)$ in the same manner as above. The rest is similar, and so we get a solution on $[T_1, 2T_1]$ with the constant energy equalled $\mathcal{H}(u_0)$. After several repetitions of the argument we construct a solution on $[0, T_0]$.

It remains to prove the uniqueness. Let $u_1, u_2 \in L^2(\Omega; C(0, T_0; H^\sigma(\mathbb{R})))$ solve Eq. (5). For $R > 0$ we introduce

$$\tau_R = \inf\left\{t \in [0, T_0] : \max_{i=1,2} \|u_i(t)\|_{H^\sigma} > R\right\}.$$

Clearly, for a.e. $\omega \in \Omega$ both u_1 and u_2 are solutions of (6) on $[0, \tau_R]$. By Proposition 1 it holds true that $u_1 = u_2$ on $[0, \tau_R]$ for a.e. $\omega \in \Omega$. Taking $R \in \mathbb{N}$ and exploiting the time-continuity of u_1, u_2 one obtains $u_1 = u_2$ on $[0, \lim_{R \to \infty} \tau_R]$ for a.e. $\omega \in \Omega$. Now from sub-additivity and Chebyshev's inequality we deduce

$$\mathbb{P}(\tau_R \geq T_0) = \mathbb{P}\left(\max_{i=1,2} \|u_i\|_{C(0,T_0;H^\sigma)} \leq R\right)$$

$$\geq 1 - \frac{1}{R^2}\mathbb{E}\left(\|u_1\|_{C(0,T_0;H^\sigma)}^2 + \|u_2\|_{C(0,T_0;H^\sigma)}^2\right) \to 1$$

as $R \to \infty$, proving $u_1 = u_2$ on $[0, T_0]$. This concludes the proof of Theorem 1.

Acknowledgments The author is grateful to the members of STUOD team for fruitful discussions and numerous helpful comments. The author acknowledges the support of the ERC EU project 856408-STUOD.

References

1. CRAIG, W., AND GROVES, M. D. Hamiltonian long-wave approximations to the water-wave problem. *Wave Motion 19*, 4 (1994), 367–389.
2. DE BOUARD, A., AND DEBUSSCHE, A. A Stochastic Nonlinear Schrödinger Equation with Multiplicative Noise. *Communications in Mathematical Physics 205*, 1 (Aug. 1999), 161–181.
3. DE BOUARD, A., AND DEBUSSCHE, A. The Stochastic Nonlinear Schrödinger Equation in H^1. *Stochastic Analysis and Applications 21*, 1 (2003), 97–126.
4. DINVAY, E., AND MÉMIN, E. Hamiltonian formulation of the stochastic surface wave problem. *Proc. R. Soc. A. 478*, (2022), 20220050. http://doi.org/10.1098/rspa.2022.0050
5. GAWARECKI, L., AND MANDREKAR, V. *Stochastic Differential Equations in Infinite Dimensions.* Springer, Berlin, Heidelberg, 2011.
6. GRAFAKOS, L. *Modern Fourier Analysis*, vol. 250. Springer, 2009.
7. LINARES, F., AND PONCE, G. *Introduction to Nonlinear Dispersive Equations.* Universitext. Springer, New York, 2015.
8. MÉMIN, E. Fluid flow dynamics under location uncertainty. *Geophysical & Astrophysical Fluid Dynamics 108*, 2 (2014), 119–146.

Observation-Based Noise Calibration: An Efficient Dynamics for the Ensemble Kalman Filter

Benjamin Dufée, Etienne Mémin, and Dan Crisan

Abstract We investigate the calibration of the stochastic noise in order to guide the realizations towards the observational data used for the assimilation. This is done in the context of the stochastic parametrization under Location Uncertainty (LU) and data assimilation. The new methodology is rigorously justified by the use of the Girsanov theorem, and yields significant improvements in the experiments carried out on the Surface Quasi Geostrophic (SQG) model, when applied to Ensemble Kalman filters. The particular test case studied here shows improvements of the peak MSE from 85% to 93%.

Keywords Stochastic parametrization · Modeling under location uncertainty · noise calibration · Ensemble Kalman filters · Square root filters

1 Introduction

Sequential data assimilation uses observational data to correct a set of realizations given by a numerical model. In the case of both high-dimensional data and model, the data assimilation methodology can be facilitated via a procedure allowing to guide the realizations towards the available observations. This is particularly helpful in high dimensions as it enables the ensemble to focus on a restricted set of the state space. That is what we intend to put forward in this paper. This work relies on a stochastic parametrization of the underlying dynamical system based on the Location Uncertainty (LU) principles, which rely on a decomposition of the Lagrangian velocity into a large-scale smooth component and a random time-uncorrelated component. In this setting, a stochastic transport operator plays the

B. Dufée (✉) · E. Mémin
Inria/Irmar, Fluminance, Campus universitaire de Beaulieu, Rennes Cedex, France
e-mail: benjamin.dufee@inria.fr; etienne.memin@inria.fr

D. Crisan
Department of Mathematics, Imperial College, London, UK
e-mail: d.crisan@imperial.ac.uk

© The Author(s) 2023
B. Chapron et al. (eds.), *Stochastic Transport in Upper Ocean Dynamics*,
Mathematics of Planet Earth 10, https://doi.org/10.1007/978-3-031-18988-3_4

role of the usual material derivative, see [1] for more details. This work aims at adding the feature of a noise specifically calibrated to play a guiding role for the realizations.

In a previous data assimilation study on the Surface Quasi Geostrophic (SQG) model, the stochastic forecast was shown to provide better results than deterministic techniques like variance inflation with perturbation on the initial condition, see [2] for details. The current study is a continuation of [2]. The noise calibration presented here further improves the results presented in [2], particularly when the system starts from poor or badly estimated initial conditions (for instance resulting from initial estimations relying on regularized inverse problems). For such initial conditions, which are generally too smooth and inaccurate, classical ensemble methods are likely to be put in difficulties. In this short paper, we will first briefly recall the principles of Location Uncertainty and how it applies to the SQG model. Then we will detail the procedure leading to the noise calibration, and finally detail and assess the numerical experiments performed.

2 The Stochastic SQG Model Under Location Uncertainty (LU)

The analysis in this paper is carried out on the 2D Surface Quasi-Geostrophic (SQG) model. The SQG equations model an idealized dynamics for surface oceanic currents. It involves many realistic non-linear features such as fronts or strong multiscale eddies (see [3, 4] for details). The deterministic SQG model couples a transport equation of the buoyancy field b, a kinematic condition and a 2D divergence-free constraint:

$$D_t b = 0 \; ; \; b = \frac{N_{strat}}{f_0} (-\Delta)^{\frac{1}{2}} \psi \; ; \; v = \nabla^{\perp} \psi, \tag{1}$$

expressed on ψ the stream function and v the velocity, where D_t is the material derivative. The kinematic condition depends on the stratification N_{strat} and the Coriolis frequency f_0.

The corresponding stochastic dynamics is derived from the Location Uncertainty (LU) principles described in [1]. The full description and numerical analysis of the LU-SQG model can be found in [5, 6]. This stochastic formalism models the impact of the small scales on the flow component that is initially smooth in time. It relies on the decomposition of the Lagrangian velocity of a fluid particle positioned at x_t in a spatial domain $\Omega \subset \mathbb{R}^2$:

$$dx_t = v(x_t, t)dt + \sigma(x_t, t)dB_t, \tag{2}$$

in terms of a resolved component v (referred to as the large-scale component in the following) and σdB_t, an unresolved highly oscillating random component, built from a (cylindrical) Wiener process B_t (ie a well-defined Brownian motion taking values in a functional space) [7]. The increments of the latter component are time-independent. Due to the lack of smoothness of the solution x_t, we rigorously derive (2) in its integral form.

The random perturbation of velocity is Gaussian and has the following distribution:

$$\sigma dB_t \sim \mathcal{N}(0, Q dt), \tag{3}$$

where Q is the covariance operator. This operator admits an orthonormal eigenfunction basis $\{\phi_n(\cdot, t)\}_{n \in N}$ with non-negative eigenvalues $(\lambda_n(t))_{n \in N}$. This generates a convenient spectral definition of the noise as

$$\sigma(x, t) dB_t = \sum_{n \in N} \sqrt{\lambda_n(t)} \phi_n(x, t) d\beta_t^n, \tag{4}$$

where the β^n are i.i.d standard one dimensional Brownian motions. From Eq. (4), the noise variance tensor a is then defined by

$$a(x, t) = \sum_{n \in N} \lambda_n(t) \phi_n(x, t) \phi_n(x, t)^T. \tag{5}$$

It can be noticed the variance tensor has the physical dimension of a viscosity (ie m^2/s). Indeed, as σdB_t is a distance, then $a(x, t) dt = \mathbb{E}[\sigma dB_t (\sigma dB_t)^T]$ is a squared distance. The procedure used to generate the orthonormal basis functions determines the spatial structure of the noise. The one used in our experiments will be presented later in this section.

While a deterministically transported tracer Θ has zero material derivative: $D_t \Theta = \partial_t \Theta + v \cdot \nabla \Theta = 0$, in the LU framework, a stochastically transported tracer cancels a related stochastic transport operator defined as:

$$\mathbf{D}_t \Theta := d_t \Theta + (v^* dt + \sigma dB_t) \cdot \nabla \Theta - \frac{1}{2} \nabla \cdot (a \nabla \Theta) dt, \tag{6}$$

where

$$d_t \Theta := \Theta(x, t + dt) - \Theta(x, t) \tag{7}$$

is the infinitesimal forward time increment of the tracer. The effective advection velocity is defined by

$$v^* = v - \frac{1}{2} \nabla \cdot a, \tag{8}$$

the term $\sigma dB_t \cdot \nabla \Theta$ is a non-Gaussian multiplicative noise corresponding to the tracer's transport by the small-scale flow, and the last term in (6) is a diffusion term, as the variance tensor a is definite positive. The expression of the stochastic transport operator comes from a generalized Itô formula (Itô-Wentzell formula), see [5] for more details.

The stochastic version of the SQG model is obtained by replacing the material derivative $D_t b$ in Eq. (1) with the stochastic transport operator $\mathbf{D}_t b$:

$$\mathbf{D}_t b = d_t b + (v^* dt + \sigma dB_t) \cdot \nabla b - \frac{1}{2} \nabla \cdot (a \nabla b) dt = 0, \qquad (9)$$

and an additional compressibility constraint on the noise:

$$\nabla \cdot \sigma dB_t = 0. \qquad (10)$$

In the case of a compressible random field, the modified advection incorporates an additional term in Eq. (8) related to the noise divergence [5]. One essential property of LU (for a divergence-free noise component) is the conservation of energy for the transported random tracer, under the same ideal boundary conditions as in the deterministic case:

$$d \int_\Omega \Theta^2(x) dx = 0, \qquad (11)$$

and, very importantly, this energy conservation property holds pathwise (i.e for any realization of the Brownian noise), see [5, 8] for details. This property highlights the strong relation between the LU-SQG version and the deterministic one.

Noise Generation The method used to generate the noise in this study relies on a data-driven method called proper orthogonal decomposition (POD) to estimate the empirical orthogonal functions in the spectral representation of Eq. (4). By a slight abuse of notation in the following, this noise will be referred to as POD noise. We give some brief details in what follows.

Considering a series of snapshots of the velocity field, this method consists in the computation of the covariance tensor around the temporal mean of the series of snapshots. Then its eigenvectors and eigenfunctions can be estimated in order to reconstruct the large-scale variability (the first "modes" or eigenfunctions), and the small-scale one (the smaller modes). In practice, this procedure is applied to coarse-grained high-resolution snapshots of deterministic simulations. The latter modes will be the ones on which the noise is decomposed. These modes are divergence-free and stationary by construction, so the global structure of the noise will not vary in time. In case of chaotic geophysical models like this one, we can also use online-computed noises as the one used in our previous work [2] which have much better uncertainty quantification, but are also much more expensive. An extension of this work to this noise is currently at work. We refer to [6] for a precise description of this procedure.

3 Girsanov Theorem and Noise Calibration

3.1 Change of Measure

Ensemble-based sequential data assimilation filters are composed of a forecasting step of the ensemble to provide a sampling of the forecast distribution, and an analysis step correcting the departure from the observations. The purpose of the proposed noise calibration is to modify the forecast distribution, taking into account the upcoming observation, in order to guide the forecast towards it. In the context of transport equations such as in the SQG model, this extra guiding term is an added drift in the noise σdB_t, which was initially built to have zero mean. Allowing σdB_t to have a non-zero mean entails a modification of the transport equation in order to rewrite it in terms of a centered noise. This is called the Girsanov transform, and it consists in a change of underlying measure so that a non-centered noise becomes centered under a new probability measure, up to a drift term accounting for this change of measure. For now, σdB_t is defined on a probability space $(\Omega, \mathcal{F}, P)$ and we define $(\mathcal{F}_t)_t$ the filtration adapted to σdB_t.

The Girsanov theorem (see [7] for details) states that if $(Y_t)_{0 \leq t \leq T}$ is a stochastic process such that:

- $(Y_t)_{0 \leq t \leq T}$ is adapted with respect to the Wiener filtration $(\mathcal{F}_t)_{0 \leq t \leq T}$.
- For the current probability measure P, we have, P-almost surely,

$$\int_0^T Y_t^2 dt < \infty.$$

- The process $(Z_t)_{0 \leq t \leq T}$ defined by

$$Z_t = \exp\left(\int_0^t Y_s dB_s - \frac{1}{2}\int_0^t Y_s^2 ds\right) \tag{12}$$

is a $\mathcal{F}_t$-martingale,

then there exists a probability measure $\tilde{P}$ under which:

- The process $(\tilde{B}_t)_{0 \leq t \leq T}$ defined by

$$\tilde{B}_t = B_t - \int_0^t Y_s ds \tag{13}$$

is a standard cylindrical Wiener process.
- The Radon-Nikodym derivative of $\tilde{P}$ with respect to P is Z_T.

Let us denote by $(\Gamma_t)_{0 \leq t \leq T}$ the drift we intend to add to the noise. With such a change of measure, let us see how Eq. (9) is modified. According to Eq. (13), we have

$$dB_t = d\tilde{B}_t + \Gamma_t dt, \tag{14}$$

so the stochastic transport operator rewrites

$$D_t b = d_t b + (v^* dt + \sigma [d\tilde{B}_t + \Gamma_t dt]) \cdot \nabla b - \frac{1}{2}\nabla \cdot (a\nabla b)dt \tag{15a}$$

$$= d_t b + (v^* dt + v_\Gamma dt + \sigma d\tilde{B}_t) \cdot \nabla b - \frac{1}{2}\nabla \cdot (a\nabla b)dt, \tag{15b}$$

where

$$v_\Gamma = \sum_{k=1}^{K} \gamma_k \phi_k \tag{16}$$

is the velocity drift entailed by the Girsanov transform and we assume that $\Gamma_t = \Gamma = (\gamma_1, \ldots, \gamma_K)$ is constant on a small time step dt, which will be the case for the discretized numerical scheme that we use.

As a result, under the probability measure $\tilde{P}$, (15) presents the same form as Eq. (9) since $\tilde{B}$ is indeed a centered cylindrical Wiener process under $\tilde{P}$, but with an added drifted advection velocity.

3.2   Computation of the Girsanov Drift

We now describe how to compute Γ in order to guide the forecast towards the next observation.

Let us start from a given time t_1 where a complete buoyancy and velocity field is available. The next observation $b^{obs}(\cdot, t_2)$ is assumed to be available at time t_2 and L numerical time steps are performed until then $(t_2 - t_1 = L\delta_t$, where δ_t is the time discretization step).

At time t_1, a rough prediction of the velocity at time t_2 can be estimated with the current velocity (which, more precisely, comes from previous stochastic iterations, but is $\mathcal{F}_{t_1}$-measurable), namely

$$b^{obs}(x + v(x, t_1)L\delta_t, t_2) := \tilde{b}(x, t_2), \tag{17}$$

that stands for the backward-registered observation with respect to the current deterministic velocity. This way the error made is

$$\Delta_t \tilde{b}(x) = \tilde{b}(x, t_2) - b(x, t_1). \tag{18}$$

So $\tilde{b}(x, t_2)$ is a value taken in a modified observation field, because b^{obs} is advected by the current velocity $v(\cdot, t_1)$. For this reason we consider that the backward-registered observation used for the calibration does not have the same nature as the raw observation used for data assimilation. It constitutes a pseudo-observation, for which we can consider that the error due to the imprecision of the backward-registration (ensuing in particular from successive bilinear interpolations) is way bigger than the observation noise, and almost uncorrelated to the latter. In the second case, only the raw observation is used for the Kalman filter, corresponding only to the observation noise. The aim is now to calibrate the current velocity by adding a Girsanov drift $v_\Gamma = \sum_{k=1}^{K} \gamma_k \phi_k$, such that the solution of the following transport equation

$$b\left(x + v(x, t_1)L\delta_t + v_\Gamma L\delta_t + \sum_{k=1}^{K}(\sqrt{\delta_t}\phi_k)(\sqrt{L\delta_t}\beta_k), t_2\right) = b(x, t_1). \qquad (19)$$

is approximated in a least square sense. In other words, we solve the following minimization problem:

$$\min_\Gamma \int_\Omega \mathbb{E}\left[b\left(x + v(x, t_1)L\delta_t + v_\Gamma L\delta_t \right.\right.$$
$$\left.\left. + \sum_{k=1}^{K}(\sqrt{\delta_t}\phi_k)(\sqrt{L\delta_t}\beta_k), t_2\right) - b(x, t_1)\right]^2 dx. \qquad (20)$$

This can be rewritten as

$$\min_\Gamma \int_\Omega \left[\Delta_t \tilde{b} + \nabla \tilde{b} \cdot v_\Gamma L\delta_t - \frac{1}{2}\nabla \tilde{b} \cdot \nabla a L\delta_t - \frac{1}{2}\nabla \cdot (a\nabla \tilde{b})L\delta_t\right]^2 dx.$$

Using the identities

$$\nabla \cdot a = \sum_{k=1}^{K}(\phi_k \cdot \nabla)\phi_k \; ; \; \nabla \cdot (a\nabla b) = \sum_{k=1}^{K}(\phi_k \cdot \nabla)(\phi_k \cdot \nabla b), \qquad (21)$$

we rewrite the minimization problem as

$$\min_\Gamma \int_\Omega \left[\Delta_t \tilde{b} + \nabla \tilde{b} \cdot \left(\sum_{k=1}^{K}\gamma_k \phi_k\right)L\delta_t - \frac{1}{2}\sum_{k=1}^{K}(\nabla \tilde{b} \cdot F_k + G_k(\tilde{b}))L\delta_t\right]^2 dx \qquad (22)$$

where

$$F_k = (\phi_k \cdot \nabla)\phi_k \; ; \; G_k(\tilde{b}) = (\phi_k \cdot \nabla)(\phi_k \cdot \nabla \tilde{b}).$$

Denoting by J the integrand, we have

$$\frac{\partial J}{\partial \gamma_i} = 2 \int_{\Omega} (\nabla \tilde{b} \cdot \phi_i) L \delta_t \left[\Delta_t \tilde{b} + \nabla \tilde{b} \cdot \left(\sum_{k=1}^{K} \gamma_k \phi_k \right) L \delta_t \right.$$

$$\left. - \frac{1}{2} \sum_{k=1}^{K} (\nabla \tilde{b} \cdot F_k + G_k(\tilde{b})) L \delta_t \right] dx. \qquad (23)$$

Finally, we add a regularization term $\alpha \|v_\Gamma\|_2^2 = \alpha \sum_{k=1}^{K} \gamma_k^2 \lambda_k$, where λ_k is the eigenvalue of the Q-eigenfunction ϕ_k in Eq. (22) to ensure the uniqueness of the solution of the proposed minimization problem, where α needs to be tuned properly. As a result, the minimization problem can be written as an inverse problem

$$A\Gamma = c \qquad (24)$$

where

$$A_{ik} := 2 \int_{\Omega} (\nabla \tilde{b} \cdot \phi_i)(\nabla \tilde{b} \cdot \phi_k) + 2\alpha \lambda_k \delta_{ik} \qquad (25a)$$

$$c_i := \int_{\Omega} (\nabla \tilde{b} \cdot \phi_i) \left[2\Delta_t \tilde{b} - \sum_{k=1}^{K} (\nabla \tilde{b} \cdot F_k + G_k(\tilde{b})) \right] dx. \qquad (25b)$$

The parameter α is a priori fixed in order to control the resulting euclidian norm of v_Γ, $\|v_\Gamma\|_2$. Large values of α lead to very small corrections (Γ tends to $(0, \ldots, 0)$ when α goes to $+\infty$) whereas small values yield very strong and noisy drifts, as we get closer to an ill-posed problem. For now, we use an empirical iterative way to tune α, we increase it until the resulting norm of v_Γ is under a given threshold.

4   Experiments

This section details the numerical experiments carried out in this work. The goal is to study the benefits brought by a noise-calibrated forecast in an up-to-date version of a localized ensemble Kalman filter. In particular we wish to observe whether or not the noise calibration brings by itself an efficient and practical improvement of the assimilation step.

Ensemble Kalman filters (see e.g. [9] for details) constitute a well-known family of data assimilation methods. They rely on an ensemble of realizations (called ensemble members) of a dynamical system $(x_n^f)_{n=1,\ldots,N}$ coming from the forecast step, and give as an output another set of members $(x_n^a)_{n=1,\ldots,N}$. Each posterior ensemble member x_n^a is obtained as a linear combination of the prior ensemble members $(x_n^f)_{n=1,\ldots,N}$ in order to minimize the distance between the ensemble and the observation in some sense.

One important assumption of the classical EnKF is to consider that the observation and model noise are uncorrelated. This observation-calibrated forecast could imply that the latter assumption no longer holds. Still, the discussion following Eq. (18) on the observation nature explains why we can consider the uncorrelation between the forecast and observation noise. If this assumption appears to be not valid, we refer to the work made in [10] to rigorously justify the introduction of an observation-dependent forecast. In this work, both Kalman and particle filter equations were rewritten in terms of the conditional expectation with respect to the underlying sequence of current and past observations. The stochastic simulations are run on a double-periodic simulation grid, G_s, of size 64×64 points and of physical size $1000\,\text{km} \times 1000\,\text{km}$, meaning that two neighbor points are approximately $15\,\text{km}$ apart. An observation is assumed to be available every day (i.e. every 600 time steps of the dynamics) on a coarser observation grid, G_o, which is a subset of G_s of size 16×16. It is generated as follows: a trajectory of buoyancy $(z_t)_t$ is run from the deterministic model (PDE) at a very fine resolution grid G_f, of size 512×512. Then a convolution-decimation procedure D is applied in order to fit to the targeted simulation grid G_s. It consists in the composition of a Gaussian filter and a decimation operator subsampling one pixel out of two. It has to be iterated three times in our case to fit the correct resolution. This is done in order to respect Shannon's theorem and to avoid spectrum folding. A projection operator P is applied from G_s to G_o, and we finally add an observation noise to get the observation

$$b^{obs}(\cdot, t) = P \circ D(z_t) + \eta_t \; ; \; \eta_t \sim \mathcal{N}(\mathbf{0}, R) \text{ and } R = r^2 I_M, \tag{26}$$

where R is the diagonal observation covariance matrix and M is the number of points on the observation grid.

Numerical Setup The simulations have been performed with a pseudo-spectral code in space (see [6] for details). The time-scheme is a fourth-order Runge-Kutta scheme for the deterministic PDE, and an Euler-Maruyama scheme for the SPDEs. We use a standard hyperviscosity model to dissipate the energy at the resolution cut-off with a hyperviscosity coefficient $\beta = (5 \times 10^{29}\,\text{m}^8.\text{s}^{-1})M_x^{-8}$, where M_x is the grid resolution [6].

The test case considered in this study is the following: an ensemble of $N = 100$ ensemble members is started from the very same initial condition at day 0, which consists in two cold vortices to the north and two warm vortices to the south. However, the amplitude of the initial vortices is underestimated compared to the initial condition used for the deterministic run (considered as the truth) by 20%, as shown in Fig. 1. We refer to [2] for a mathematical expression of this field.

In this experiment, we study the differences of efficiency of the localized Ensemble Square Root Filter (an up-to-date version of the Ensemble Kalman filter, see for instance [11] for details of the square root filters (ESRF) and [12] for a description of the observation covariance localization procedure) with both noise-calibrated forecast and classical stochastic simulations. We also refer to [13] for the

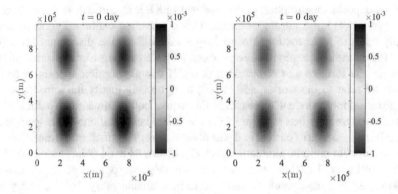

Fig. 1 Initial conditions for the truth (on the left) and for each stochastic run (on the right, common to all ensemble members). We enforce an underestimation of the amplitude of the initial vortices of 20%

extension of the square root filter for additive forecast noise based on covariance transformation, where the advantages of additional model error in the forecast step are shown.

In both cases, starting from the underestimated initial condition, the stochastic dynamics is simulated using the POD noise with $K = 10$ modes. An observation is provided each day (i.e. every 600 time steps of the SPDE), with an observation error covariance set to $r = 10^{-5}$ in (26), which corresponds to a weak (but not negligible, 1% of the maximum amplitude in the initial buoyancy field) noise on the observation. The localization radius is set to l_{obs} here, where $l_{obs} \simeq 60\,\text{km}$ denotes the distance between two neighboring observational sites, as it provided the best results for both cases.

The typical behaviour of the vortices, at least at the beginning of the simulation, is to spin with no translation of the cores. In our case, the true vortices will spin much faster than those in the biased stochastic runs. The goal of calibration is then to speed these vortices up in order to get them closer to the truth.

The forecast is calibrated at each time step of the SPDE, using the upcoming observation to do it. Multiple parameters were tried for the regularization parameter α, or alternatively for the upper bound allowed for the L^2-norm of the Girsanov drift v_Γ. Figure 2 compares the MSE along time for all the range of parameters tested here, with also the same experiment without noise calibration. For this latter, the LESRF has a difficult task, as it tries to find linear combinations of the prior ensemble members, which all have an underestimated velocity, to get closer to the observation. This is a general issue for ensemble methods (as well as for particle filters), which are not able and designed to correct the bias if this correction is not made in the forecast. By contrast, the LU calibration offers an additional degree of freedom to guide the ensemble towards the observation. This procedure significantly improves the results in terms of MSE. At day 13, when the MSE is maximal for the usual case, we observe an improvement from 85% to 93% depending on the

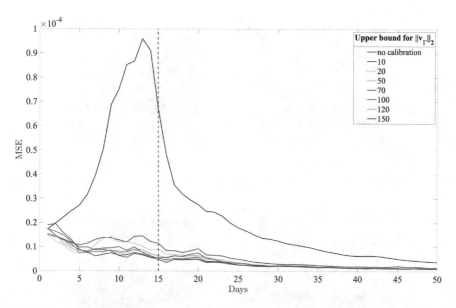

Fig. 2 Comparison of MSE along time between the non calibrated forecast (in black) and all the different parameters tested here for the noise calibration. The snapshots shown in Fig. 3 are taken at day 15 (black dashed line)

parameters tested. The case of the underestimation is an example, but we expect this procedure to be efficient in any situation in which all ensemble members have a similar problem of bias, bad amplitude estimation, artefacts, unsymmetrical features, etc. With a reasonably small ensemble size, which is generally the case in practice, this is likely to occur if the initial conditions have such features.

As explained previously, the regularization term α controls the amplitude of the allowed correction drift. In our experiments, all parameters tested yield significant improvements compared to the classical case, still a good trade-off seems to be found with a control of $||v_\Gamma||_2$ between 70 and 150. Starting from 150, we observe higher MSE in the very first days, certainly due to a lack of constraint on the inverse problem. In addition to the MSE results, we show in Fig. 3 a more visual example of what calibration does. At day 15, the configuration of the truth is that all four vortices are horizontal. Without calibration (first row), the vortices are slanted because of the initial underestimation of the velocity. The velocity field has not been properly corrected. On the other hand, the LU calibration offers a more reliable prediction, as we recovered the global shape of the vortices, with additional spread around the mean.

Finally, we show in Fig. 4 an insight of how the Girsanov correction v_Γ behaves in time. As the structure of the noise is stationary, so is the structure of v_Γ because it relies on the same modes as the noise. What is interesting is the evolution of the amplitude of this field, which decreases in time, meaning that most of the calibration

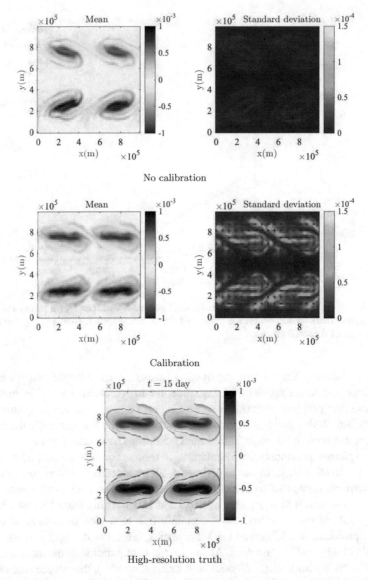

Fig. 3 Comparison between the ensemble mean (left) and the ensemble standard deviation (right) maps, with and without calibration, at day 15 with the high-resolution truth

work is done in the very first days of simulation, and once the forecast manages to get closer to the truth, the need for calibration is less crucial and the Girsanov correction gets weaker.

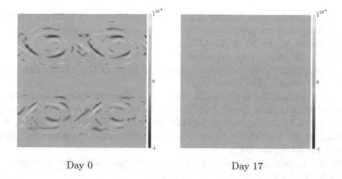

Day 0   Day 17

Fig. 4 Vorticity of the Girsanov drift v_Γ computed for one ensemble member at the first time step after the initial condition (left) and at the first time step after day 17 (right)

5 Conclusion

The findings of this paper show the ability of a data-driven noise calibration procedure to improve significantly the assimilation by EnKF of a system initialized with an underestimated initial condition.

As already mentioned in Sect. 2, we intend to extend this setting to non-stationary noises, as they were shown to be associated to a better quantification of the uncertainty (see [6] for details). Regarding computational effort, the calibration procedure is intrinsically paralellizable ensemble-wise, and the techniques used are close to optical flow estimation procedures, for which efficient solutions exist. The tuning step of α is the more expensive step for now, for which more sophisticated methods could be envisaged.

References

1. E. Mémin. Fluid flow dynamics under location uncertainty. *Geophysical & Astrophysical Fluid Dynamics*, 108(2):119–146, 2014.
2. B. Dufée, E. Mémin and D. Crisan Stochastic parametrization: an alternative to inflation in EnKF. *Quarterly Journal of the Royal Meteorological Society*, doi:10.1002/qj.4247 2022
3. P. Constantin, Q. Nie, and N. Schörghofer. Front formation in an active scalar equation. *Physical Review E*, 60(3):2858, 1999.
4. G. Lapeyre and P. Klein. Dynamics of the upper oceanic layers in terms of surface quasigeostrophy theory. *Journal of physical oceanography*, 36(2):165–176, 2006.
5. V. Resseguier, E. Mémin, and B. Chapron. Geophysical flows under location uncertainty, Part I Random transport and general models. *Geophys. & Astro. Fluid Dyn.*, 111(3):149–176, 2017a.
6. V. Resseguier, L. Li, G. Jouan, P. Derian, E. Mémin, and B. Chapron. New trends in ensemble forecast strategy: uncertainty quantification for coarse-grid computational fluid dynamics. *Archives of Computational Methods in Engineering*, pages 1886–1784, 2020a.
7. G. Da Prato and J. Zabczyk. *Stochastic equations in infinite dimensions*. Cambridge University Press, 1992.

8. W. Bauer, P. Chandramouli, B. Chapron, L. Li, and E. Mémin. Deciphering the role of small-scale inhomogeneity on geophysical flow structuration: a stochastic approach. *Journal of Physical Oceanography*, 50(4):983–1003, 2020a.

9. G. Evensen. Sequential data assimilation with a nonlinear quasi-geostrophic model using monte carlo methods to forecast error statistics. *Journal of Geophysical Research: Oceans*, 99(C5):10143–10162, 1994.

10. E. Arnaud, E. Mémin, and B. Cernuschi. Conditional Filters for Image Sequence Based Tracking – Application to Point Tracking *IEEE transactions on image processing : a publication of the IEEE Signal Processing Society*, 14(1):63–79, doi:10.1109/TIP.2004.838707, 2005

11. J.S. Whitaker and T.M. Hamill. Ensemble data assimilation without perturbed observations. *Monthly Weather Review*, 2002.

12. P. Sakov and L. Bertino. Relation between two common localisation methods for the enkf. *Computational Geosciences*, 15(2):225–237, 2011.

13. P.N. Raanes, A. Carrassi, and L. Bertino. Extending the Square Root Method to Account for Additive Forecast Noise in Ensemble Methods *Monthly Weather Review*, 143(10):3857–3873, 2015

A Two-Step Numerical Scheme in Time for Surface Quasi Geostrophic Equations Under Location Uncertainty

Camilla Fiorini, Pierre-Marie Boulvard, Long Li, and Etienne Mémin

Abstract In this work we consider the surface quasi-geostrophic (SQG) system under location uncertainty (LU) and propose a Milstein-type scheme for these equations, which is then used in a multi-step method. The SQG system considered here consists of one stochastic partial differential equation, which models the stochastic transport of the buoyancy, and a linear operator linking the velocity and the buoyancy. In the LU setting, the Euler-Maruyama scheme converges with weak order 1 and strong order 0.5. Our aim is to develop higher order schemes in time, based on a Milstein-type scheme in a multi-step framework. First we compared different kinds of Milstein schemes. The scheme with the best performance is then included in the two-step scheme. Finally, we show how our two-step scheme decreases the error in comparison to other multi-step schemes.

1 Introduction

The main aim of the modelling under location uncertainty (LU) consists in simulating on coarse meshes an enriched system mimicking a high resolution deterministic chaotic dynamics. Such LU models allow one to recover phenomena such as backscattering, dissipation and reorganisation on very coarse meshes. Furthermore, it provides a natural framework for uncertainty quantification analysis [14]. The LU framework, first introduced in [11], is based on the decomposition of the Lagrangian velocity into two components: a large-scale smooth component and

C. Fiorini (✉)
Laboratoire M2N, Conservatoire National des Arts et Métiers, Paris, France
e-mail: camilla.fiorini@lecnam.net

P.-M. Boulvard
Inria Paris, Equipe ANGE, Paris, France

Inria Rennes - Bretagne Atlantique, Équipe FLUMINANCE, Rennes, France

L. Li · E. Mémin
Inria Rennes - Bretagne Atlantique, Équipe FLUMINANCE, Rennes, France

© The Author(s) 2023
B. Chapron et al. (eds.), *Stochastic Transport in Upper Ocean Dynamics*,
Mathematics of Planet Earth 10, https://doi.org/10.1007/978-3-031-18988-3_5

a small-scale fast oscillating one. This decomposition leads to a stochastic transport operator, and one can, in turn, develop the stochastic version of classical fluid-dynamics systems derived from the Navier–Stokes equations. SQG in particular consists of one stochastic partial differential equation (SPDE), which models the stochastic transport of the buoyancy, and a linear operator relating the velocity and the buoyancy:

$$
\begin{cases}
db_t = \frac{1}{2}\nabla \cdot (a\nabla b_t)dt - v^* \cdot \nabla b_t dt - \nabla b_t \cdot \sigma d\mathbf{B}_t, \\
b_t = N(-\Delta)^{1/2}\psi, \\
\mathbf{u} = \nabla^\perp \psi,
\end{cases}
\tag{1}
$$

where b_t is the buoyancy at time t, $\mathbf{u}$ the large-scale smooth velocity, N a constant depending on the vertical oscillation frequency of the buoyancy and a Coriolis parameter, $\mathbf{B}$ a Wiener process, ψ the stream function and $v^* = u - \frac{1}{2}\nabla \cdot a + \sigma\nabla \cdot \sigma$ is a corrected velocity associated with the effect of the noise inhomogeneity on the advected variables. The spatial correlations of the noise are given through an integral kernel operator σ (here assumed deterministic and symmetric for sake of simplicity), and the variance matrix, a, given by the matrix kernel of the operator $\sigma\sigma$ provides a local measure of the noise strength. For more details on the derivation of this system, see [10, 13]. In the rest of this work we will mainly focus on the first equation, and the last two will be condensed in $\mathbf{u} = \mathcal{H}(b)$. Concerning the modelling of the noise, we use the equivalent convenient spectral definition:

$$
\sigma d\mathbf{B}_t = \sum_m \varphi^m d\beta_t^m,
$$

where $\beta^m = \beta^m(t)$ are independent one-dimensional standard Brownian motions and $\varphi^m = [\varphi_x^m, \varphi_y^m]^T (\mathbf{x})$ are basis functions. The number of terms involved in the sum is in theory infinite, but in numerical application a truncation is considered. In the definition of the numerical schemes we will thus assume that it is a finite sum. For the computation of the basis functions, two strategies are possible: an offline strategy, where they are defined from the eigenfunctions of an empirical covariance tensor built from high-resolution data as described in [10, 13]; of strategies, where the functions are updated during the simulation and in this case they are a function of the buoyancy b. With this representation, the variance tensor reads:

$$
a = \sum_m \varphi^m (\varphi^m)^T.
$$

2   Numerical Schemes

In this section we derive a two-step numerical scheme in time for the SQG system under LU (SQG-LU). We compare this scheme to other multi-step schemes for the SPDE, in particular the ones developed in [5] and [4], and show how our scheme improves the precision. Concerning discretisation in space, standard spectral methods are used: the linear terms are treated in the Fourier space, whilst the nonlinear terms are discretised in the physical space.

The derivation of the time scheme consists of two steps: first, we derive a class of Milstein schemes for SQG-LU and we empirically verify their convergence, then a two-step scheme is proposed.

2.1 Derivation of a Milstein Scheme

To design the Milstein schemes, we consider the integral form of the SPDE in (1), namely

$$b_t = b_{t_0} + \int_{t_0}^{t} \left(\frac{1}{2} \nabla \cdot (a \nabla b_s) - v^* \cdot \nabla b_s \right) ds - \int_{t_0}^{t} \sum_m \nabla b_s \cdot \varphi^m d\beta_s^m, \qquad (2)$$

and we can define the following functions:

$$f(b_t, t) = \frac{1}{2} \nabla \cdot (a \nabla b_t) - v^* \cdot \nabla b_t \qquad \text{and} \qquad g^m(b_t, t) = -\nabla b_t \cdot \varphi^m. \qquad (3)$$

We can now use the functional extension of the Itô formula [3] for both f and g to write their differential forms:

$$f(b_t, t) = f(b_{t_0}, t_0) + \int_{t_0}^{t} \frac{\partial f}{\partial s}(b_s, s)ds + \int_{t_0}^{t} \frac{\partial f}{\partial b}(b_s, s)db_s$$
$$+ \frac{1}{2} \int_{t_0}^{t} \frac{\partial^2 f}{\partial b^2}(b_s, s)d\langle b, b \rangle_s \qquad (4)$$

$$g^m(b_t, t) = g^m(b_{t_0}, t_0) + \int_{t_0}^{t} \frac{\partial g^m}{\partial s}(b_s, s)ds + \int_{t_0}^{t} \frac{\partial g^m}{\partial b}(b_s, s)db_s$$
$$+ \frac{1}{2} \int_{t_0}^{t} \frac{\partial^2 g^m}{\partial b^2}(b_s, s)d\langle b, b \rangle_s \qquad (5)$$

We remark that, since the basis φ^m is constant in time then so is a and the functions f and g^m do not depend explicitly on time, therefore $\partial f / \partial t = \partial g^m / \partial t = 0$.

Concerning the first derivatives with respect to b, it has to be interpreted as a Fréchet derivative. The Fréchet derivative of an operator F is the bounded linear operator $DF(\bar{x})$ which satisfies the following relation:

$$\lim_{\|h\| \to 0} \frac{\|F(\bar{x} + h) - F(\bar{x}) - DF(\bar{x})h\|}{\|h\|} = 0, \qquad (6)$$

which implies that for a linear operator $DF(\bar{x})h = F(h)$. We start for g and use the fact that ∇ is a linear operator:

$$\frac{\partial g}{\partial b}(\bar{b})b = -\nabla b \cdot \boldsymbol{\varphi}^m - \nabla b \cdot \frac{\partial \boldsymbol{\varphi}^m}{\partial b}. \tag{7}$$

If the basis is computed offline, $\boldsymbol{\varphi}^m$ does not depend on b and therefore the second term in (7) is zero. If the basis is computed online and $\boldsymbol{\varphi}^m$ does depend on b, we can rewrite the second term of the sum by components and, using the chain rule, one has:

$$\nabla b \cdot \frac{\partial \boldsymbol{\varphi}^m}{\partial b} = \frac{\partial b}{\partial x}\frac{\partial \varphi_x^m}{\partial b} + \frac{\partial b}{\partial y}\frac{\partial \varphi_y^m}{\partial b} = \nabla \cdot \boldsymbol{\varphi}^m. \tag{8}$$

For the second term of f, i.e. $\boldsymbol{v}^* \cdot \nabla b$, the same considerations are valid. To compute the derivative of the first term of f, we remark that it is a composition and product of three operators, two of which are linear. We can define:

$$F_1(\boldsymbol{h}) = \frac{1}{2}\nabla \cdot \boldsymbol{h}, \quad F_2(b) = \boldsymbol{a}(b), \quad F_3(b) = \nabla b. \tag{9}$$

Using the chain rule and the linearity of F_1 and F_3 one has:

$$\begin{aligned}
D\Big(F_1\big(F_2(b)F_3(b)\big)\Big)b &= DF_1\big(F_2(b)F_3(b)\big)\big(DF_2(b)F_3(b) + F_2(b)DF_3(b)\big)b \\
&= F_1\big(F_3(b)DF_2(b)b + F_2(b)F_3(b)\big) \\
&= \frac{1}{2}\nabla \cdot \left(\frac{\partial \boldsymbol{a}}{\partial b}\nabla b + \boldsymbol{a}\nabla b\right).
\end{aligned} \tag{10}$$

Finally, with the same considerations used above, we remark that we can write $(\partial \boldsymbol{a}/\partial b)\nabla b = \nabla \cdot \boldsymbol{a}$. Therefore:

$$\frac{\partial f}{\partial b}(\bar{b})b = f(b) + \frac{1}{2}\nabla \cdot \nabla \cdot \boldsymbol{a} - \nabla \cdot \boldsymbol{v}^*, \quad \frac{\partial g^m}{\partial b}(\bar{b})b = g^m(b) - \nabla \cdot \boldsymbol{\varphi}^m. \tag{11}$$

As for the Itô covariation bracket, one has:

$$\langle b, b\rangle_t = \Big\langle \int_{t_0}^{\cdot} \sum_m g^m(b_s, s)\mathrm{d}\beta_s^m, \int_{t_0}^{\cdot} \sum_k g^k(b_\tau, \tau)\mathrm{d}\beta_\tau^k\Big\rangle_t = \int_{t_0}^{t}\left(\sum_m g^m(b_s, s)\right)^2 \mathrm{d}s$$

We now suppose to be in either one of the following cases:

- the basis functions $\boldsymbol{\varphi}^m$ (and therefore $\boldsymbol{a}$) do not depend on b and $\nabla \cdot \boldsymbol{v}^* = 0$,
- the basis functions $\boldsymbol{\varphi}^m$ depend on b but are such that $\nabla \cdot \boldsymbol{v}^* = \nabla \cdot \nabla \cdot \boldsymbol{a} = \nabla \cdot \boldsymbol{\sigma} = \nabla \cdot \boldsymbol{\varphi}^m = 0$.

It can be noticed that the first case corresponds to a noise defined from external high-resolution data (and thus that does not depend on the solution) while the second case boils down to impose an incompressibility condition constraint on the large scale component, $\nabla \cdot \boldsymbol{u} = 0$, that is indeed often considered in practice with particular scaling of the noise [1, 2]. With these assumptions, we have then:

$$\frac{\partial f}{\partial b} = \frac{\partial^2 f}{\partial b^2} = f, \quad \frac{\partial g^m}{\partial b} = \frac{\partial^2 g^m}{\partial b^2} = g^m. \tag{12}$$

We can now replace all these expressions into (4) and (5), and then (4) and (5) into (2). Keeping only the terms of order one or lower, we obtain:

$$b_t = b_{t_0} + f(b_{t_0})\Delta t + \sum_m g^m(b_{t_0})\Delta\beta^m + \int_{t_0}^t \int_{t_0}^s \sum_{m,k} g^m(g^k(b_\tau))\mathrm{d}\beta_\tau^k \mathrm{d}\beta_s^m, \tag{13}$$

where $\Delta t = t - t_0$ and $\Delta\beta^m = \beta_t^m - \beta_{t_0}^m$. We define the following quantities:

$$G^{m,k} := g^m(g^k(b_{t_0})), \quad I^{m,k} := \int_{t_0}^t \int_{t_0}^s \mathrm{d}\beta_\tau^k \mathrm{d}\beta_s^m,$$

then the double iterated Itô integral in (13) can be approximated as follows:

$$\sum_{m,k} G^{m,k} I^{m,k} = \sum_{m,k} G^{m,k} \frac{I^{m,k} + I^{k,m}}{2} + G^{m,k} \frac{I^{m,k} - I^{k,m}}{2}.$$

The first symmetric term can be computed analytically from Itô integration by part formulae, $I^{m,k} + I^{k,m} = \Delta\beta^m \Delta\beta^k - \delta_{m,k}\Delta t$, however the second antisymmetric term $(I^{m,k} - I^{k,m})/2 =: A_{t_0,t}^{m,k}$ cannot and it is known as the Lévy area.

2.1.1 Lévy Area Simulation

In this subsection, we briefly introduce the methods we used to simulate the Lévy area. More details can be found in [6, 8], where these methods were proposed. The first method to simulate the Lévy area will be referred to as the weak approximation in the rest of this work: in this method, we simulate a random variable that has the same moments as the Lévy area. The second method, which will be referred to as the conditional method, is a recursive method: the time interval (t_0, t) is recursively split into two subintervals of the same length, and the two following relations are used:

$$A_{t_0,t}^{m,k} = A_{t_0,u}^{m,k} + A_{u,t}^{m,k} + \frac{1}{2}\left((\beta_u^m - \beta_{t_0}^m)(\beta_t^k - \beta_u^k) - (\beta_u^k - \beta_{t_0}^k)(\beta_t^m - \beta_u^m)\right) \tag{14}$$

$$\mathbb{E}[A_{t_0,t}|\mathbf{B}_t - \mathbf{B}_{t_0}] = 0.$$

For more details on these two methods, see [7]. Finally, we consider a third approach, where we neglect the Lévy area. We remark that this approach is exact if $G^{m,k} = G^{k,m}$, which is not the case here.

2.2 Multi-Step Schemes

We next propose a two-step scheme in which the Milstein method is used as the prediction step and the Euler method is adopted as the correction step, it reads:

$$
\begin{cases}
b_t^* = b_{t_0} + f(b_{t_0}, \boldsymbol{u}_{t_0})\Delta t + \sum_m g^m(b_{t_0})\Delta\beta^m + \sum_{m,k} G^{m,k}\left(S_{t_0,t}^{m,k} + \tilde{A}_{t_0,t}^{m,k}\right) \\
\boldsymbol{u}_t^* = \mathcal{H}(b_t^*) \\
b_t = \frac{1}{2}b_{t_0} + \frac{1}{2}\left(b_t^* + f(b_t^*, \boldsymbol{u}_t^*)\Delta t + \sum_m g^m(b_t^*)\Delta\beta^m\right)
\end{cases}
$$
(15)

where $S_{t_0,t}^{m,k} := (\Delta\beta^m \Delta\beta^k - \delta_{m,k}\Delta t)/2$ and $\tilde{A}_{t_0,t}^{m,k}$ is one of the approximations of the Lévy area described in the previous subsection. This scheme will be referred to as SRK2-EM (EM stands for Euler-Milstein not for Euler-Maruyama) in the rest of the paper.

In the next section, we first analyse the results of the Milstein schemes with the different Lévy area approximations in order to select the best one. Then, we compare our multi-step scheme to two other multi-step schemes developed in [5] and [4]. We briefly recall them here. The first one, based on a third order Runge-Kutta scheme, (SSPRK3) [5], is:

$$
\begin{cases}
b^{(1)} = b_{t_0} + f_s(b_{t_0}, \boldsymbol{u}_{t_0})\Delta t + \sum_m g^m(b_{t_0})\Delta\beta^m \\
\boldsymbol{u}^{(1)} = \mathcal{H}(b^{(1)}) \\
b^{(2)} = \frac{3}{4}b_{t_0} + \frac{1}{4}\left(b^{(1)} + f_s(b^{(1)}, \boldsymbol{u}^{(1)})\Delta t + \sum_m g^m(b^{(1)})\Delta\beta^m\right) \\
\boldsymbol{u}^{(2)} = \mathcal{H}(b^{(2)}) \\
b_t = \frac{1}{3}b_{t_0} + \frac{2}{3}\left(b^{(2)} + f_s(b^{(2)}, \boldsymbol{u}^{(2)})\Delta t + \sum_m g^m(b^{(2)})\Delta\beta^m\right)
\end{cases}
$$
(16)

where $f_s = f - \nabla\cdot(a\nabla b)/2$ denotes the modified drift under Stratonovich integral. The second one, relies on Euler-Heun method [4] equally for Stratonovich integral, reads:

$$
\begin{cases}
b^{(1)} = b_{t_0} + f_s(b_{t_0}, \boldsymbol{u}_{t_0})\Delta t + \sum_m g^m(b_{t_0})\Delta \beta^m \\
\boldsymbol{u}^{(1)} = \mathcal{H}(b^{(1)}) \\
b_t = \frac{1}{2}b_{t_0} + \frac{1}{2}\left(b^{(1)} + f_s(b^{(1)}, \boldsymbol{u}^{(1)})\Delta t + \sum_m g^m(b^{(1)})\Delta \beta^m \right)
\end{cases}
\tag{17}
$$

3 Numerical Results

In this section we show some numerical results. First, the effect of the different approximations of the Lévy area is studied on the Milstein scheme. Then, the multi-step scheme is assessed and compared to the ones already proposed in the literature. We focus on two variations of one specific test case plotted in Fig. 1: the initial condition (left) consists of two warm elliptical anticyclones on the bottom of the domain and two cold elliptical cyclones on the top. After one day under moderate noise (centre), the four structures have rotated of approximately 45°. After one day under strong noise (right) the nonlinearity of the dynamic is more noticeable. One can find all the configuration details used for these simulations in Chapter 6 of [10] for the moderate noise configuration. For the strong noise, all the basis functions φ^m are multiplied by a factor 10.

We will use the following abbreviations for the different numerical schemes

- Euler: Euler-Maruyama scheme.
- Milstein-0: Milstein scheme without the Lévy area.
- Milstein-weak: Milstein scheme with the weak approximation of the Lévy area.
- Milstein-cond-n: Milstein scheme with the conditional approximation of the Lévy area. Here n stands for the number of times the interval is recursively split (cf. (14)).
- SRK2-EM: scheme (15) with $\tilde{A}_{t_0,t}^{m,k} = 0$.
- SSPRK3: scheme (16).
- Heun: scheme (17).

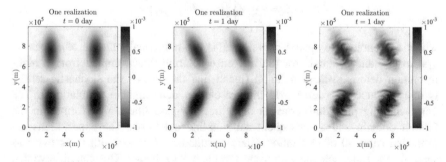

Fig. 1 Euler-Maruyama simulation of system (1) on a 128×128 spatial grid

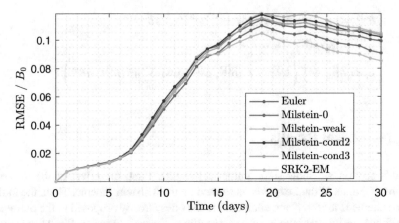

Fig. 2 RMSE (normalised by the amplitude of buoyancy $B_0 = 10^{-3}$ m/s^2) of different schemes during 30 days of simulation under moderate noise

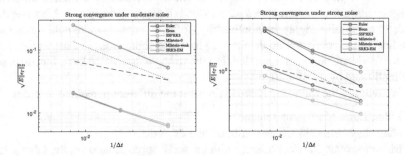

Fig. 3 Convergence of different schemes under weak and strong noise. Order 1 in dotted black, order 0.5 in dashed black

In Figs. 2 and 3 one can see the difference among the Euler-Maruyama scheme and all the Milstein schemes proposed. In Fig. 2 we plot for each scheme for a period of 30 day the root mean squared error (RMSE), defined as:

$$\text{RMSE} = \frac{1}{|\Omega|} \mathbb{E}\Big[\big\|b_h - b\big\|_{L^2(\Omega)}^2\Big]^{1/2}, \tag{18}$$

where Ω denotes the spatial domain, b_h is the numerical solution of stochastic system (1), and b stands for the reference solution downsampled from a high-resolution deterministic simulation (recall that the aim of the stochastic setting is to reproduce on coarse grid high-resolution deterministic simulations). The downsampling procedure consists of a first low-pass filtering performed in the Fourier domain and a subsequent subsampling operation. The expectations are estimated from 30 of realization. These results are obtained with a Δt twice as small for the Euler scheme with respect to the other schemes. One can observe that Milstein-0 performs slightly better than the other Milstein schemes.

In Fig. 3, we show the rate of strong convergence γ of all the schemes discussed, under weak and strong noise. Since the exact solution is unknown, we use the following method [15] to estimate γ, for a sufficiently small Δt:

$$\gamma \simeq \log_2\left(\frac{e_1}{e_2}\right), \text{ with } e_i := \mathbb{E}\left[\left\|b_h\left(T, \frac{\Delta t}{2^{i-1}}\right) - b_h\left(T, \frac{\Delta t}{2^i}\right)\right\|_{L^2(\Omega)}^2\right]^{1/2},$$

where $b_h(T, \Delta t)$ is the numerical solution at the final time T obtained with a time step Δt. It is important to underline that in order for this method to work, the Brownian trajectories must be fixed. We applied this method for time steps $30, 60, 120, 240$, hence obtaining two estimates for γ. Is is important to remark that the value of the time steps is given in seconds and the time-scale of the studied phenomenon is of the order of one day. For reference, the CFL condition for this problem at the initial time would give a time step around $300\,\text{s}$. The smallest time step we considered to obtain this estimate is ten times smaller than this. As one can see from Fig. 3, under weak noise all the one-step schemes provide almost identical results and all the multi-step schemes are very similar. It is hard to distinguish among the different numerical schemes proposed. In particular, for the considered span of time steps, the error of the Euler scheme under moderate noise displays a linear trend and the prevailing convergence order in this case is one. The reason of that is explained in Appendix.

Under strong noise, it is easier to see the differences among the schemes. Milstein-weak is a slight improvement on the Euler-Maruyama, but its rate of convergence is far from 1. Milstein-0 has the highest rate of convergence among all the schemes.

In conclusion, Milstein-0 seem to perform better than the other Milstein schemes. Furthermore, it is less computationally demanding. For these reasons, we built our two-step scheme based on Milstein-0.

In Fig. 3 we also compare the multi-step schemes mentioned above: they all have a similar behaviour, with a rate of convergence $0.5 \leq \gamma \leq 1$, but a much smaller error when compared to the one-step schemes. In particular, the two-step scheme proposed in this work (SRK2-EM in the figures) yields the smallest error of all for this test case. The SRK2-EM schemes also yields the smallest RMSE (cf. Fig. 2).

4   Conclusion and Perspectives

The Milstein schemes analysed in this work improve the numerical results, in particular when used in a multi-step framework. The Lévy area does not seem to play a key role in these test cases, which allows us to drastically reduce the computational costs. It must be pointed out that under weak noise, all the schemes tested provide very similar results. Some ongoing and future work include the understanding of the (non) importance of the Lévy area and whether this is related to the test case, the equations, or other factors.

Appendix: Convergence of Euler-Maruyama Scheme Under Moderate Noise

To study the behaviour of our system under moderate noise, we use the formalism of [12]; in particular, we write our system in the following generic form:

$$dX_t = a(x, t)dt + \epsilon b(x, t)dW_t + \epsilon^2 c(x, t)dt, \quad t \in [0, T] \tag{19}$$

with a, b, c, being jointly L^2-measurable in (x, t), Lipschitz, bounded linear-growth functions in x.

Let Y^δ be an Euler-Maruyama integration scheme for X. with integration step δ. Then we may prove in a similar fashion to theorem 4.5.4 in [9] that:

1. $\mathbb{E}[X_t]^2 \le C, \quad \forall t \in [0, T]$
2. $\mathbb{E}\left[|X_{t+\delta} - Y^\delta_{t+\delta}| \big| X_{t+\delta} = x\right] \le K(x)(\delta + \sqrt{\epsilon}\sqrt{\delta} + \circ(\delta)).$

Using this and the Lipschitziannity of the coefficients in (19), we may prove a result, to some extent similar to theorem 2.1 in [12], namely that

$$\mathbb{E}\left[\sup_{t_0 \le t \le T} |X_t - Y^\delta_t| \Big| X_{t_0} = x\right] \le K'(x)(\delta + \sqrt{\epsilon}\sqrt{\delta} + \circ(\delta)). \tag{20}$$

In light of this estimate, we may interpret the convergence rate displayed in Fig. 3 as a case where δ is not small enough when compared to ϵ so that $\sqrt{\epsilon}\sqrt{\delta}$ does not necessarily prevail over δ which is evidenced by the linear rate of convergence.

References

1. W. Bauer, P. Chandramouli, B. Chapron, L. Li, and E. Mémin. Deciphering the role of small-scale inhomogeneity on geophysical flow structuration: a stochastic approach. *Journal of Physical Oceanography*, 50(4):983–1003, 2020.
2. R. Brecht, L. Li, W. Bauer, and E. Mémin. Rotating shallow water flow under location uncertainty with a structure-preserving discretization. *J. of Advances in Modeling of Earth Systems*, 13(12), 2021.
3. R. Cont and D. Fournie. A functional extension of the Itō formula. *Comptes Rendus Mathematique*, 348(1–2):57–61, 2010.
4. C. Cotter, D. Crisan, D. Holm, W. Pan, and I. Shevchenko. Modelling uncertainty using stochastic transport noise in a 2-layer quasi-geostrophic model. *Foundations of Data Science*, 2(2):173, 2020.
5. C. Cotter, D. Crisan, D. D. Holm, W. Pan, and I. Shevchenko. Numerically modeling stochastic lie transport in fluid dynamics. *Multiscale Modeling & Simulation*, 17(1):192–232, 2019.
6. G. Flint and T. Lyons. Pathwise approximation of sdes by coupling piecewise abelian rough paths. *arXiv preprint arXiv:1505.01298*, 2015.
7. J. Foster. Lévy area simulation. https://github.com/james-m-foster/levy-area-simulation/blob/master/levy_area_simulation.ipynb, 2019.

8. T. L. J. Foster and H. Oberhauser. An optimal polynomial approximation of Brownian motion. *SIAM Journal on Numerical Analysis*, 58(3):1393–1421, 2020.
9. P. E. Kloeden and E. Platen. *Numerical solution of stochastic differential equations.* Springer-Verlag Inc, Berlin; New York, 1995.
10. L. Li. *Stochastic modeling and numerical simulation of ocean dynamics.* PhD thesis, Université Rennes 1, 2021.
11. E. Mémin. Fluid flow dynamics under location uncertainty. *Geophysical & Astrophysical Fluid Dynamics*, 108(2):119–146, 2014.
12. G. N. Milstein and M. V. Tretyakov. Numerical methods in the weak sense for stochastic differential equations with small noise. *SIAM J. Numer. Anal.*, 34(6):2142–2167.
13. V. Resseguier, L. Li, G. Jouan, P. Derian, E. Mémin, and B. Chapron. New trends in ensemble forecast strategy: uncertainty quantification for coarse-grid computational fluid dynamics. *Archives of Computational Methods in Engineering*, 28(1):1886–1784, 2020.
14. V. Resseguier, A. M. Picard, E. Mémin, and B. Chapron. Quantifying truncation-related uncertainties in unsteady fluid dynamics reduced order models. *SIAM/ASA Journal on Uncertainty Quantification*, 9(3):1152–1183, 2021.
15. K. Schmitz A. and W. T. Shaw. Measure order of convergence without an exact solution, Euler vs Milstein scheme. *International Journal of Pure and Applied Mathematics*, 24(3):365–381, 2005.

The Dissipation Properties of Transport Noise

Franco Flandoli and Eliseo Luongo

Abstract The aim of this work is to present, in a compact way, the latest results about the dissipation properties of transport noise in fluid mechanics. Starting from the reasons why transport noise is natural in a passive scalar equation for the heat diffusion and transport, several results about enhanced dissipation due to the noise are presented. Rigorous statements are matched with numerical experiments in order to understand that the sufficient conditions stated are not yet optimal but give a first useful indication.

Keywords Dissipation by noise · Turbulence · Eddy diffusion · Vortex patch · Transport noise · Dirichlet boundary condition

1 Introduction

In the last four years, a new understanding of heat diffusion in a turbulent fluid modeled by white noise has been developed. This model has the interesting feature of describing properly the dissipation properties of a turbulent fluid. The equation for the heat diffusion and transport, with a heat source q, is

$$\partial_t \theta + u \cdot \nabla \theta = \kappa \Delta \theta + q \tag{1}$$

where $\theta = \theta(t, x)$ is the temperature, κ is the diffusion constant and $u = u(t, x)$ is the velocity field of the fluid. The turbulent fluid is a priori described by a random field, Gaussian and white in time, with covariance structure given a priori (hence the temperature is a passive scalar). In this review we consider the following description for u:

F. Flandoli (✉) · E. Luongo
Scuola Normale Superiore, Pisa, Italy
e-mail: franco.flandoli@sns.it

$$u(t, x) = \sum_{k \in K} \sigma_k(x) \frac{dW_t^k}{dt} \tag{2}$$

where σ_k are divergence free vector fields satisfying no slip-boundary conditions and W_t^k are independent Brownian motions on a filtered probability space $(\Omega, \mathcal{F}, (\mathcal{F}_t)_{t \geq 0}, \mathbb{P})$; for simplicity, assume K is a finite set, but the case of a countable set can be studied without troubles at the price of additional summability assumptions. Some rigorous justification for describing the velocity of a turbulent fluid by Eq. (2) are available in Sect. 2.2. Here we want just give some ideas. Let us denote by $u^{\nu, \varepsilon}$ the solution in a domain with boundary D of the SPDE

$$\begin{cases} \partial_t u^{\nu, \varepsilon} + \nabla p^{\nu, \varepsilon} &= \nu \Delta u^{\nu, \varepsilon} - \frac{1}{\varepsilon} u^{\nu, \varepsilon} + \frac{1}{\varepsilon} \sum_{k \in K} \sigma_k \partial_t dW_t^k \\ \operatorname{div}(u^{\nu, \varepsilon}) &= 0 \\ u|_{\partial D} &= 0, \end{cases} \tag{3}$$

where the terms $-\frac{1}{\varepsilon} u^{\nu, \varepsilon} + \frac{1}{\varepsilon} \sum_{k \in K} \sigma_k \partial_t dW_t^k$ describe the roughness of the boundary as stated in Sect. 2.2. Let, moreover, $W_t^{\nu, \varepsilon} = \int_0^t u^{\nu, \varepsilon}(s)\, ds$, then it can be proven than

$$\lim_{\varepsilon \to 0} \mathbb{E} \left[\sup_{t \in [0, T]} \| W_t^{\nu, \varepsilon} - \sum_{k \in K} \sigma_k W_t^k \|_{L^2(D)}^2 \right] = 0,$$

see for example [6].

The correct interpretation of Eq. (1) when u has the form (2) is the Stratonovich equation

$$d\theta + \sum_{k \in K} \sigma_k \cdot \nabla\theta \circ dW_t^k = (\kappa \Delta\theta + q)\, dt \tag{4}$$

or equivalently the Itô equation with corrector $\mathcal{L}\theta$ given by the second order differential operator (7) below:

$$d\theta + \sum_{k \in K} \sigma_k \cdot \nabla\theta dW_t^k = (\kappa \Delta\theta + \mathcal{L}\theta + q)\, dt. \tag{5}$$

There are some motivations for the analysis of Eq. (4) based on the idea to extend to SPDE the remarkable principle of Wong-Zakai [20], see for example [2, 3, 14, 15, 18, 19, 16].

Assuming that the external source q and the initial temperature θ_0 are deterministic, under suitable mild assumptions the deterministic function

$$\Theta(t, x) = \mathbb{E}[\theta(t, x)]$$

is the solution of the deterministic parabolic equation

$$\partial_t \Theta = (\kappa \Delta + \mathcal{L}) \Theta + q \tag{6}$$

where $\mathbb{E}$ denotes the mathematical expectation on $(\Omega, \mathcal{F}, \mathbb{P})$. The main results in the last years are quantitative estimates on the difference $\theta - \Theta$, some convergence properties of the solution of Eq. (5) to the stationary solution of Eq. (6) and the enhanced dissipative properties of the second order differential operator $\kappa \Delta + \mathcal{L}$, see [7, 9, 10]. These kinds of results explained properly the dissipation properties of transport noise and are the core of this review article.

In Sect. 2 we will present some motivations for the analysis of Eq. (4) as a good model for the heat diffusion in a turbulent fluid and we will introduce the main notations. In Sect. 3 we will present the main results, referring to [7, 9, 10] for some rigorous proofs. Lastly, in Sect. 4 we will present some cases where the coefficients σ_k introduce more dissipation in the model with respect to the theoretical predictions made by the rigorous sufficient conditions, exploiting real computations or numerical simulations following the ideas of [7, 10].

Remark 1 In this review we only considered the effects of the transport noise on passive scalars. Actually, some results can be stated also for the scalar vorticity of the fluid itself, in two space dimensions. We refer to [8, 11, 12] for further readings. The case of the influence on vector fields is much more difficult and still to be understood.

2 Well-Posedness and Motivations

2.1 Notations and Definitions

In this review we will denote by D a 2D domain with boundary, either a smooth bounded open set or an infinite 2D channel, namely $\mathbb{R} \times (-1, 1)$. We write the coordinates using the notation

$$x = (x_1, z) \in D.$$

Let Z be a separable Hilbert space, denote by $L^2(\mathcal{F}_{t_0}, Z)$ the space of square integrable random variables with values in Z, measurable with respect to $\mathcal{F}_{t_0}$. Moreover, denote by $C_{\mathcal{F}}([0, T]; Z)$ the space of continuous adapted processes $(X_t)_{t \in [0, T]}$ with values in Z such that

$$\mathbb{E} \left[\sup_{t \in [0, T]} \|X_t\|_Z^2 \right] < \infty$$

and by $L^2_{\mathcal{F}}(0, T; Z)$ the space of progressively measurable processes $(X_t)_{t \in [0,T]}$ with values in Z such that

$$\mathbb{E}\left[\int_0^T \|X_t\|_Z^2 \, dt\right] < \infty.$$

Denote by $L^2(D)$ and $W^{k,2}(D)$ the usual Lebesgue and Sobolev spaces and by $W_0^{k,2}(D)$ the closure in $W^{k,2}(D)$ of smooth compact support functions. Set $H = L^2(D)$, $V = W_0^{1,2}(D)$, $D(A) = W^{2,2}(D) \cap V$. We denote by $\langle \cdot, \cdot \rangle$ and $\|\cdot\|$ the inner product and the norm in H respectively.

Assume that K is a finite set and $\sigma_k \in \left(D(A) \cap C_b^\infty(D)\right)^2$, $\nabla \cdot \sigma_k = 0, k \in K$ (less is sufficient but we do not stress this level of generality). Define the matrix-valued function

$$Q(x, y) = \sum_{k \in K} \sigma_k(x) \otimes \sigma_k(y).$$

If we denote by $W(t, x)$ the vector valued random field

$$W(t, x) = \sum_{k \in K} \sigma_k(x) W_t^k$$

(the velocity field u given by (2) is the distributional time derivative of W) then we see that $Q(x, y)$ is the space-covariance of $W(1, x)$:

$$Q(x, y) = \mathbb{E}\left[W(1, x) \otimes W(1, y)\right].$$

The matrix-function $Q(x, x)$ is elliptic:

$$\sum_{i,j=1}^d Q_{ij}(x, x) \xi_i \xi_j = \mathbb{E}\left[|W(t, x) \cdot \xi|^2\right] \geq 0$$

for all $\xi = (\xi_1, \ldots, \xi_d) \in \mathbb{R}^d$. Associated to it define the bounded linear operator

$$Q : L^2(D; \mathbb{R}^2) \to L^2(D; \mathbb{R}^2), \qquad (Qv)(x) = \int_D Q(x, y) v(y) \, dy$$

and the quantities:

$$\tilde{q}(x) := \min_{\xi \neq 0} \frac{\xi^T Q(x, x) \xi}{|\xi|^2},$$

$$\varepsilon_Q := \|Q^{1/2}\|^2_{L^2(D;\mathbb{R}^2) \to L^2(D;\mathbb{R}^2)}.$$

Consider the divergence form elliptic operator $\mathcal{L}$ defined as

$$(\mathcal{L}\theta)(x) = \frac{1}{2} \sum_{i,j=1}^{d} \partial_i \left(Q_{ij}(x,x) \partial_j \theta(x) \right) \tag{7}$$

for $\theta \in W^{2,2}(D)$. Define the linear operator $A : D(A) \subset H \to H$ as

$$A\theta = (\kappa\Delta + \mathcal{L})\theta.$$

It is the infinitesimal generator of an analytic semigroup of negative type, see [1, 4, 13, 17], that we denote by e^{tA}, $t \geq 0$. Moreover, if D is bounded, we denote by $\kappa\lambda$ the first eigenvalue of $-\kappa\Delta$ and by $\kappa\lambda_{\mathcal{L}}$ the first eigenvalue of $-(\kappa\Delta + \mathcal{L})$.

Definition 1 Given $\theta_0 \in L^2(\mathcal{F}_0, H)$ and $q \in L^2(0, T; H)$, a stochastic process

$$\theta \in C_{\mathcal{F}}([0, T]; H) \cap L^2_{\mathcal{F}}(0, T; V)$$

is a mild solution of Eq. (5) if the following identity holds

$$\theta(t) = e^{tA}\theta_0 + \int_0^t e^{(t-s)A} q(s)\, ds - \sum_{k \in K} \int_0^t e^{(t-s)A} \sigma_k \cdot \nabla\theta(s)\, dW_s^k$$

for every $t \in [0, T]$, $\mathbb{P}$-a.s.

Theorem 1 *For every $\theta_0 \in L^2(\mathcal{F}_0, H)$ and $q \in L^2(0, T; H)$ there exists a unique θ mild solution of Eq. (5). Moreover θ depends continuously on θ_0 and q.*

Definition 2 Given $\theta_0 \in L^2(\mathcal{F}_0, H)$ and $q \in L^2(0, T; H)$, we say that a stochastic process θ is a weak solution of Eq. (5) if

$$\theta \in C_{\mathcal{F}}([0, T]; H) \cap L^2_{\mathcal{F}}(0, T; V)$$

and for every $\phi \in D(A)$, we have

$$\langle \theta(t), \phi \rangle = \langle \theta_0, \phi \rangle + \int_0^t \langle \theta(s), A\phi \rangle\, ds + \int_0^t \langle q(s), \phi \rangle$$
$$+ \sum_{k \in K} \int_0^t \langle \theta(s), \sigma_k \cdot \nabla\phi \rangle\, dW_s^k$$

for every $t \in [0, T]$, $\mathbb{P}$ − a.s.

Theorem 2 *θ is a weak solution of problem (5) if and only if is a mild solution of problem (5). Moreover the Itô formula*

$$\|\theta(t)\|^2 - \|\theta(0)\|^2 = 2\int_0^t \langle \theta(s), q(s) \rangle\, ds + \sum_{k \in K} \int_0^t \|\sigma_k \cdot \nabla\theta(s)\|^2\, ds$$
$$-2\int_0^t \langle (-A)^{\frac{1}{2}}\theta(s), (-A)^{\frac{1}{2}}\theta(s) \rangle\, ds$$

holds.

These results are classical and can be found in [5, 10] together with several generalizations.

2.2 Motivations

In this section we want to give some heuristics to accept Eq. (4) as a correct model for heat diffusion in a turbulent fluid. In the domain D we have a fluid with velocity u (pressure p, constant density = 1) and the heat θ. Both u and θ are equal to zero on ∂D:

$$u|_{\partial D} = 0$$

$$\theta|_{\partial D} = 0.$$

The condition $u|_{\partial D} = 0$ provokes several interesting technical questions. The equations are

$$\partial_t u + u \cdot \nabla u + \nabla p = f$$
$$\nabla \cdot u = 0$$
$$\partial_t \theta + u \cdot \nabla \theta = \kappa \Delta \theta + q \qquad (8)$$
$$u|_{t=0} = u_0$$
$$\theta|_{t=0} = \theta_0.$$

where f and q take care of interaction with external sources. In particular, physical boundaries are never completely smooth. Hence, the external source f want to model the effects of the roughness of the boundary and its influence to the velocity of the fluid. The instability of the flow at the boundary, originating vortices, is very strong, hence the frequency and intensity of creation of vortices at the boundary strongly suffers from the imprecision of the description of the true boundary. Replacing the true details of the boundary by a random mechanism of vorticity production would increase the realism of the model. Emergence of vortices near obstacles is commonly observed and we content ourselves with an ad hoc inclusion of this fact into the equations. Assume the velocity field at time t is $u(t, x)$. Assume that, as a consequence of an instability near the boundary, a modification occurs and in a very short time we have a field $u(t + \Delta t, x)$ which is not just equal to the smooth evolution of $u(t, x)$. We may assume that at some time t we have a jump:

$$u(t + \Delta t, x) = u(t, x) + \sigma(x)$$

where $\sigma(x)$ is presumably localized in space and corresponds to a vortex structure. After these preliminary comments we can accept to model the roughness via a friction term of intensity $-\frac{u}{\varepsilon}$ and a term of jump described by

$$W_N(t, x) = \sum_{k \in K} \frac{\sigma_k(x)}{N} \frac{N^{k,1}_{N^2 t/\varepsilon^2} - N^{k,2}_{N^2 t/\varepsilon^2}}{\sqrt{2}},$$

where $N_t^{\cdot \cdot}$ are independent Poisson processes. More on this topic can be found in [6]. Applying a Donsker invariance principle to the stochastic process $W_N(t, x)$, it converges in law to the gaussian process

$$W_t(x) = \frac{1}{\varepsilon} \sum_{k \in K} \sigma_k(x) W_t^k,$$

where W_t^k are independent Brownian motions. Parameterizing the solutions of system (8) by ε we arrive to the following stochastic coupled system

$$\begin{aligned}
\partial_t u^\varepsilon + u^\varepsilon \cdot \nabla u^\varepsilon + \nabla p^\varepsilon &= -\frac{1}{\varepsilon}(u^\varepsilon - \partial_t W) \\
\nabla \cdot u^\varepsilon &= 0 \\
\partial_t \theta^\varepsilon + u^\varepsilon \cdot \nabla \theta^\varepsilon &= \kappa \Delta \theta^\varepsilon + q \\
u|_{t=0} &= u_0 \\
\theta|_{t=0} &= \theta_0.
\end{aligned} \tag{9}$$

The last step for moving from system (9) to Eq. (4) is trying to understand the behavior of system (9) letting $\varepsilon \to 0$ and it based on a result proved in [12] in the case of the 2D torus and under analysis in the case of general 2D domains with boundary. Thus just for the last sentence of this subsection we assume the $D = \mathbb{T}^2 := \mathbb{R}^2/(2\pi \mathbb{Z}^2)$.

Theorem 3 *Under previous assumptions on q and σ, if moreover:*

- *the coefficients σ_k are zero-mean and there exists $l \geq 1$ such that*
  $\sigma_k \in W^{l,\infty} \; \forall k \in K$;
- $\forall x \in \mathbb{T}^2$ *it holds* $\sum_{k \in K} ((K * \sigma_k) \cdot \nabla \sigma_k)(x) = 0$;
- $q \in L^1([0, T]; L^\infty(\mathbb{T}^2))$;
- $\theta_0 \in L^\infty(\mathbb{T}^2)$,

then for every $f \in L^1(\mathbb{T}^2)$

$$\mathbb{E}\left[\left| \int_{\mathbb{T}^2} (\theta_t^\varepsilon - \theta_t)(x) f(x) \, dx \right| \right] \to 0 \qquad as \; \varepsilon \to 0$$

for every fixed $t \in [0, T]$ and in $L^p([0, T])$ for every finite p. Moreover, if $q \in L^1([0, T]; Lip(\mathbb{T}^2))$ then the previous convergence holds uniformly for $t \in [0, T]$ and $f \in Lip(\mathbb{T}^2)$ with $[f]_{Lip(\mathbb{T}^2)} \leq 1$ and $\|f\|_{L^\infty(\mathbb{T}^2)} \leq 1$.

3   Main Results

The results related to the analysis of these equations can be classified in three
categories:

1. Convergence of the solution of Eq. (5) to some quantities related to Eq. (6).
2. Quantification of the dissipation of the function $\mathbb{E}\left[\|\theta(t)\|^2\right]$.
3. Enhanced dissipative properties of the second order differential operator $\kappa\Delta + \mathcal{L}$.

Remark 2 Even if Q is a covariance operator, the third question is far to be trivial.
In fact we assumed that $\sigma_k|_{\partial D} = 0$. Thus the operator $\mathcal{L}$ degenerates at the boundary.

We will treat all the three problems above, sometimes specializing our general
framework.

Theorem 4 *Assume D is a bounded domain.*

1. If $\theta_0 \in L^2(\mathcal{F}_0, H)$, $q \equiv 0$. Then, $\forall \phi \in L^\infty(D)$,

$$\mathbb{E}\left[\langle\phi, \theta(t) - \Theta(t)\rangle^2\right] \le \frac{\varepsilon_Q}{\kappa}\mathbb{E}\left[\|\theta_0\|^2\right]\|\phi\|_{L^\infty(D)}.$$

2. Moreover, if $\theta_0 \ge 0$

$$\mathbb{E}\left[\|\theta(t)\|^2\right] \le \left(\frac{\varepsilon_Q}{\kappa} + 2|D|e^{-\kappa\lambda_{\mathcal{L}}t}\right)\mathbb{E}\left[\|\theta_0\|^2\right].$$

Remark 3 A result similar to the first item can be proved also in the case of D
infinite channel and $q \ne 0$ adapting the proof of Theorem 7 in [10] to such finite
time case.

Thanks to previous theorem is evident that the dissipation properties of the solution
of the stochastic Eq. (5) are influenced obviously by the first eigenvalue of the
operator $\mathcal{L}$ but also by the operatorial norm of $\mathbb{Q}^{1/2}$. Thus, our next step will be
state state some sufficient conditions in order to have ε_Q very small and $\kappa\lambda_{\mathcal{L}} \gg \kappa\lambda$.
For $\delta > 0$ fixed, let us define

$$D_\delta := \{x \in D : \; dist(x, \partial D) > \delta\}.$$

Then the following theorems hold.

Theorem 5 *Assume that the family of coefficients $(\sigma_k(\cdot))_{k \in K}$ has the following
approximate orthogonality property: there exists a finite number $M \in \mathbb{N}$ and a
partition $K = K_1 \cup \ldots \cup K_M$ such that*

$$\langle\sigma_k, \sigma_{k'}\rangle = 0 \text{ for all } k, k' \in K_i$$

for all $i = 1, \ldots, M$. Then

$$\varepsilon_Q \leq M \sup_{k \in K} \|\sigma_k\|^2.$$

Theorem 6 *Assuming that $\tilde{q}(x) \geq \sigma^2$ in D_δ, then for any $\kappa > 0$ fixed*

$$\lim_{(\sigma, \delta) \to (+\infty, 0)} \kappa \lambda_{\mathcal{L}} = +\infty.$$

Theorem 7 *There exists a constant $C_D > 0$ such that*

$$\kappa \lambda_{\mathcal{L}} \geq C_D \min\left(\sigma^2, \frac{\kappa}{\delta}\right)$$

for every Q such that

$$\tilde{q}(x) \geq \sigma^2 \text{ in } D_\delta.$$

When D is the unit ball, asymptotically as $\delta \to 0$, one can take $C_D = 1$ and

$$\kappa \lambda_{\mathcal{L}} \geq \frac{2\kappa}{\kappa + \delta\sigma^2}\sigma^2.$$

From the last two theorems we understand that the dissipation properties enhance if ε_Q is very small and $\tilde{q}(x)$ is very large except for a small boundary layer around ∂D. Obviously ε_Q is related to the operatorial norm of $\mathbb{Q}^{1/2}$ and thus, loosely speaking, is related to the operatorial norm of $\mathbb{Q}$. Instead $\tilde{q}(x)$ is related to the trace of $\mathbb{Q}$, i.e.

$$Tr(\mathbb{Q}) = \int_D Tr\, Q(x, x)\, dx.$$

Consequently we want that the operatorial norm of $\mathbb{Q}$ is small and the trace of $\mathbb{Q}$ is arbitrarily large and, possibly, infinity. Hence the existence of such operators Q which increase the dissipativity properties of the equation is not surprising. The last issue related to this topic is the presentation of an operator Q which has a fluid dynamics interpretation and satisfies previous property. This definition for general domain D is a bit implicit. Thus in the last section we will present some more explicit computations.

Let us fix a parameter ℓ such that $0 < \ell \leq \delta$, consider a smooth probability density function $\Psi : \mathbb{R}^2 \to \mathbb{R}$ with compact support in $\overline{B(0, 1)}$ and let us denote by $K(x, y)$ the Biot-Savart kernel in D. We recall that a point vortex in x_0 has vorticity δ_{x_0} and smoothing it by $\Psi_\ell(x) := \frac{1}{\ell^2}\Psi\left(\frac{x}{\ell}\right)$, then it has vorticity $\frac{1}{\ell^2}\Psi\left(\frac{x - x_0}{\ell}\right)$.

Now let us consider a random variable X_0 distributed uniformly on $D_{2\delta}$, a real random variable Γ_0 such that

$$\mathbb{E}[\Gamma_0] = 0, \qquad \varepsilon_0^2 := \mathbb{E}\left[\Gamma_0^2\right] < \infty$$

and set

$$u(x) = \Gamma_0 \int_D K(x, y)\,\theta_\ell(y - X_0)dy$$

$$=: \Gamma_0 K_\ell(x, X_0).$$

If we consider in Eq. (5) the Brownian motion $W(t, x)$, with covariance operator

$$Q_\ell(x, y) = \varepsilon_0^2 \mathbb{E}[K_\ell(x, X_0) \otimes K_\ell(y, X_0)]$$

one has

$$v^T Q_\ell(x, x)\, v = \varepsilon_0^2 \mathbb{E}\left[|K_\ell(x, X_0) \cdot v|^2\right] \qquad \text{for } v \in \mathbb{R}^2$$

$$\langle \mathbb{Q}_\ell w, w \rangle = \varepsilon_0^2 \mathbb{E}\left[\left(\int_D w(x) \cdot K_\ell(x, X_0)dx\right)^2\right] \qquad \text{for } w \in L^2(D; \mathbb{R}^2).$$

Inside the previous identities there is the key to have $v^T Q(x, x)\, v$ large and $\langle \mathbb{Q}w, w \rangle$ small. Moreover, the law of u on the space of divergence free square integrable vector fields with null normal trace, heuristically, is a Poisson Point Process generating smoothed point vortices (and the associated velocity field) in random positions of D. Thus this kind of noise is reasonable for model what we expect from the heuristic analysis described in Sect. 2.

Theorem 8

– *There exists a constant $C > 0$ such that*

$$\langle \mathbb{Q}_\ell v, v \rangle \leq C \varepsilon_0^2 \|v\|_H^2$$

for every $v \in H$ and $\ell > 0$.
– *For every $x \in D$, let $q_\ell(x) \geq 0$ be the largest number such that*

$$v^T Q_\ell(x, x)\, v \geq q_\ell(x)\, |v|^2$$

for all $v \in \mathbb{R}^2$. Then

$$\lim_{\ell \to \infty} \inf_{x \in D_{2\delta}} q_\ell(x) = +\infty.$$

In the last result of this section the presence of the external source q is crucial. Moreover we assume that q is independent of time and introduce the stationary solution of Eq. (6)

$$\Theta_{st} := -A^{-1}q.$$

In fact we want to study the convergence of the solution of the stochastic Eq. (5) to Θ_{st}.
Set

$$C_\infty(\theta_0, q) := \sup_{t \geq 0} \mathbb{E}\left[\|\theta(t)\|_\infty^2\right].$$

Theorem 9 *If $\theta_0 \in L^2(\mathcal{F}_0, D(A))$ and $q \in D(A)$, then*

- $C_\infty(\theta_0, q) < \infty$.
- *For every $\phi \in H$*

$$\limsup_{t \to \infty} \mathbb{E}\left[|\langle\theta(t) - \Theta_{st}, \phi\rangle|^2\right] \leq \frac{\varepsilon_Q}{\kappa}\|\phi\|^2 C_\infty(\theta_0, q).$$

In order to be of interest for applications, this theorem requires two conditions:

(1) that ε_Q is small.
(2) that Θ_{st} is significantly affected by the noise.

Obviously if $\kappa\lambda_L \gg \kappa\lambda$ then Θ_{st} is significantly affected by the noise. Thus we reconduct ourselves to the previous framework already treated. In Sect. 4 we will show a concrete example where this phenomenon appears.

4 Explicit Computations

Theorem 8 is not completely suitable for numerical simulations because the definition of $K(x, y)$ is not explicitly available for every domain smooth and bounded. In this section we will present an explicit construction with a fluid dynamics interpretation, again based on vortex structures, which satisfies both ε_Q arbitrarily small and $\tilde{q}$ arbitrarily large outside a boundary layer. Moreover we will show numerically that, even relaxing the conditions in this construction, the noise influences the behavior of the stationary solution.

4.1 Explicit Construction

We will construct a noise of the form $\Gamma \sum_{k \in K} u_k(x) \, dW_t^k$ with

$$u_k(x) = w_r(x - x_k), \qquad w_r(x) = r^{-1} w\left(\frac{x}{r}\right)$$

for suitable r and w. Thus, the covariance of this noise is

$$Q(x, y) = \Gamma^2 \sum_{z \in \Lambda_N} w_r(x - z) \otimes w_r(y - z).$$

We need to choose x_k, called the "centers" of the vortex blobs, and a suitable vector field w. The vector field w must satisfy several conditions:

1. w is smooth and $\nabla \cdot w = 0$;
2. w has compact support contained in $\overline{B(0, 1)}$;
3. w is close to $\frac{1}{2\pi} \frac{x^\perp}{|x|^2}$ near $x = 0$.

The first two properties are useful in order to have that the u_j's model the velocity of an incompressible fluid at rest. The third one is close to our idea of vortex structures.

Now we choose the centers. For a fixed $\delta > 0$, we choose a positive integer N such that $\frac{1}{N} \leq \delta$. Then we consider the set Λ_N of all points of D_δ having coordinates of the form $\left(\frac{k}{N}, \frac{h}{N}\right)$ with $k, h \in \mathbb{Z}$. Thanks to this choice we have

$$\min_{z_1 \neq z_2 \in \Lambda_N} |z_1 - z_2| = \frac{1}{N}, \qquad \min_{z \in \Lambda_N} d(z, \partial D) \geq \delta.$$

We choose another positive integer M and we decompose the set Λ_N as the disjoint union of the sets

$$\Lambda_N = \bigcup_{(k_0, h_0) \in \{0, 1, \dots, M-1\}^2} \Lambda_N^{(M, k_0, h_0)}$$

where $\left(\frac{k}{N}, \frac{h}{N}\right) \in \Lambda_N^{(M, k_0, h_0)}$ if $k = Mn + k_0$, $h = Mm + h_0$, with $n, m \in \mathbb{Z}$. In this way, we have

$$\min_{z_1 \neq z_2 \in \Lambda_N^{(M, k_0, h_0)}} |z_1 - z_2| = \frac{M}{N}$$

for each $(k_0, h_0) \in \{0, 1, \dots, M - 1\}^2$. We have introduced M and the sets $\Lambda_N^{(M, k_0, h_0)}$ in order to have that each couple of u_j and u_k in the same class have disjoint supports for r small enough and this is sufficient for our estimates, because it implies that the vector fields are "almost" orthogonal in the sense of Theorem 5.

In order to have the supports disjoint for elements of $\Lambda_N^{(M,k_0,h_0)}$ and the action of the noise covers the full set $D_{2\delta}$ we ask $r \leq \frac{M}{2N}$. Now we can focus on the vector field w. In order of being divergence free we set $w = \nabla^\perp \psi$. Thus, we look for a smooth function ψ on $\mathbb{R}^2$, compactly supported in $\overline{B\,(0, 1)}$, close to $\frac{1}{2\pi} \log |x|$ near $x = 0$. A possible construction is the following one:

$$\psi(x) = \int_{\mathbb{R}^2} \psi_0(x - y) f_\varepsilon(y)\, dy$$

where f_ε is a mollifier with support in $B(0, \varepsilon)$ and ψ_0 is a $C^\infty(\mathbb{R}^2 \setminus \{0\})$ radial function such that

$$\psi_0(x) = \frac{\log |x|}{2\pi} \quad \text{for } |x| \leq \frac{1}{3} \quad \text{and } \psi_0(x) = 0 \quad \text{for } |x| > \frac{2}{3}.$$

Moreover, it can be proved that w defined above satisfies

$$\|w\|^2 \leq C \log \frac{1}{\varepsilon}, \qquad \|w_r\|^2 = \|w\|^2.$$

Thanks to these relations we can obtain, easily, an estimate of ε_Q

$$\int\int v(x)^T\, Q(x, y)\, v(y)\, dxdy = \Gamma^2 \sum_{z \in \Lambda_N} \left(\int w_r(x - z) \cdot v(x)\, dx \right)^2$$

$$= \|w\|\, \Gamma^2 \sum_{(k_0,h_0) \in \{0,1,\dots,M-1\}^2}\, \sum_{z \in \Lambda_N^{(M,k_0,h_0)}} \left(\int \frac{w_r(x - z)}{\|w\|_{L^2}} \cdot v(x)\, dx \right)^2$$

$$\leq M^2\, \|w\|^2\, \Gamma^2\, \|v\|^2.$$

Thus, taking $\varepsilon = \frac{1}{N}$ we get

$$\varepsilon_Q \leq M^2 \Gamma^2 C \log N$$

which is small if, given N, Γ is small enough.

For what concern the analysis of a lower bound for $\tilde{q}(x)$ in $D_{2\delta}$, the computations are a bit more involving and we refer to [7] for a complete discussion which is out of our scope. We just claim that if

$$r \geq \frac{12}{N}, \qquad M > 24, \qquad N \text{ is large enough}$$

then

$$\tilde{q}(x) \geq \frac{\Gamma^2 N}{16\pi} \quad \text{in } D_{2\delta}.$$

4.2 Numerical Simulation

Summing up the results of previous subsection, we have seen that if

$$r \leq \frac{M}{2N}, \qquad r \leq \delta, \qquad r \geq \frac{12}{N}, \qquad \varepsilon = \frac{1}{N} \leq \frac{1}{6}, \qquad N \text{ is large enough}$$

we have

$$\varepsilon_Q \leq M^2 \Gamma^2 C \log N, \qquad \tilde{q}(x) \geq \frac{\Gamma^2 N}{16\pi} \quad \text{in } D_{2\delta}.$$

These conditions are strong from the numerical point of view: the cardinality of K must be very large and a finite but not small M is required. Certain supports have to overlap so that the noise acts everywhere. However in [10] it has been shown, numerically, that these conditions are overabundant and much less is required to see the influence of the noise on the solution, namely that Θ_{st} differs significantly from the parabolic profile even for relatively modest sets K and for $M = 1$. In this subsection we are working in an infinite 2D channel, suspend the requirement that q, Θ have to decay at infinity, although not strictly covered by the theory described in Sect. 2.1. We assume that the function $q(x)$ is equal to a constant q.

For numerical reasons we consider the problem in the bounded domain

$$\tilde{D} = (\tan(-1.54), \tan(1.54)) \times (-0.1, 0.1).$$

In order to have that the σ_k's model a fluid at rest, we can take

$$r \leq max_{k \in K} d(\partial \tilde{D}, x_k) \text{ and } \varepsilon < \frac{1}{6}.$$

These are the real constraints on the parameters of our numerical simulation. The other parameters Γ, K, $\{x_k\}_{k \in K}$ can be chosen more arbitrarily in order to have satisfactory results.

Differently from [10], here the vortex structures have not been chosen on a grid equally spaced in both directions. In particular the points thicken in the x_1 direction. We have chosen 2 points in the z direction between -0.05 and 0.05 and for what concern the x_1 direction we have chosen 2 points between 0 and 0.2, 4 points between 0.2 and 0.4 and 8 points between 0.4 and 0.6. In order to improve the smoothness of the solution, avoiding a shock in the number of vortices, we prefer to consider some few vortices for $x_1 > 0.6$. They only slightly affect the behavior of our solution in the critical region of interest $x_1 < 0.5$. Thus we consider 4 points between 0.6 and 0.8 and 2 points between 0.8 and 1. Obviously we avoid repetition of the vortices. In conclusion we have 34 vortices. Moreover, we take $r = 0.05$, $\varepsilon = 0.1$ and $\Gamma = 0.03$. The other parameters of the problem are $\kappa = 0.05$ and $q \equiv 1$. In this way the quantity M and N are not well defined and the impact

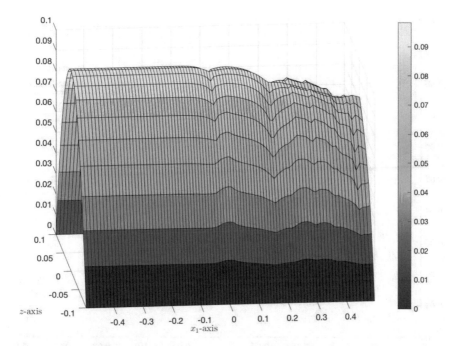

Fig. 1 Solution in the critical region

of the operator $\mathcal{L}$ is related to a small portion of the domain $\tilde{D}$, however we can completely appreciate how it changes the profile of the solution.

Figures 1 and 2 illustrate the modification of the profile, from the standard parabolic one of free diffusion in a steady medium, to the case of turbulent decay. Even if we use just a really reduced number of vortices we can observe a significant decay modification of the profile due to turbulence where vortices thicken.

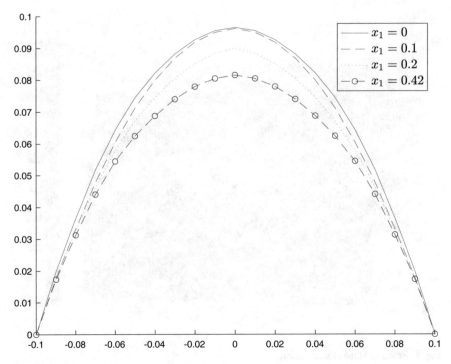

Fig. 2 Profiles at different values of x_1

References

1. Agmon, S.: Lectures on elliptic boundary value problems, vol. 369. American Mathematical Soc. (2010)
2. Brzeźniak, Z., Capiński, M., Flandoli, F.: Approximation for diffusion in random fields. Stochastic Analysis and Applications **8**(3), 293–313 (1990)
3. Brzeźniak, Z., Flandoli, F.: Almost sure approximation of wong-zakai type for stochastic partial differential equations. Stochastic processes and their applications **55**(2), 329–358 (1995)
4. Da Prato, G., Zabczyk, J.: Stochastic equations in infinite dimensions. Cambridge university press (1992)
5. Flandoli, F.: Regularity Theory and Stochastic Flows for Parabolic SPDEs, vol. 9. Gordon and Breach Publishers (1995)
6. Flandoli, F., Luongo, E.: Stochastic Partial Differential Equations in Fluid Mechanics To appear in Lecture Notes in Mathematics, Springer.
7. Flandoli, F., Galeati, L., Luo, D.: Eddy heat exchange at the boundary under white noise turbulence. Philosophical Transactions of the Royal Society A **380**(2219), 20210096 (2022)
8. Flandoli, F., Galeati, L., Luo, D.: Quantitative convergence rates for scaling limit of spdes with transport noise. arXiv preprint arXiv:2104.01740 (2021)
9. Flandoli, F., Huang, R.: Noise based on vortex structures in 2d and 3d (In preparation)
10. Flandoli, F., Luongo, E.: Heat diffusion in a channel under white noise modeling of turbulence. Mathematics in Engineering **4**(4), 1–21 (2022)
11. Flandoli, F., Pappalettera, U.: 2d euler equations with stratonovich transport noise as a large-scale stochastic model reduction. Journal of Nonlinear Science **31**(1), 1–38 (2021)

12. Flandoli, F., Pappalettera, U.: From additive to transport noise in 2d fluid dynamics. Stochastics and Partial Differential Equations: Analysis and Computations pp. 1–41 (2022)
13. Grisvard, P.: Commutativité de deux foncteurs d interpolation et applications. Journal de mathématiques pures et appliquées **45**(2), 143 (1966)
14. Gyöngy, I.: On the approximation of stochastic partial differential equations i. Stochastics: An International Journal of Probability and Stochastic Processes **25**(2), 59–85 (1988)
15. Gyöngy, I.: On the approximation of stochastic partial differential equations ii. Stochastics: An International Journal of Probability and Stochastic Processes **26**(3), 129–164 (1989)
16. Pappalettera, U.: Quantitative mixing and dissipation enhancement property of ornstein-uhlenbeck flow. Communications in Partial Differential Equations pp. 1–32 (2022).
17. Pazy, A.: Semigroups of linear operators and applications to partial differential equations, vol. 44. Springer Science & Business Media (2012)
18. Tessitore, G., Zabczyk, J.: Wong-zakai approximations of stochastic evolution equations. Journal of Evolution Equations **6**(4), 621–655 (2006)
19. Twardowska, K.: Approximation theorems of wong-zakai type for stochastic differential equations in infinite dimensions (1993)
20. Wong, E., Zakai, M.: On the convergence of ordinary integrals to stochastic integrals. The Annals of Mathematical Statistics **36**(5), 1560–1564 (1965)

Existence and Uniqueness of Maximal Solutions to a 3D Navier-Stokes Equation with Stochastic Lie Transport

Daniel Goodair

Abstract We present here a criterion to conclude that an abstract SPDE possesses a unique maximal strong solution, which we apply to a three dimensional Stochastic Navier-Stokes Equation. Motivated by the work of Kato and Lai we ask that there is a comparable result here in the stochastic case whilst facilitating a variety of noise structures such as additive, multiplicative and transport. In particular our criterion is designed to fit viscous fluid dynamics models with Stochastic Advection by Lie Transport (SALT) as introduced in Holm (Proc R Soc A: Math Phys Eng Sci 471(2176):20140963, 2015). Our application to the Incompressible Navier-Stokes equation matches the existence and uniqueness result of the deterministic theory. This short work summarises the results and announces two papers (Crisan et al., Existence and uniqueness of maximal strong solutions to nonlinear SPDEs with applications to viscous fluid models, in preparation; Crisan and Goodair, Analytical properties of a 3D stochastic Navier-Stokes equation, 2022, in preparation) which give the full details for the abstract well-posedness arguments and application to the Navier-Stokes Equation respectively.

Keywords Stochastic transport · SPDE · Navier-Stokes · Well-posedness

1 Introduction

The theoretical analysis of fluid models perturbed by transport noise has been in significant demand since the release of the seminal works [16] and [17]. In the papers Holm and Mémin establish a new class of stochastic equations driven by transport noise which serve as much improved fluid dynamics models by adding uncertainty in the transport of the fluid parcels to reflect the unresolved scales. Here we consider the SALT [16] Navier-Stokes Equation given by

D. Goodair (✉)
Imperial College London, London, England, UK
e-mail: daniel.goodair16@imperial.ac.uk
https://www.imperial.ac.uk/people/daniel.goodair16

© The Author(s) 2023
B. Chapron et al. (eds.), *Stochastic Transport in Upper Ocean Dynamics*,
Mathematics of Planet Earth 10, https://doi.org/10.1007/978-3-031-18988-3_7

$$u_t - u_0 + \int_0^t \mathcal{L}_{u_s} u_s \, ds - \int_0^t \Delta u_s \, ds + \int_0^t B u_s \circ d\mathcal{W}_s + \int_0^t \nabla \rho_s ds = 0 \qquad (1)$$

and supplemented with the divergence-free (incompressibility) and zero-average conditions on the three dimensional torus $\mathbb{T}^3$. The equation is presented here in velocity form where u represents the fluid velocity, ρ the pressure, $\mathcal{L}$ is the mapping corresponding to the nonlinear term, $\mathcal{W}$ is a cylindrical Brownian Motion and B is the relevant transport operator defined with respect to a collection of functions (ξ_i) which physically represent spatial correlations. The explicit meaning of these conditions and the definitions of the operators involved are given at the beginning of Sect. 2.2. These (ξ_i) can be determined at coarse-grain resolutions from finely resolved numerical simulations, and mathematically are derived as eigenvectors of a velocity-velocity correlation matrix (see [3, 4, 5]). The corresponding stochastic Euler equation was derived in [12] and the viscous term plays no additional role in the stochastic derivation (without loss of generality we set the viscosity coefficient to be 1).

There has been limited progress in proving well-posedness for this class of equations: Crisan, Flandoli and Holm [5] have shown local existence and uniqueness for the 3D Euler Equation on the torus, whilst Crisan and Lang [9, 11, 10] demonstrated the same result for the Euler, Rotating Shallow Water and Great Lake Equations on the torus once more. Whilst this represents a strong start in the theoretical analysis (alongside works for SPDEs with general transport noise e.g. [2, 1]), the modelling literature continues to expand in both the deterministic fluid models (see for example Figure 2 of [8] and the analysis therein) and method of stochastic perturbation (for example we may soon look to introduce nonlinearity and time dependence in the (ξ_i)). The significance of an abstract approach to the well-posedness question is clear, and whilst we discuss here only an application to SALT Navier-Stokes [16, 12] the hope is that other stochastic viscous fluid models can be similarly solved by simply checking the required assumptions. We state our equation in the form

$$\Psi_t = \Psi_0 + \int_0^t \mathcal{A}(s, \Psi_s) ds + \int_0^t \mathcal{G}(s, \Psi_s) d\mathcal{W}_s \qquad (2)$$

for operators $\mathcal{A}$ and $\mathcal{G}$ to be elucidated in due course. The most notable contribution to the well-posedness theory for an abstract nonlinear SPDE is from [13]. Here the authors prove the existence of a unique maximal solution to their abstract equation and apply this to the three dimensional primitive equations with a Lipschitz type multiplicative noise. The class of equations which we are concerned with include a differential operator in the noise term, preventing us from applying this framework. Moreover the assumptions on their operator $\mathcal{A}$ are quite explicit in terms of the sum of the standard fluid nonlinear term and a linear operator, which we don't restrict ourselves to. Overall our assumptions are much more general and allow for a straightforwards application to a wider class of SPDEs. Another relevant piece here is the work of Glatt-Holtz and Ziane [14] whom show the same existence and

$$u_t - u_0 + \int_0^t \mathcal{L}_{u_s} u_s \, ds - \int_0^t \Delta u_s \, ds + \int_0^t B u_s \circ dW_s + \int_0^t \nabla \rho_s ds = 0 \quad (1)$$

and supplemented with the divergence-free (incompressibility) and zero-average conditions on the three dimensional torus $\mathbb{T}^3$. The equation is presented here in velocity form where u represents the fluid velocity, ρ the pressure, $\mathcal{L}$ is the mapping corresponding to the nonlinear term, $\mathcal{W}$ is a cylindrical Brownian Motion and B is the relevant transport operator defined with respect to a collection of functions (ξ_i) which physically represent spatial correlations. The explicit meaning of these conditions and the definitions of the operators involved are given at the beginning of Sect. 2.2. These (ξ_i) can be determined at coarse-grain resolutions from finely resolved numerical simulations, and mathematically are derived as eigenvectors of a velocity-velocity correlation matrix (see [3, 4, 5]). The corresponding stochastic Euler equation was derived in [12] and the viscous term plays no additional role in the stochastic derivation (without loss of generality we set the viscosity coefficient to be 1).

There has been limited progress in proving well-posedness for this class of equations: Crisan, Flandoli and Holm [5] have shown local existence and uniqueness for the 3D Euler Equation on the torus, whilst Crisan and Lang [9, 11, 10] demonstrated the same result for the Euler, Rotating Shallow Water and Great Lake Equations on the torus once more. Whilst this represents a strong start in the theoretical analysis (alongside works for SPDEs with general transport noise e.g. [2, 1]), the modelling literature continues to expand in both the deterministic fluid models (see for example Figure 2 of [8] and the analysis therein) and method of stochastic perturbation (for example we may soon look to introduce nonlinearity and time dependence in the (ξ_i)). The significance of an abstract approach to the well-posedness question is clear, and whilst we discuss here only an application to SALT Navier-Stokes [16, 12] the hope is that other stochastic viscous fluid models can be similarly solved by simply checking the required assumptions. We state our equation in the form

$$\boldsymbol{\Psi}_t = \boldsymbol{\Psi}_0 + \int_0^t \mathcal{A}(s, \boldsymbol{\Psi}_s) ds + \int_0^t \mathcal{G}(s, \boldsymbol{\Psi}_s) dW_s \quad (2)$$

for operators $\mathcal{A}$ and $\mathcal{G}$ to be elucidated in due course. The most notable contribution to the well-posedness theory for an abstract nonlinear SPDE is from [13]. Here the authors prove the existence of a unique maximal solution to their abstract equation and apply this to the three dimensional primitive equations with a Lipschitz type multiplicative noise. The class of equations which we are concerned with include a differential operator in the noise term, preventing us from applying this framework. Moreover the assumptions on their operator $\mathcal{A}$ are quite explicit in terms of the sum of the standard fluid nonlinear term and a linear operator, which we don't restrict ourselves to. Overall our assumptions are much more general and allow for a straightforwards application to a wider class of SPDEs. Another relevant piece here is the work of Glatt-Holtz and Ziane [14] whom show the same existence and

Existence and Uniqueness of Maximal Solutions to a 3D Navier-Stokes Equation with Stochastic Lie Transport

Daniel Goodair

Abstract We present here a criterion to conclude that an abstract SPDE possesses a unique maximal strong solution, which we apply to a three dimensional Stochastic Navier-Stokes Equation. Motivated by the work of Kato and Lai we ask that there is a comparable result here in the stochastic case whilst facilitating a variety of noise structures such as additive, multiplicative and transport. In particular our criterion is designed to fit viscous fluid dynamics models with Stochastic Advection by Lie Transport (SALT) as introduced in Holm (Proc R Soc A: Math Phys Eng Sci 471(2176):20140963, 2015). Our application to the Incompressible Navier-Stokes equation matches the existence and uniqueness result of the deterministic theory. This short work summarises the results and announces two papers (Crisan et al., Existence and uniqueness of maximal strong solutions to nonlinear SPDEs with applications to viscous fluid models, in preparation; Crisan and Goodair, Analytical properties of a 3D stochastic Navier-Stokes equation, 2022, in preparation) which give the full details for the abstract well-posedness arguments and application to the Navier-Stokes Equation respectively.

Keywords Stochastic transport · SPDE · Navier-Stokes · Well-posedness

1 Introduction

The theoretical analysis of fluid models perturbed by transport noise has been in significant demand since the release of the seminal works [16] and [17]. In the papers Holm and Mémin establish a new class of stochastic equations driven by transport noise which serve as much improved fluid dynamics models by adding uncertainty in the transport of the fluid parcels to reflect the unresolved scales. Here we consider the SALT [16] Navier-Stokes Equation given by

D. Goodair (✉)
Imperial College London, London, England, UK
e-mail: daniel.goodair16@imperial.ac.uk
https://www.imperial.ac.uk/people/daniel.goodair16

© The Author(s) 2023
B. Chapron et al. (eds.), *Stochastic Transport in Upper Ocean Dynamics*,
Mathematics of Planet Earth 10, https://doi.org/10.1007/978-3-031-18988-3_7

uniqueness for the incompressible $3D$ Navier-Stokes with again a Lipschitz noise term. Though we cannot apply this method in the presence of our transport noise we look to adapt this argument to fit not just our Navier-Stokes equation but the wider class of stochastic viscous fluid models and SPDEs beyond. The impact of the boundary is fundamental to the equation and the approach of Glatt-Holtz and Ziane copes with the arising issues by working in the right function spaces; we recognised the importance of this in establishing an abstract framework which we hope to apply to such stochastic transport equations on the bounded domain as well.

This short summary work contains three more sections: in the subsequent one we properly define our Stochastic Navier-Stokes equation through the operators involved, the relevant function spaces, the notions of solution and main results. Following this we concretely define our abstract formulation and notion of solution, giving the assumptions that we require and the main results for the abstract equation. These assumptions are then all that needs to be checked to conclude the relevant existence and uniqueness for the proposed SPDE. In the final section we discuss the key steps behind proving these results; in the spirit of this as a summary work announcing our results we do not give a complete proof, though all such arguments are to be found in [7]. We then address how our Navier-Stokes equation fits the context of the abstract formulation, though once more we do not give a thorough justification that the operators of our equation satisfy the required assumptions, with this precise treatment to come in [6].

2 SALT Navier-Stokes and Results

As alluded to in this section we formally introduce Eq. (1) and state the main results.

2.1 Preliminaries from Stochastic Analysis

Throughout the paper we work with a fixed filtered probability space $(\Omega, \mathcal{F}, (\mathcal{F}_t), \mathbb{P})$ satisfying the usual conditions of completeness and right continuity. We take $\mathcal{W}$ to be a cylindrical Brownian Motion over some Hilbert Space $\mathfrak{U}$ with orthonormal basis (e_i). The choice of $\mathfrak{U}$ and the subsequent basis play no role in the analysis. Recall ([15, Subsection 1.4]) that $\mathcal{W}$ admits the representation $\mathcal{W}_t = \sum_{i=1}^{\infty} e_i W_t^i$ as a limit in $L^2(\Omega; \mathfrak{U}')$ whereby the (W^i) are a collection of i.i.d. standard real valued Brownian Motions and $\mathfrak{U}'$ is an enlargement of the Hilbert Space $\mathfrak{U}$ such that the embedding $J : \mathfrak{U} \to \mathfrak{U}'$ is Hilbert-Schmidt and $\mathcal{W}$ is a JJ^*-cylindrical Brownian Motion over $\mathfrak{U}'$. Given a process $F : [0, T] \times \Omega \to \mathscr{L}^2(\mathfrak{U}; \mathscr{H})$ progressively measurable and such that $F \in L^2\left(\Omega \times [0, T]; \mathscr{L}^2(\mathfrak{U}; \mathscr{H})\right)$, for any $0 \leq t \leq T$ we understand the stochastic integral

$$\int_0^t F_s dW_s$$

to be the infinite sum

$$\sum_{i=1}^{\infty} \int_0^t F_s(e_i) dW_s^i$$

taken in $L^2(\Omega; \mathcal{H})$. We can extend this notion to processes F which are such that $F(\omega) \in L^2\left([0, T]; \mathscr{L}^2(\mathfrak{U}; \mathcal{H})\right)$ for $\mathbb{P} - a.e.$ ω via the traditional localisation procedure. In this case the stochastic integral is a local martingale in $\mathcal{H}$. A complete, direct construction of this integral, a treatment of its properties and the fundamentals of stochastic calculus in infinite dimensions can be found in [15, Section 1].

2.2 SALT Navier-Stokes Equation

We present Eq. (1) on the three dimensional torus $\mathbb{T}^3$ (noting that all results hold on $\mathbb{T}^2$), and detail now the operators involved alongside the function spaces which define the equations. The operator $\mathcal{L}$ is defined for sufficiently regular functions $\phi, \psi : \mathbb{T}^3 \to \mathbb{R}^3$ by

$$\mathcal{L}_\phi \psi := \sum_{j=1}^{3} \phi^j \partial_j \psi$$

where $\phi^j : \mathbb{T}^3 \to \mathbb{R}$ is the jth coordinate mapping of ϕ and $\partial_j \psi$ is defined by its kth coordinate mapping $(\partial_j \psi)^k = \partial_j \psi^k$. The operator B is defined as a linear operator on $\mathfrak{U}$ (introduced in Sect. 2.1) by its action on the basis vectors $B(e_i, \cdot) := B_i(\cdot)$ by

$$B_i = \mathcal{L}_{\xi_i} + \mathcal{T}_{\xi_i}$$

for $\mathcal{L}$ as above and

$$\mathcal{T}_\phi \psi := \sum_{j=1}^{3} \psi^j \nabla \phi^j.$$

A complete discussion of how B is then defined on $\mathfrak{U}$ is given in [15, Subsection 2.2]. We embed the divergence-free and zero-average conditions into the relevant function spaces and simply define our solutions as belonging to these spaces. To be explicit, by a divergence-free function we mean a $\phi \in W^{1,2}(\mathbb{T}^3; \mathbb{R}^3)$ such that

3.1 Assumption Set 1

We work with a quartet of continuously embedded Hilbert Spaces

$$V \hookrightarrow H \hookrightarrow U \hookrightarrow X$$

and the operators

$$\mathcal{A} : [0, \infty) \times V \to U,$$

$$\mathcal{G} : [0, \infty) \times V \to \mathscr{L}^2(\mathfrak{U}; H).$$

We ask that there is a continuous bilinear form $\langle \cdot, \cdot \rangle_{X \times H} : X \times H \to \mathbb{R}$ such that for $\phi \in U$ and $\psi \in H$,

$$\langle \phi, \psi \rangle_{X \times H} = \langle \phi, \psi \rangle_U. \tag{8}$$

Moreover the continuity and bilinearity ensures that there exists some constant c whereby for all such ϕ, ψ,

$$|\langle \phi, \psi \rangle_{X \times H}| \leq c \|\phi\|_X \|\psi\|_H. \tag{9}$$

As we look to use a Galerkin Scheme to solve our equation, we introduce now a sequence of spaces (V_n) contained in V given by $V_n := \text{span}\{a_1, \ldots, a_n\}$ for (a_n) an orthogonal basis in U. Defining $\mathcal{P}_n$ to be the orthogonal projection onto V_n in X, we shall also assume that the restriction of $\mathcal{P}_n$ to U is an orthogonal projection in U and that the sequence of these projections is uniformly bounded on H: that is, that there exists some constant c independent of n such that for all $\phi \in H$,

$$\|\mathcal{P}_n \phi\|_H^2 \leq c \|\phi\|_H^2. \tag{10}$$

We also require the existence of a real valued sequence (μ_n) with $\mu_n \to \infty$, which is such that for any $\phi \in U$ and $\psi \in H$,

$$\|(I - \mathcal{P}_n)\phi\|_X \leq \frac{1}{\mu_n} \|\phi\|_U, \tag{11}$$

$$\|(I - \mathcal{P}_n)\psi\|_U \leq \frac{1}{\mu_n} \|\psi\|_H \tag{12}$$

where I represents the identity operator in the corresponding spaces. These assumptions are of course supplemented by a series of assumptions on the operators. We shall use general notation c_t to represent a function $c. : [0, \infty) \to \mathbb{R}$ bounded on $[0, T]$ for any $T > 0$, evaluated at the time t. Moreover we define functions K,

$$\sum_{j=1}^{3} \partial_j \phi^j = 0$$

and by zero-average we ask for a $\psi \in L^2(\mathbb{T}^3; \mathbb{R}^3)$ with the property

$$\int_{\mathbb{T}^3} \psi \, d\lambda = 0$$

for λ the Lebesgue measure on $\mathbb{T}^3$. We introduce the space $L_\sigma^2(\mathbb{T}^3; \mathbb{R}^3)$ as the subspace of $L^2(\mathbb{T}^3; \mathbb{R}^3)$ consisting of zero-average functions which are 'weakly divergence-free'; see [18] Definition 2.1 for the precise construction. $W_\sigma^{1,2}(\mathbb{T}^3; \mathbb{R}^3)$ is then defined as the subspace of $W^{1,2}(\mathbb{T}^3; \mathbb{R}^3)$ consisting of zero-average divergence-free functions, and $W_\sigma^{2,2}(\mathbb{T}^3; \mathbb{R}^3) := W^{2,2}(\mathbb{T}^3; \mathbb{R}^3) \cap W_\sigma^{1,2}(\mathbb{T}^3; \mathbb{R}^3)$.

As is standard in the treatment of the incompressible Navier-Stokes Equation we consider a projected version to eliminate the pressure term and facilitate us working in the above spaces. Note that ρ does not come with an evolution equation and is simply chosen to ensure the incompressibility condition. The idea is to solve the projected equation and then append a pressure to it, see [18]. To this end we introduce the standard Leray Projector $\mathcal{P}$ defined as the orthogonal projection in $L^2(\mathbb{T}^3; \mathbb{R}^3)$ onto $L_\sigma^2(\mathbb{T}^3; \mathbb{R}^3)$. As we look to project equation (1) as discussed, we ought to address the Stratonovich integral. We look to convert this term into an Itô integral to enable our analysis, but the resulting converted and projected equation should not depend on the order in which the projection and conversion occur. To this end we assume that the (ξ_i) are such that $\xi_i \in W_\sigma^{1,2}(\mathbb{T}^3; \mathbb{R}^3) \cap W^{3,\infty}(\mathbb{T}^3; \mathbb{R}^3)$ and satisfy the bound

$$\sum_{i=1}^{\infty} \|\xi_i\|_{W^{3,\infty}}^2 < \infty. \tag{3}$$

The significance of the bound (3) will be revisited, but for now we note that as each ξ_i is divergence-free then each B_i satisfies the property that $\mathcal{P} B_i$ is equal to $\mathcal{P} B_i \mathcal{P}$ on $W^{1,2}(\mathbb{T}^3; \mathbb{R}^3)$ which ensures that the projection and conversion commute. Our new equation is then

$$u_t - u_0 + \int_0^t \mathcal{P} \mathcal{L}_{u_s} u_s \, ds + \int_0^t A u_s ds$$
$$- \frac{1}{2} \sum_{i=1}^{\infty} \int_0^t \mathcal{P} B_i^2 u_s ds + \sum_{i=1}^{\infty} \int_0^t \mathcal{P} B_i u_s dW_s^i = 0 \tag{4}$$

where $A := -\mathcal{P}\Delta$ is known as the Stokes Operator. Details of the Itô-Stratonovich conversion can be found in [15, Subsection 2.3]. We shall use the Stokes operator to define inner products with which we equip our function spaces. Recall from [18] Theorem 2.24 for example that there exists a collection of functions (a_k), $a_k \in W_\sigma^{1,2}(\mathbb{T}^3; \mathbb{R}^3) \cap C^\infty(\mathbb{T}^3; \mathbb{R}^3)$ such that the (a_k) are eigenfunctions of A, are an orthonormal basis in $L^2_\sigma(\mathbb{T}^3; \mathbb{R}^3)$ and an orthogonal basis in $W^{1,2}_\sigma(\mathbb{T}^3; \mathbb{R}^3)$ considered as Hilbert Spaces with standard $L^2(\mathbb{T}^3; \mathbb{R}^3)$, $W^{1,2}(\mathbb{T}^3; \mathbb{R}^3)$ inner products. The corresponding eigenvalues (λ_k) are strictly positive and approach infinity as $k \to \infty$. Thus any $\phi \in W^{1,2}_\sigma(\mathbb{T}^3; \mathbb{R}^3)$ admits the representation

$$\phi = \sum_{k=1}^{\infty} \phi_k a_k$$

so for $m \in \mathbb{N}$ we can define $A^{m/2}$ by

$$A^{m/2} : \phi \mapsto \sum_{k=1}^{\infty} \lambda_k^{m/2} \phi_k a_k$$

which is a well defined element of $L^2_\sigma(\mathbb{T}^3; \mathbb{R}^3)$ on any ϕ such that

$$\sum_{k=1}^{\infty} \lambda_k^m \phi_k^2 < \infty. \tag{5}$$

For ϕ, ψ with the property (5) then the bilinear form

$$\langle \phi, \psi \rangle_m := \langle A^{m/2}\phi, A^{m/2}\psi \rangle$$

is well defined. For $m = 1, 2$ this is an inner product on the spaces $W^{1,2}_\sigma(\mathbb{T}^3; \mathbb{R}^3)$, $W^{2,2}_\sigma(\mathbb{T}^3; \mathbb{R}^3)$ respectively which is equivalent to the standard $W^{1,2}(\mathbb{T}^3; \mathbb{R}^3)$, $W^{2,2}(\mathbb{T}^3; \mathbb{R}^3)$ inner product. Of course $\langle \cdot, \cdot \rangle_3$ is well defined on $\bigcup_{k=1}^{\infty} \text{span}\{a_1, \ldots, a_k\}$ and so we define $W^{3,2}_\sigma(\mathbb{T}^3; \mathbb{R}^3)$ as the completion of $\bigcup_{k=1}^{\infty} \text{span}\{a_1, \ldots, a_k\}$ in this inner product. We consider $W^{m,2}_\sigma(\mathbb{T}^3; \mathbb{R}^3)$ as a Hilbert Space equipped with the $\langle \cdot, \cdot \rangle_m$ inner product, and define our solution to the equation (4) relative to these spaces.

2.3   Notions of Solution and Results

We frame this definition for an $\mathcal{F}_0$−measurable $u_0 : \Omega \to W^{1,2}_\sigma(\mathbb{T}^3; \mathbb{R}^3)$. Here and throughout we use the notation $\mathbf{1}$ for the indicator function.

Definition 1 A pair (u, τ) where τ is a $\mathbb{P} − a.s.$ positive stopping time and u is a process such that for $\mathbb{P} − a.e.$ ω, $u_.(\omega) \in C\left([0, T]; W^{1,2}_\sigma(\mathbb{T}^3; \mathbb{R}^3)\right)$ and $u_.(\omega)\mathbf{1}_{.\leq\tau(\omega)} \in L^2\left([0, T]; W^{2,2}_\sigma(\mathbb{T}^3; \mathbb{R}^3)\right)$ for all $T > 0$ with $u.\mathbf{1}_{.\leq\tau}$ progressively measurable in $W^{2,2}_\sigma(\mathbb{T}^3; \mathbb{R}^3)$, is said to be a local strong solution of the equation (2) if the identity

$$u_t - u_0 + \int_0^{t\wedge\tau} \mathcal{P}\mathcal{L}_{u_s} u_s\, ds + \int_0^{t\wedge\tau} A u_s ds$$
$$- \frac{1}{2}\sum_{i=1}^{\infty}\int_0^{t\wedge\tau} \mathcal{P}B_i^2 u_s ds + \sum_{i=1}^{\infty}\int_0^{t\wedge\tau} \mathcal{P}B_i u_s dW_s^i = 0 \tag{6}$$

holds $\mathbb{P} − a.s.$ in $L^2_\sigma(\mathbb{T}^3; \mathbb{R}^3)$ for all $t \geq 0$.

We shall address why this definition makes sense in the abstract setting in Sect. 3.3, before then translating this abstract framework back to our Navier-Stokes Equation.

Definition 2 A pair (u, Θ) such that there exists a sequence of stopping times (θ_j) which are $\mathbb{P} − a.s.$ monotone increasing and convergent to Θ, whereby $(u_{.\wedge\theta_j}, \theta_j)$ is a local strong solution of the equation (4) for each j, is said to be a maximal strong solution of the equation (4) if for any other pair (v, Γ) with this property then $\Theta \leq \Gamma$ $\mathbb{P} − a.s.$ implies $\Theta = \Gamma$ $\mathbb{P} − a.s.$

Definition 3 A maximal strong solution (u, Θ) of the equation (4) is said to be unique if for any other such solution (v, Γ), then $\Theta = \Gamma$ $\mathbb{P} − a.s.$ and for all $t \in [0, \Theta)$,

$$\mathbb{P}(\{\omega \in \Omega : u_t(\omega) = v_t(\omega)\}) = 1.$$

We can now state the main result of the paper.

Theorem 1 *For any given $\mathcal{F}_0−$ measurable $u_0 : \Omega \to W^{1,2}_\sigma(\mathbb{T}^3; \mathbb{R}^3)$, there exists a unique maximal strong solution (u, Θ) of the equation (4). Moreover at $\mathbb{P} − a.e.$ ω for which $\Theta(\omega) < \infty$, we have that*

$$\sup_{r\in[0,\Theta(\omega))} \|u_r(\omega)\|_1^2 + \int_0^{\Theta(\omega)} \|u_r(\omega)\|_2^2 dr = \infty. \tag{7}$$

3   Abstract Framework and Results

We now establish the abstract framework through which we arrive at Theorem 1. This involves giving two sets of assumptions before exploring the abstract method with the assumptions in place, and then in Sect. 4.2 discussing how (4) fits into this framework. These assumption sets pertain to two different notions of solution (both strong in the probabilistic sense but related to different spaces), the reason for which will be illustrated in Sect. 4. We give these as two distinct sets of assumptions in the event that an equation fits the first set of assumptions but not the second, such that we would still be able to conclude that some type of solution exists for the equation.

$\tilde{K}$ relative to some non-negative constants $p, \tilde{p}, q, \tilde{q}$. We use a generic notation to define the functions $K : U \to \mathbb{R}$, $K : U \times U \to \mathbb{R}$, $\tilde{K} : H \to \mathbb{R}$ and $\tilde{K} : H \times H \to \mathbb{R}$ by

$$K(\phi) := 1 + \|\phi\|_U^p,$$

$$K(\phi, \psi) := 1 + \|\phi\|_U^p + \|\psi\|_U^q,$$

$$\tilde{K}(\phi) := K(\phi) + \|\phi\|_H^{\tilde{p}},$$

$$\tilde{K}(\phi, \psi) := K(\phi, \psi) + \|\phi\|_H^{\tilde{p}} + \|\psi\|_H^{\tilde{q}}$$

Distinct use of the function K will depend on different constants but in no meaningful way in our applications, hence no explicit reference to them shall be made. In the case of $\tilde{K}$, when $\tilde{p}, \tilde{q} = 2$ then we shall denote the general $\tilde{K}$ by $\tilde{K}_2$. In this case no further assumptions are made on the p, q. That is, $\tilde{K}_2$ has the general representation

$$\tilde{K}_2(\phi, \psi) = K(\phi, \psi) + \|\phi\|_H^2 + \|\psi\|_H^2 \tag{13}$$

and similarly as a function of one variable.

We state the assumptions for arbitrary elements $\phi, \psi \in V$, $\phi^n \in V_n$ and $t \in [0, \infty)$, and a fixed $\kappa > 0$. Understanding $\mathcal{G}$ as an operator $\mathcal{G} : [0, \infty) \times V \times \mathfrak{U} \to H$, we introduce the notation $\mathcal{G}_i(\cdot, \cdot) := \mathcal{G}(\cdot, \cdot, e_i)$.

Assumption 1 *For any $T > 0$, $\mathcal{A} : [0, T] \times V \to U$ and $\mathcal{G} : [0, T] \times V \to \mathscr{L}^2(\mathfrak{U}; H)$ are measurable.*

Remark 1 Measurability here and throughout the paper is defined with respect to the Borel Sigma Algebra on the relevant Hilbert Spaces.

Assumption 2

$$\|\mathcal{A}(t, \phi)\|_U^2 + \sum_{i=1}^{\infty} \|\mathcal{G}_i(t, \phi)\|_H^2 \leq c_t K(\phi) \left[1 + \|\phi\|_V^2\right], \tag{14}$$

$$\|\mathcal{A}(t, \phi) - \mathcal{A}(t, \psi)\|_X \leq c_t \left[K(\phi, \psi) + \|\phi\|_V + \|\psi\|_V\right] \|\phi - \psi\|_H, \tag{15}$$

$$\sum_{i=1}^{\infty} \|\mathcal{G}_i(t, \phi) - \mathcal{G}_i(t, \psi)\|_X \leq c_t K(\phi, \psi) \|\phi - \psi\|_H \tag{16}$$

Assumption 3

$$2\langle \mathcal{P}_n \mathcal{A}(t, \boldsymbol{\phi}^n), \boldsymbol{\phi}^n \rangle_H + \sum_{i=1}^{\infty} \|\mathcal{P}_n \mathcal{G}_i(t, \boldsymbol{\phi}^n)\|_H^2 \leq$$

$$c_t \tilde{K}_2(\boldsymbol{\phi}^n) \left[1 + \|\boldsymbol{\phi}^n\|_H^2 \right] - \kappa \|\boldsymbol{\phi}^n\|_V^2, \tag{17}$$

$$\sum_{i=1}^{\infty} \langle \mathcal{P}_n \mathcal{G}_i(t, \boldsymbol{\phi}^n), \boldsymbol{\phi}^n \rangle_H^2 \leq c_t \tilde{K}_2(\boldsymbol{\phi}^n) \left[1 + \|\boldsymbol{\phi}^n\|_H^4 \right]. \tag{18}$$

Assumption 4

$$2\langle \mathcal{A}(t, \boldsymbol{\phi}) - \mathcal{A}(t, \boldsymbol{\psi}), \boldsymbol{\phi} - \boldsymbol{\psi} \rangle_U + \sum_{i=1}^{\infty} \|\mathcal{G}_i(t, \boldsymbol{\phi}) - \mathcal{G}_i(t, \boldsymbol{\psi})\|_U^2$$

$$\leq c_t \tilde{K}_2(\boldsymbol{\phi}, \boldsymbol{\psi}) \|\boldsymbol{\phi} - \boldsymbol{\psi}\|_U^2 - \kappa \|\boldsymbol{\phi} - \boldsymbol{\psi}\|_H^2, \tag{19}$$

$$\sum_{i=1}^{\infty} \langle \mathcal{G}_i(t, \boldsymbol{\phi}) - \mathcal{G}_i(t, \boldsymbol{\psi}), \boldsymbol{\phi} - \boldsymbol{\psi} \rangle_U^2 \leq c_t \tilde{K}_2(\boldsymbol{\phi}, \boldsymbol{\psi}) \|\boldsymbol{\phi} - \boldsymbol{\psi}\|_U^4 \tag{20}$$

Assumption 5

$$2\langle \mathcal{A}(t, \boldsymbol{\phi}), \boldsymbol{\phi} \rangle_U + \sum_{i=1}^{\infty} \|\mathcal{G}_i(t, \boldsymbol{\phi})\|_U^2 \leq c_t K(\boldsymbol{\phi}) \left[1 + \|\boldsymbol{\phi}\|_H^2 \right], \tag{21}$$

$$\sum_{i=1}^{\infty} \langle \mathcal{G}_i(t, \boldsymbol{\phi}), \boldsymbol{\phi} \rangle_U^2 \leq c_t K(\boldsymbol{\phi}) \left[1 + \|\boldsymbol{\phi}\|_H^4 \right]. \tag{22}$$

3.2 Assumption Set 2

These assumptions are only checked in addition to Assumption Set 1 and so take place in the same framework. We state the assumptions now for arbitrary elements $\boldsymbol{\phi}, \boldsymbol{\psi} \in H$ and $t \in [0, \infty)$, and continue to use the $c, K, \tilde{K}, \kappa$ notation of Assumption Set 1.

Assumption 6 *For any $T > 0$, $\mathcal{A} : [0, T] \times H \to X$ is measurable, and whenever $\boldsymbol{\Phi}$ is a progressively measurable process in H we have that $\mathcal{G}(\cdot, \boldsymbol{\Phi}_\cdot)$ is progressively measurable in $\mathscr{L}^2(\mathfrak{U}; U)$.*

Assumption 7

$$\|\mathcal{A}(t, \boldsymbol{\phi})\|_X^2 + \sum_{i=1}^{\infty} \|\mathcal{G}_i(t, \boldsymbol{\phi})\|_U^2 \le c_t K(\boldsymbol{\phi}) \left[1 + \|\boldsymbol{\phi}\|_H^2 \right], \tag{23}$$

$$\|\mathcal{A}(t, \boldsymbol{\phi}) - \mathcal{A}(t, \boldsymbol{\psi})\|_X \le c_t \left[K(\phi, \psi) + \|\phi\|_H + \|\psi\|_H \right] \|\phi - \psi\|_H \tag{24}$$

Assumption 8

$$2\langle \mathcal{A}(t, \boldsymbol{\phi}) - \mathcal{A}(t, \boldsymbol{\psi}), \boldsymbol{\phi} - \boldsymbol{\psi} \rangle_X + \sum_{i=1}^{\infty} \|\mathcal{G}_i(t, \boldsymbol{\phi}) - \mathcal{G}_i(t, \boldsymbol{\psi})\|_X^2 \le$$

$$c_t \tilde{K}_2(\boldsymbol{\phi}, \boldsymbol{\psi}) \|\boldsymbol{\phi} - \boldsymbol{\psi}\|_X^2, \tag{25}$$

$$\sum_{i=1}^{\infty} \langle \mathcal{G}_i(t, \boldsymbol{\phi}) - \mathcal{G}_i(t, \boldsymbol{\psi}), \boldsymbol{\phi} - \boldsymbol{\psi} \rangle_X^2 \le$$

$$c_t \tilde{K}_2(\boldsymbol{\phi}, \boldsymbol{\psi}) \|\boldsymbol{\phi} - \boldsymbol{\psi}\|_X^4 \tag{26}$$

We in fact state Assumption 9 for $\phi \in V$ and some $\kappa > 0$, making this a stronger assumption than 5.

Assumption 9 *With the stricter requirement that* $\phi \in V$ *then*

$$2\langle \mathcal{A}(t, \boldsymbol{\phi}), \boldsymbol{\phi} \rangle_U + \sum_{i=1}^{\infty} \|\mathcal{G}_i(t, \boldsymbol{\phi})\|_U^2 \le c_t K(\boldsymbol{\phi}) - \kappa \|\boldsymbol{\phi}\|_H^2, \tag{27}$$

$$\sum_{i=1}^{\infty} \langle \mathcal{G}_i(t, \boldsymbol{\phi}), \boldsymbol{\phi} \rangle_U^2 \le c_t K(\boldsymbol{\phi}). \tag{28}$$

3.3 Notions of Solution and Results

Here we define the two different notions of solution, which we call V-valued solutions and H-valued solutions. The corresponding definitions of uniqueness and maximality are given in one for both notions of solution. We frame the definition of the V-valued solutions for an initial condition $\boldsymbol{\Psi}_0 : \Omega \to H$ which is an $\mathcal{F}_0$-measurable mapping, and for the H-valued solutions a $\boldsymbol{\Psi}_0 : \Omega \to U$ which is likewise $\mathcal{F}_0$-measurable.

Definition 4 A pair $(\boldsymbol{\Psi}, \tau)$ where τ is a $\mathbb{P} - a.s.$ positive stopping time and $\boldsymbol{\Psi}$ is a process such that for $\mathbb{P} - a.e.$ ω, $\boldsymbol{\Psi}_{\cdot}(\omega) \in C([0, T]; H)$ and $\boldsymbol{\Psi}_{\cdot}(\omega) \mathbf{1}_{\cdot \le \tau(\omega)} \in$

$L^2([0, T]; V)$ for all $T > 0$ with $\boldsymbol{\Psi}.\mathbf{1}._{\leq\tau}$ progressively measurable in V, is said to be a V-valued local strong solution of the equation (2) if the identity

$$\boldsymbol{\Psi}_t = \boldsymbol{\Psi}_0 + \int_0^{t\wedge\tau} \mathcal{A}(s, \boldsymbol{\Psi}_s)ds + \int_0^{t\wedge\tau} \mathcal{G}(s, \boldsymbol{\Psi}_s)d\mathcal{W}_s \tag{29}$$

holds $\mathbb{P} - a.s.$ in U for all $t \geq 0$.

Remark 2 If $(\boldsymbol{\Psi}, \tau)$ is a V-valued local strong solution of the equation (2), then $\boldsymbol{\Psi}. = \boldsymbol{\Psi}._{\wedge\tau}$.

Remark 3 The progressive measurability condition on $\boldsymbol{\Psi}.\mathbf{1}._{\leq\tau}$ may look a little suspect as $\boldsymbol{\Psi}_0$ itself may only belong to H and not V making it impossible for $\boldsymbol{\Psi}.\mathbf{1}._{\leq\tau}$ to be even adapted in V. We are mildly abusing notation here; what we really ask is that there exists a process $\boldsymbol{\Phi}$ which is progressively measurable in V and such that $\boldsymbol{\Phi}. = \boldsymbol{\Psi}.\mathbf{1}._{\leq\tau}$ almost surely over the product space $\Omega \times [0, \infty)$ with product measure $\mathbb{P} \times \lambda$ for λ the Lebesgue measure on $[0, \infty)$.

Remark 4 If Assumption 1 and (14) hold, then the time integral is well defined in U and the stochastic integral is well defined as a local martingale in H.

Definition 5 A pair $(\boldsymbol{\Psi}, \tau)$ where τ is a $\mathbb{P} - a.s.$ positive stopping time and $\boldsymbol{\Psi}$ is a process such that for $\mathbb{P} - a.e.$ ω, $\boldsymbol{\Psi}.(\omega) \in C([0, T]; U)$ and $\boldsymbol{\Psi}.(\omega)\mathbf{1}._{\leq\tau(\omega)} \in L^2([0, T]; H)$ for all $T > 0$ with $\boldsymbol{\Psi}.\mathbf{1}._{\leq\tau}$ progressively measurable in H, is said to be an H-valued local strong solution of the equation (2) if the identity

$$\boldsymbol{\Psi}_t = \boldsymbol{\Psi}_0 + \int_0^{t\wedge\tau} \mathcal{A}(s, \boldsymbol{\Psi}_s)ds + \int_0^{t\wedge\tau} \mathcal{G}(s, \boldsymbol{\Psi}_s)d\mathcal{W}_s \tag{30}$$

holds $\mathbb{P} - a.s.$ in X for all $t \geq 0$.

Remark 5 The analogy to Remarks 2, 3 hold for the H-valued solutions.

Remark 6 If Assumption 6 and (23) hold, then the time integral is well defined in X and the stochastic integral is well defined as a local martingale in U.

In the following we use $V; H$ to mean V or H respectively.

Definition 6 A pair $(\boldsymbol{\Psi}, \Theta)$ such that there exists a sequence of stopping times (θ_j) which are $\mathbb{P} - a.s.$ monotone increasing and convergent to Θ, whereby $(\boldsymbol{\Psi}._{\wedge\theta_j}, \theta_j)$ is a $(V; H)$–valued local strong solution of the equation (2) for each j, is said to be a $(V; H)$–valued maximal strong solution of the equation (2) if for any other pair $(\boldsymbol{\Phi}, \Gamma)$ with this property then $\Theta \leq \Gamma\, \mathbb{P} - a.s.$ implies $\Theta = \Gamma\, \mathbb{P} - a.s.$

Definition 7 A $(V; H)$-valued maximal strong solution $(\mathbf{\Psi}, \Theta)$ of the equation (2) is said to be unique if for any other such solution $(\mathbf{\Phi}, \Gamma)$, then $\Theta = \Gamma \; \mathbb{P} - a.s.$ and for all $t \in [0, \Theta)$,

$$\mathbb{P}\left(\{\omega \in \Omega : \mathbf{\Psi}_t(\omega) = \mathbf{\Phi}_t(\omega)\}\right) = 1.$$

Theorem 2 *Suppose that Assumption Set 1 holds. Then for any given $\mathcal{F}_0$-measurable $\mathbf{\Psi}_0 : \Omega \to H$, there exists a unique V-valued maximal strong solution $(\mathbf{\Psi}, \Theta)$ of the equation (2). Moreover at $\mathbb{P} - a.e.$ ω for which $\Theta(\omega) < \infty$, we have that*

$$\sup_{r \in [0, \Theta(\omega))} \|\mathbf{\Psi}_r(\omega)\|_H^2 + \int_0^{\Theta(\omega)} \|\mathbf{\Psi}_r(\omega)\|_V^2 dr = \infty. \tag{31}$$

Theorem 3 *Suppose that Assumption Set 1 and 2 hold. Then for any given $\mathcal{F}_0$-measurable $\mathbf{\Psi}_0 : \Omega \to U$, there exists a unique H-valued maximal strong solution $(\mathbf{\Psi}, \Theta)$ of the equation (2). Moreover at $\mathbb{P} - a.e.$ ω for which $\Theta(\omega) < \infty$, we have that*

$$\sup_{r \in [0, \Theta(\omega))} \|\mathbf{\Psi}_r(\omega)\|_U^2 + \int_0^{\Theta(\omega)} \|\mathbf{\Psi}_r(\omega)\|_H^2 dr = \infty. \tag{32}$$

4 Abstract Solution Method and Application

In this final section we give the main steps of the proofs of Theorems 2 and 3, followed by a brief exposition of how our SALT Navier-Stokes Equation fits into this framework.

4.1 Abstract Solution Method

Proof (Theorem 2) We suppose that Assumption Set 1 holds and address the question first for an initial condition $\mathbf{\Psi}_0$ which is such that for $\mathbb{P} - a.e.$ ω,

$$\|\mathbf{\Psi}_0(\omega)\|_H^2 \leq M' \tag{33}$$

for some constant M'. We work with this bounded initial condition in the first instance as we shall use local solutions up to first hitting times given in terms of the initial condition, so this boundedness translates to boundedness of the relevant

process up until these times. As directed in Sect. 3.1 we are to use a Galerkin
Scheme, whereby we consider the equations

$$\Psi_t^n = \Psi_0^n + \int_0^t \mathcal{P}_n \mathcal{A}(s, \Psi_s^n)ds + \int_0^t \mathcal{P}_n \mathcal{G}(s, \Psi_s^n)d\mathcal{W}_s \tag{34}$$

with notation $\mathcal{P}_n \mathcal{G}(\cdot, \cdot, e_i) := \mathcal{P}_n \mathcal{G}_i(\cdot, \cdot)$. A local strong solution of this equation is
defined as a pair (Ψ^n, τ) where τ is a $\mathbb{P} - a.s.$ positive stopping time and Ψ^n is
an adapted process in V_n such that for $\mathbb{P} - a.e.$ ω, $\Psi_\cdot^n(\omega) \in C([0, T]; V_n)$ for all
$T > 0$, and the identity

$$\Psi_t^n = \Psi_0^n + \int_0^{t \wedge \tau} \mathcal{P}_n \mathcal{A}(s, \Psi_s^n)ds + \int_0^{t \wedge \tau} \mathcal{P}_n \mathcal{G}(s, \Psi_s^n)d\mathcal{W}_s \tag{35}$$

holds $\mathbb{P} - a.s.$ in V_n for all $t \geq 0$. We can conclude that for any fixed $t > 0$ and
$M > 1$, a local strong solution $(\Psi^n, \tau_n^{M,t})$ of (34) exists for the stopping time $\tau_n^{M,t}$
defined by

$$\tau_n^{M,t} := t \wedge \inf \left\{ s \geq 0 : \sup_{r \in [0,s]} \|\Psi_r^n\|_U^2 + \int_0^s \|\Psi_r^n\|_H^2 dr \geq M + \|\Psi_0^n\|_U^2 \right\}. \tag{36}$$

This conclusion is reached thanks to Assumption 2, through standard theory in the
finite dimensional Hilbert Space V_n though some care must be taken for the infinite
dimensional Brownian Motion. Understanding that

$$\|\Psi_0^n(\omega)\|_H^2 \leq c\|\Psi_0(\omega)\|_H^2 \leq cM' \tag{37}$$

coming from (10) and (33), it is clear that

$$\|\Psi_0^n(\omega)\|_U^2 \leq \tilde{M} \tag{38}$$

for some $\tilde{M}$ clearly still independent of n and ω. Thus we see the bound

$$\sup_{r \in [0, \tau_n^{M,t}(\omega)]} \|\Psi_r^n(\omega)\|_U^2 + \int_0^{\tau_n^{M,t}(\omega)} \|\Psi_s^n(\omega)\|_H^2 ds \leq M + \tilde{M} \tag{39}$$

holds true for every n and $\mathbb{P} - a.e.$ ω. This boundedness plays a significant role
in our analysis and demonstrates the importance of starting from this bounded
initial condition in the first instance. The motivation for choosing these stopping
times comes from the work of Glatt-Holtz and Ziane in the referenced paper [14].
The authors prove an abstract result which is the central theorem of the paper,
which we simply restate in the Appendix as Theorem 4. In the original paper, the
authors use the traditional Galerkin Scheme for Navier-Stokes (given by the basis

of eigenfunctions of the Stokes Operator) and apply this theorem directly with the spaces $\mathcal{H}_1 := W_\sigma^{2,2}(\mathbb{T}^3; \mathbb{R}^3)$, $\mathcal{H}_2 := W_\sigma^{1,2}(\mathbb{T}^3; \mathbb{R}^3)$. We have to take a slight detour from this method in the case of transport noise due to the condition (47). Translating this to our framework through $\mathcal{H}_1 = H$ and $\mathcal{H}_2 = U$, the idea in showing this condition is to apply the Itô Formula in U to the difference process $\boldsymbol{\Psi}^n - \boldsymbol{\Psi}^m$. When we simplify down the term arising from the quadratic variation of the stochastic integral, we must control

$$\sum_{i=1}^\infty \|[I - \mathcal{P}_m]\mathcal{G}_i(s, \boldsymbol{\Psi}_s^m)\|_U^2$$

which we would do via (12) and (10) to bound the above by

$$\sum_{i=1}^\infty \frac{1}{\mu_m} \|\mathcal{G}_i(s, \boldsymbol{\Psi}_s^m)\|_H^2.$$

In order to send this to zero as $m \to \infty$ we use some uniform boundedness of the term $\sum_{i=1}^\infty \|\mathcal{G}_i(s, \boldsymbol{\Psi}_s^m)\|_H^2$ which in the case of a Lipschitz operator as in the original paper is immediate from (39). Where $\mathcal{G}_i$ is a differential operator we must obtain uniform boundedness of the solutions $(\boldsymbol{\Psi}^n)$ in a higher norm, hence the need for our space V (which in the context of our SALT Navier-Stokes, would then be $W_\sigma^{3,2}(\mathbb{T}^3; \mathbb{R}^3)$). For this reason we must introduce another step to the proof, whereby we show that there exists constants $C, \tilde{C}$ dependent on M, M', t but independent of n such that for the local strong solution $(\boldsymbol{\Psi}^n, \tau_n^{M,t})$ of (34),

$$\mathbb{E} \sup_{r \in [0, \tau_n^{M,t}]} \|\boldsymbol{\Psi}_r^n\|_H^2 + \mathbb{E} \int_0^{\tau_n^{M,t}} \|\boldsymbol{\Psi}_s^n\|_V^2 ds \leq C \left[\mathbb{E}\left(\|\boldsymbol{\Psi}_0^n\|_H^2 \right) + 1 \right] \tag{40}$$

and in particular

$$\mathbb{E} \sup_{r \in [0, \tau_n^{M,t}]} \|\boldsymbol{\Psi}_r^n\|_H^2 + \mathbb{E} \int_0^{\tau_n^{M,t}} \|\boldsymbol{\Psi}_s^n\|_V^2 ds \leq \tilde{C}. \tag{41}$$

This result is proven by considering V_n as a Hilbert Space with H inner product, applying the Itô Formula in this context and using Assumption 3. Equation (41) then follows from (40) due to (10) so we see the significance of starting from an initial condition bounded in H and not just U (or at least, square integrable in H). From Assumption 4, along with the requirement that each $\mathcal{P}_n$ is an orthogonal projection in X and U and the conditions (8),(11),(12), we deduce that for any $m < n$ and $\lambda_m := \min\{\mu_m, \mu_m^2\}$,

$$2\langle \mathcal{P}_n \mathcal{A}(t, \boldsymbol{\phi}) - \mathcal{P}_m \mathcal{A}(t, \boldsymbol{\psi}), \boldsymbol{\phi} - \boldsymbol{\psi} \rangle_U + \sum_{i=1}^{\infty} \| \mathcal{P}_n \mathcal{G}_i(t, \boldsymbol{\phi}) - \mathcal{P}_m \mathcal{G}_i(t, \boldsymbol{\psi}) \|_U^2$$

$$\leq c_t \tilde{K}_2(\boldsymbol{\phi}, \boldsymbol{\psi}) \| \boldsymbol{\phi} - \boldsymbol{\psi} \|_U^2 - \frac{\kappa}{2} \| \boldsymbol{\phi} - \boldsymbol{\psi} \|_H^2 + \frac{c_t}{\lambda_m} K(\boldsymbol{\phi}, \boldsymbol{\psi}) \left[1 + \| \boldsymbol{\phi} \|_V^2 + \| \boldsymbol{\psi} \|_V^2 \right],$$

$$\sum_{i=1}^{\infty} \langle \mathcal{P}_n \mathcal{G}_i(t, \boldsymbol{\phi}) - \mathcal{P}_m \mathcal{G}_i(t, \boldsymbol{\psi}), \boldsymbol{\phi} - \boldsymbol{\psi} \rangle_U^2$$

$$\leq c_t \tilde{K}_2(\boldsymbol{\phi}, \boldsymbol{\psi}) \| \boldsymbol{\phi} - \boldsymbol{\psi} \|_U^4 + \frac{c_t}{\lambda_m} K(\boldsymbol{\phi}, \boldsymbol{\psi}) \left[1 + \| \boldsymbol{\psi} \|_V^2 \right].$$

Along with (41) these bounds allow us to conclude that

$$\lim_{m \to \infty} \sup_{n \geq m} \left[\mathbb{E} \sup_{r \in [0, \tau_m^{M,t} \wedge \tau_n^{M,t}]} \| \boldsymbol{\Psi}_r^n - \boldsymbol{\Psi}_r^m \|_U^2 \right.$$

$$\left. + \mathbb{E} \int_0^{\tau_m^{M,t} \wedge \tau_n^{M,t}} \| \boldsymbol{\Psi}_s^n - \boldsymbol{\Psi}_s^m \|_H^2 ds \right] = 0 \qquad (42)$$

again via an application of the Itô Formula for V_n considered as a Hilbert Space with U inner product, on the difference process $\boldsymbol{\Psi}^n - \boldsymbol{\Psi}^m$. With similar ideas and the Assumption 5, we infer that

$$\lim_{S \to 0} \sup_{n \in \mathbb{N}} \mathbb{P} \left(\left\{ \sup_{r \in [0, \tau_n^{M,t} \wedge S]} \| \boldsymbol{\Psi}_r^n \|_U^2 \right. \right.$$

$$\left. \left. + \int_0^{\tau_n^{M,t} \wedge S} \| \boldsymbol{\Psi}_r^n \|_H^2 dr \geq M - 1 + \| \boldsymbol{\Psi}_0^n \|_U^2 \right\} \right) = 0. \qquad (43)$$

We then apply Theorem 4 for $\mathcal{H}_1 = H$, $\mathcal{H}_2 = U$ and claim that the resulting pair $(\boldsymbol{\Psi}, \tau_\infty^{M,t})$ satisfies the additional properties that:

- $\boldsymbol{\Psi}.\mathbf{1}_{\cdot \leq \tau_\infty^{M,t}}$ is progressively measurable in V;
- For $\mathbb{P} - a.e.\ \omega$, $\boldsymbol{\Psi}(\omega) \in C([0, T]; H)$ and $\boldsymbol{\Psi}.(\omega)\mathbf{1}_{\cdot \leq \tau_\infty^{M,t}(\omega)} \in L^2([0, T]; V)$ for all $T > 0$;
- $\boldsymbol{\Psi}^{n_l} \to \boldsymbol{\Psi}$ holds in the sense that

$$\mathbb{E} \left[\sup_{r \in [0, \tau_\infty^{M,t}]} \| \boldsymbol{\Psi}_r^{n_l} - \boldsymbol{\Psi}_r \|_U^2 + \int_0^{\tau_\infty^{M,t}} \| \boldsymbol{\Psi}_r^{n_l} - \boldsymbol{\Psi}_r \|_H^2 dr \right] \longrightarrow 0. \qquad (44)$$

Indeed the first two are true from using the uniform boundedness (41) and taking weakly convergent subsequences in the appropriate spaces, then using uniqueness of limits in the weak topology and the embeddings $V \hookrightarrow H \hookrightarrow U$ to identify this limit with $\boldsymbol{\Psi}$. The weak convergence preserves the measurability and so the progressive measurability of each $\boldsymbol{\Psi}^n$ (from the continuity and adaptedness in V_n) is what gives the result here. The final item is then a simple application of the dominated convergence theorem. To conclude that $(\boldsymbol{\Psi}, \tau_\infty^{M,i})$ is a V-valued local strong solution it only remains to show the identity (29), which is done by taking limits of the corresponding terms in (35) and applying (15), (16) alongside the already used assumptions on the $(\mathcal{P}_n)$. We take the limit in X and argue that the identity being satisfied in X is sufficient to conclude the satisfaction of the identity in U, given that all integrals can be constructed in U from the regularity of the solution.

We have now shown the existence of a V−valued local strong solution but for the bounded initial condition (33). We then show a uniqueness result for such solutions, which is: suppose that $(\boldsymbol{\Psi}^1, \tau_1)$ and $(\boldsymbol{\Psi}^2, \tau_2)$ are two V−valued local strong solutions of the equation (2) for a given initial condition $\boldsymbol{\Psi}_0$. Then for all $s \in [0, \infty)$,

$$\mathbb{P}\left(\left\{\omega \in \Omega : \boldsymbol{\Psi}^1_{s \wedge \tau_1(\omega) \wedge \tau_2(\omega)}(\omega) = \boldsymbol{\Psi}^2_{s \wedge \tau_1(\omega) \wedge \tau_2(\omega)}(\omega)\right\}\right) = 1.$$

This is proven through applying Assumption 4 in the context of an Itô Formula in U of the difference process of any two solutions. With this uniqueness in place we then conclude the results of Theorem 2 but still for the bounded initial condition, via similar arguments to those used in [14]. To pass to a general initial condition we consider a sequence of such maximal strong solutions $(\boldsymbol{\Psi}^k, \Theta^k)$ corresponding to the bounded initial conditions $(\boldsymbol{\Psi}_0 \mathbf{1}_{k \leq \|\boldsymbol{\Psi}_0\|_H \leq k+1})$ and use the maximality on these pieces to show that the pair $(\boldsymbol{\Psi}, \Theta)$ defined at each time $t \in [0, T]$ and $\omega \in \Omega$ by

$$\boldsymbol{\Psi}_t(\omega) := \sum_{k=1}^{\infty} \boldsymbol{\Psi}^k_t(\omega)\mathbf{1}_{k \leq \|\boldsymbol{\Psi}_0(\omega)\|_H < k+1}, \quad \Theta(\omega) := \sum_{k=1}^{\infty} \Theta^k(\omega)\mathbf{1}_{k \leq \|\boldsymbol{\Psi}_0(\omega)\|_H < k+1}$$

is our desired solution for the initial condition $\boldsymbol{\Psi}_0$ (where the limit for $\boldsymbol{\Psi}$ is in reality just a finite sum). It is clear that for any ω, there exists a k such that $(\boldsymbol{\Psi}(\omega), \Theta(\omega)) = (\boldsymbol{\Psi}^k(\omega), \Theta^k(\omega))$ so the property (31) follows from the same property in the case of the bounded initial condition. This rounds off our discussion for the proof of Theorem 2.

In the case where Assumption Set 2 holds, we then look to use the V-valued local strong solutions to obtain an H-valued local strong solution but now just for a U-valued initial condition. At this juncture it is well worth addressing the question of why we consider these distinct types of solution; that is if we wanted an H-valued local strong solution then why not restate Assumption Set 1 for the spaces V as H, H as U and U as X? The reason lies in the application to our stochastic Navier-Stokes equation, which would then not satisfy the required assumption. This will be discussed more explicitly in Sect. 4.2.

Proof (Theorem 3) The idea now is to apply this existence result to the sequence of initial conditions $(\mathcal{P}_n \boldsymbol{\Psi}_0)$, and apply the same Theorem 4 argument to the corresponding sequence of solutions. From here we now need to suppose that Assumption Set 2 holds in addition to Assumption Set 1. In the same manner we start again from a bounded $\boldsymbol{\Psi}_0$, this time such that

$$\|\boldsymbol{\Psi}_0(\omega)\|_U^2 \leq \tilde{M}. \tag{45}$$

We could immediately apply Theorem 2 for each initial condition $\mathcal{P}_n \boldsymbol{\Psi}_0$, though we want to apply Theorem 4 for the same spaces $\mathcal{H}_1 = H$ and $\mathcal{H}_2 = U$. Recall that we could not do this immediately for a U-valued initial condition and the sequence of Galerkin solutions due to gaining a suitable control on the noise term arising from the difference of the projections. In the present scenario we consider solutions to the unprojected (2) and so we are not burdened with this difficulty. An application of Theorem 4 would rely on us being able to conclude that each maximal solution $(\boldsymbol{\Psi}^n, \Theta^n)$ corresponding to the initial condition $\mathcal{P}_n \boldsymbol{\Psi}_0$ exists up until the stopping time (36) (where the $\boldsymbol{\Psi}^n$ notation has now shifted to the above). This is not immediate from Theorem 2, though we can use similar maximality arguments to extend these solutions to $\tau_n^{M,t}$ at the cost of some regularity. Indeed for these extended solutions we have only the regularity of the H-valued solution but with the additional benefit that $\boldsymbol{\Psi}_t(\omega) \mathbf{1}_{\cdot \leq \tau_n^{M,t}(\omega)} \in V$ almost everywhere on the product space $\Omega \times [0, \infty)$. This facilitates the use of Assumption 4 in order to show the Cauchy property (42), but only via first using an Itô Formula with the bilinear form $\langle \cdot, \cdot \rangle_{X \times H}$. We must make this step as the identity for these extended solutions is only satisfied in X hence we cannot use the U inner product. The stochastic integral though can be constructed in U following from Remark 5, and the regularity $\boldsymbol{\Psi}_t(\omega) \mathbf{1}_{\cdot \leq \tau_n^{M,t}(\omega)} \in V$ allows us to call upon the property (8) so that we can apply Assumption 4. Without the uniform boundedness (41) for these solutions we need Assumption 9 instead of just 5 to deduce (43). The conclusion of the proof of Theorem 3 then follows identically to that of 2, now using Assumption 8 for the uniqueness part and (24) to show the convergence of the time integral term when justifying that the limiting pair $(\boldsymbol{\Psi}, \tau_\infty^{M,t})$ obtained from Theorem 4 is an H-valued local strong solution.

4.2 SALT Navier-Stokes in the Abstract Framework

We now briefly comment on the application of this abstract framework to Eq. (4) in order to conclude the paper. In the previous subsection we have already established the identification of the spaces

$$V := W_\sigma^{3,2}(\mathbb{T}^3; \mathbb{R}^3), \, H := W_\sigma^{2,2}(\mathbb{T}^3; \mathbb{R}^3), \, U := W_\sigma^{1,2}(\mathbb{T}^3; \mathbb{R}^3), \, X := L_\sigma^2(\mathbb{T}^3; \mathbb{R}^3)$$

at which point we address the question posed in that subsection as to why we need to make this effort with first the $V-$valued solutions before showing the existence for the $H-$valued ones. That is, why would Assumption Set 1 not hold if we were to shift the spaces from V to H, H to U and U to X (with some modifications of the reference to X in Assumption Set 1)? One clear answer is in the treatment of the nonlinear term for (17): for $H = W_\sigma^{2,2}(\mathbb{T}^3; \mathbb{R}^3)$ we have the algebra property of the Sobolev Space which affords us a bound

$$\|\mathcal{L}_{\phi^n}\phi^n\|_2 \leq c\|\phi^n\|_2\|\phi^n\|_3$$

using the equivalence of the $\|\cdot\|_2$ and the standard $W^{2,2}$ one. In the $W^{1,2}$ norm we do not have the same luxury and so this nonlinear term cannot be bounded just in terms of the $W^{1,2}$ and $W^{2,2}$ norms as would be required. It is worth noting the significance of using the $\langle \cdot, \cdot \rangle_2$ inner product here, as in the same assumption this facilitates the 'integration by parts' property for the Stokes Operator in order to gain the additional control we require (i.e. the $-\kappa\|\phi^n\|_V^2$ term). There is some additional care required then to control the noise terms in these inner products, but this is facilitated by using the same standard cancellation argument that

$$\langle \mathcal{L}_{\xi_i}\phi, \phi \rangle_{L^2} = 0 \tag{46}$$

for $\phi \in W^{1,2}(\mathbb{T}^3; \mathbb{R}^3)$, as well as appreciating that the commutator $[\Delta, B_i]$ is of second order and commuting through the B_i with Δ until we reduce to a term of the form (46). The control (3) allows the ξ_i to be effectively ignored in many of these computations, by just pulling them out with the supremum. We refer once more to [6] for the complete details. Of course it is Theorem 3 which is what translates into our main Theorem of the paper (1), though it is also worth noting that having showed Theorem 2 in this context then we can also say something about the retained regularity of our solutions coming from a more regular initial condition. To really make this point we'd have to say that the maximal times for the different notions of solution were in fact the same, and this is to be addressed in [6].

Appendix

Here we state [14, Lemma 5.1].

Theorem 4 *Let $\mathcal{H}_1 \subset \mathcal{H}_2$ be Hilbert Spaces with continuous embedding, and (Ψ^n) be a sequence of processes such that for $\mathbb{P} - a.e.$ ω, $\Psi^n(\omega) \in C([0, T]; \mathcal{H}_2) \cap L^2([0, T]; \mathcal{H}_1)$ which is a Banach Space with norm*

$$\|\psi\|_{X(T)} := \left(\sup_{r \in [0,T]} \|\psi_r\|_{\mathcal{H}_2}^2 + \int_0^T \|\psi_r\|_{\mathcal{H}_1}^2 dr \right)^{\frac{1}{2}}.$$

For some fixed M > 1 and t > 0 define the stopping times

$$\tau_n^{M,t}(\omega) := t \wedge \inf\left\{s \geq 0 : \|\boldsymbol{\Psi}^n(\omega)\|_{X(s)}^2 \geq M + \|\boldsymbol{\Psi}_0^n(\omega)\|_{\mathcal{H}_2}^2\right\}$$

and suppose that

$$\lim_{m \to \infty} \sup_{n \geq m} \mathbb{E}\|\boldsymbol{\Psi}^n - \boldsymbol{\Psi}^m\|_{X(\tau_m^{M,t} \wedge \tau_n^{M,t})}^2 = 0 \qquad (47)$$

and

$$\lim_{S \to 0} \sup_{n \in \mathbb{N}} \mathbb{P}\left(\left\{\|\boldsymbol{\Psi}^n\|_{X(\tau_n^{M,t} \wedge S)}^2 \geq M - 1 + \|\boldsymbol{\Psi}_0^n\|_{\mathcal{H}_2}^2\right\}\right) = 0.$$

Then there exists a stopping time $\tau_\infty^{M,t}$, a subsequence $(\boldsymbol{\Psi}^{n_l})$ and process $\boldsymbol{\Psi} = \boldsymbol{\Psi}_{\cdot \wedge \tau_\infty^{M,t}}$ such that:

- $\mathbb{P}\left(\left\{0 < \tau_\infty^{M,t} \leq \tau_{n_l}^{M,t}\right\}\right) = 1;$
- *For $\mathbb{P} - a.e.\ \omega$, $\boldsymbol{\Psi}(\omega) \in C\left([0, \tau_\infty^{M,t}(\omega)]; \mathcal{H}_2\right) \cap L^2\left([0, \tau_\infty^{M,t}(\omega)]; \mathcal{H}_1\right);$*
- *For $\mathbb{P} - a.e.\ \omega$, $\boldsymbol{\Psi}^{n_l}(\omega) \to \boldsymbol{\Psi}(\omega)$ in*
  $\left(C\left([0, \tau_\infty^{M,t}(\omega)]; \mathcal{H}_2\right) \cap L^2\left([0, \tau_\infty^{M,t}(\omega)]; \mathcal{H}_1\right), \|\cdot\|_{X(\tau_\infty^{M,t}(\omega))}\right).$

References

1. Alonso-Orán, D. and Bethencourt de León, A., 2020. On the well-posedness of stochastic Boussinesq equations with transport noise. Journal of Nonlinear Science, 30(1), pp.175–224.
2. Brzeźniak, Z. and Slavik, J., 2021. Well-posedness of the 3D stochastic primitive equations with multiplicative and transport noise. Journal of Differential Equations, 296, pp.617–676.
3. Cotter, C., Crisan, D., Holm, D., Pan, W. and Shevchenko, I., 2020. Modelling uncertainty using stochastic transport noise in a 2-layer quasi-geostrophic model. Foundations of Data Science, 2(2), p.173.
4. Cotter, C., Crisan, D., Holm, D.D., Pan, W. and Shevchenko, I., 2019. Numerically modeling stochastic Lie transport in fluid dynamics. Multiscale Modeling and Simulation, 17(1), pp.192–232.
5. Crisan, D., Flandoli, F. and Holm, D.D., 2019. Solution properties of a 3D stochastic Euler fluid equation. Journal of Nonlinear Science, 29(3), pp.813–870.
6. Crisan, D., Goodair, D. 2022. Analytical Properties of a 3D Stochastic Navier-Stokes Equation. In preparation.
7. Crisan, D., Goodair, D., Lang, O., Mensah, P.R., 2022. Existence and Uniqueness of Maximal Strong Solutions to Nonlinear SPDEs with Applications to Viscous Fluid Models. In preparation.
8. Crisan, D., Holm, D.D., Luesink, E., Mensah, P.R. and Pan, W., 2021. Theoretical and computational analysis of the thermal quasi-geostrophic model. arXiv preprint arXiv:2106.14850.
9. Crisan, D. and Lang, O., 2021. Well-posedness Properties for a Stochastic Rotating Shallow Water Model. arXiv preprint arXiv:2107.06601.

10. Crisan, D. and Lang, O., 2021. Local Well-Posedness for the Great Lake Equation with Transport Noise. REV. ROUMAINE MATH. PURES APPL, 66(1), pp.131–155.
11. Crisan, D. and Lang, O., 2022. Well-posedness for a stochastic 2D Euler equation with transport noise. Stochastics and Partial Differential Equations: Analysis and Computations, pp.1–48.
12. Crisan, D., and Street, O.D. 2021. Semi-martingale driven variational principles. Proceedings of the Royal Society A, 477(2247), p.20200957.
13. Debussche, A., Glatt-Holtz, N. and Temam, R., 2011. Local martingale and pathwise solutions for an abstract fluids model. Physica D: Nonlinear Phenomena, 240(14-15), pp.1123–1144.
14. Glatt-Holtz, N. and Ziane, M., 2009. Strong pathwise solutions of the stochastic Navier-Stokes system. Advances in Differential Equations, 14(5/6), pp.567–600.
15. Goodair, D., 2022. Stochastic Calculus in Infinite Dimensions and SPDEs. arXiv preprint arXiv:2203.17206.
16. Holm, D.D., 2015. Variational principles for stochastic fluid dynamics. Proceedings of the Royal Society A: Mathematical, Physical and Engineering Sciences, 471(2176), p.20140963.
17. Mémin, E., 2014. Fluid flow dynamics under location uncertainty. Geophysical and Astrophysical Fluid Dynamics, 108(2), pp.119–146.
18. Robinson, J.C., Rodrigo, J.L. and Sadowski, W., 2016. The three-dimensional Navier–Stokes equations: Classical theory (Vol. 157). Cambridge university press.

Coupling of Waves to Sea Surface Currents Via Horizontal Density Gradients

Darryl D. Holm, Ruiao Hu, and Oliver D. Street

Abstract The mathematical models and numerical simulations reported here are motivated by satellite observations of horizontal gradients of sea surface temperature and salinity that are closely coordinated with the slowly varying envelope of the rapidly oscillating waves. This coordination of gradients of fluid material properties with wave envelopes tends to occur when strong horizontal buoyancy gradients are present. The nonlinear models of this coordinated movement presented here may provide future opportunities for the optimal design of satellite imagery that could simultaneously capture the dynamics of both waves and currents directly.

The model derived here appears in two levels of approximation: first for rapidly oscillating waves, and then for their slowly varying envelope (SVE) approximation obtained by using the WKB approach. The WKB wave-current-buoyancy interaction model derived here for a free surface with significant horizontal buoyancy gradients indicates that the mechanism for the emergence of these correlations is the ponderomotive force of the slowly varying envelope of rapidly oscillating waves acting on the surface currents via the horizontal buoyancy gradient. In this model, the buoyancy gradient appears explicitly in the WKB wave momentum, which in turn generates density-weighted potential vorticity whenever the buoyancy gradient is not aligned with the wave-envelope gradient.

Keywords Nonlinear water waves · Free surface fluid dynamics · Geometric mechanics

Supplementary Information The online version contains supplementary material available at https://doi.org/10.1007/978-3-031-18988-3_8

D. D. Holm · R. Hu · O. D. Street (✉)
Department of Mathematics, Imperial College London, London, UK
e-mail: d.holm@imperial.ac.uk; ruiao.hu15@imperial.ac.uk; o.street18@imperial.ac.uk

© The Author(s) 2023
B. Chapron et al. (eds.), *Stochastic Transport in Upper Ocean Dynamics*,
Mathematics of Planet Earth 10, https://doi.org/10.1007/978-3-031-18988-3_8

109

1   Introduction

1.1   Submesoscale Sea Surface Dynamics

Capabilities in sea surface observation have been improving rapidly during the past two decades [1]. In particular, new high-resolution satellite observation capabilities are revealing sea surface features seen for the first time at *submesoscale* spatial scales of 100 m–10 km and time scales of hours to weeks. Invariably, the new satellite imagery reveals a plethora of coupled dynamical surface phenomena, including currents, spiral filaments, flotsam patterns, jets and fronts, some of which are detected indirectly through gradients of sea surface temperature, salinity or colour, in addition to the imagery [5, 10, 13, 20, 26].

The new capabilities in sea surface observation are still developing. For example, the impending Surface Water Ocean Topography (SWOT) mission will map the ocean surface mesoscale sea surface height field, as well as a large fraction of the associated submesoscale field, including buoyancy fronts [17]. A sample of this type of submesoscale data taken from [5] is shown in Figs. 1 and 2.

The coming new age of higher-resolution upper ocean observations will present a formidable array of challenges for the next generation in data management, computational simulation and mathematical modelling. This paper will offer a mathematical modelling framework that is flexible enough to admit uncertainty

Fig. 1 Wave activity in the submesoscale ocean is dynamically complex, as illustrated in this figure showing the zoomed image of a submesoscale sea surface elevation, seen in Envisar MERIS glitter observations. This image shows the wave elevation tracking a cyclonic eddy visible in the sea surface glitter observations. The pixel resolution is 250 m. This glitter image demonstrates the complex, highly-coordinated dynamical forms taken in wave-current interaction on the submesoscale sea surface. In particular, notice the instabilities developing in the eddy's outer boundary. Image courtesy of B. Chapron

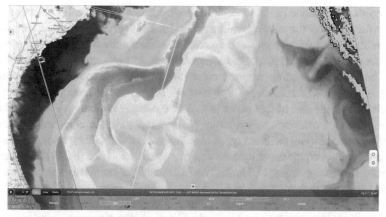

(a) Sea surface temperature near the Gulf Stream, on April 1st 2010, from the Envisat AATSR measurements.

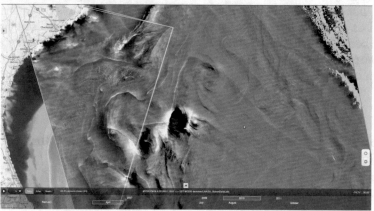

(b) Sea surface glitter contrasts near the Gulf Stream, on April 1st 2010, from the Envisar MERIS observations.

Fig. 2 Comparison of the two images above demonstrates the emergent coherence between sea surface temperature and the glitter patterns visible from satellite imagery. The thermal fronts visible are dynamic, and sea surface roughness is most obvious along the strongest fronts. Discussions of the interpretation of sun glitter measurements are given in [5, 20, 26]. Images courtesy of B. Chapron

quantification through stochastic modelling and analysis, applied in concert with high-resolution observations, computational simulations, and stochastic data assimilation for large data sets. This framework involves decomposing the surface motion into a two-dimensional horizontal flow map representing transport by the current acting on a one-dimensional vertical flow map representing wave-like motion of the elevation. This composition-of-maps modelling framework is described and applied to model sea-surface dynamics in two deterministic examples in Sect. 2 of the present paper.

Emergent Coherence (EC) Combining high-resolution thermal data (buoyancy) with glitter data for the wave elevation as in Fig. 2 has recently revealed yet another interesting feature of submesoscale dynamics. Namely, the observed submesoscale data show extremely high correlations of wave, current and thermal properties [5]. This emergent spatial-temporal coherence of dynamic and thermal properties presents a significant challenge for dynamical submesoscale modelling. Accepting this challenge, the aim of this paper is to derive a mathematical model of nonlinear sea surface dynamics whose solutions also demonstrate the emergent coherence observed in combining different types of submesoscale data. This paper derives new *two-dimensional* equations that show the emergent coherence (EC) seen in the sea surface features appearing in Fig. 2. The EC behaviour produced by the equations derived here are demonstrated in Fig. 3 which shows a snapshot of the coherence of buoyancy and wave amplitude distributions in the dynamics of divergence-free two-dimensional flow acting on free surface vertical elevation wave features moving under gravity. In the model equations, the horizontal buoyancy gradients mediate the interactions between the vertical elevation waves and the horizontal currents. The equations of motion represent the current as a time-dependent, area-preserving map of the horizontal plane into itself and the waves as the composition of the horizontal flow map with a time-dependent vertical elevation map. Thus, the model involves a dynamical composition of maps (CoM).

2 Submesoscale Thermal Wave-Current Dynamics on a Free Surface

2.1 Surface Waves as Symmetry-Breaking Features of Local Force Imbalances

Waves are propagating symmetry-breaking features that signify the response to a local imbalance of forces. Thus, from the viewpoint of satellite oceanography, observations of waves—defined as propagating sea surface elevation features— signify processes at the surface or below the surface whose presence introduces forces that locally break the symmetry of the surface. The sea surface would otherwise follow the stable global gravitational balance of the geoid, which we regard here as being spherical. Thus, waves arise from a spatially local imbalance of forces in the neighbourhood of a stable equilibrium. The propagating feature of relevance here is the wave elevation, measured as the local departure of the surface level in the direction normal to its equilibrium mean level. The symmetry broken here is the invariance of the sea surface under spatial translations tangent to the equilibrium surface level, also known as the local *horizontal* direction. Hence, from the viewpoint of satellite oceanography, sea surface waves are observed as local vertical displacements of the otherwise horizontal motion of the ocean currents on

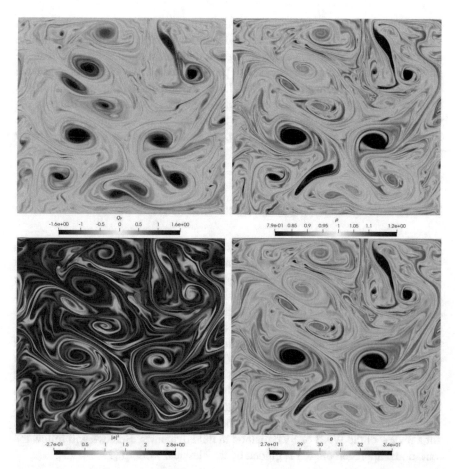

Fig. 3 This is a 512^2 snapshot of the CoM equations in the SVE approximation in the potential vorticity form in (45). The four panels display the following distributions, modified potential vorticity Q-PV in (43) (top left), buoyancy (top right), square of the wave amplitude (bottom left) and wave phase (bottom right) in the numerical simulation of the dynamics of divergence-free flow on a free surface moving under gravity. The simulation began with a spin-up period with zero wave amplitude. After the spin-up period, as explained in Sect. 3, a checker-board pattern of finite wave amplitude with *zero phase* was introduced and the simulation was resumed. The 'mixing' of these wave patterns eventually brought them into coherence with the spatial distributions of thermal properties and potential vorticity. These features show an emergent coherence in patterns similar to those seen in the corresponding high-resolution satellite data in Fig. 2

the sea surface. From the mathematical modelling viewpoint, sea surface waves are local vertical oscillations of the horizontal surface that are carried along by the horizontal current flow, envisaged as a smooth invertible time-dependent map of the horizontal surface into itself. This is the composition of maps (CoM) modelling approach for describing the dynamics of horizontal fluid flows (currents) acting on oscillating vertical elevations (waves). Since the surface current velocity, its advected material properties and the wave elevation are all that can be observed in satellite oceanography, the task in three-dimensional ocean modelling for satellite oceanography devolves into determining the dynamical surface features that are produced by the three-dimensional flow processes below the surface arising from e.g., bathymetry, stratification, rotation, Langmuir circulation, and thermal effects such as frontogenesis. The dynamics of the surface signatures of these three-dimensional flow processes, as well as the effects of air-sea interactions on the surface, need to be interpreted in order to understand what satellite oceanography observes.

2.2   A Tale of Two Maps: Currents and Waves

Story Line Waves on the surface of the ocean are modelled here as a composition of two smooth invertible maps describing the temporal evolution and advection of two degrees of dynamical freedom interacting at widely separated space-time scales. In this composition of maps (CoM) approach, the waves are regarded as local vertical disturbances that rapidly oscillate as they are swept along by the broad, slowly changing horizontal currents. Thus, the slow current motion is a Lagrangian coordinate for the rapid wave oscillations. This wide separation in space-time scales invokes the classical WKB description. The standard WKB approach seeks a rapidly oscillating wave packet solution whose phase-averaged amplitude possesses a slowly varying envelope (SVE) spatially. The WKB method is often applied via a variational principle because in a variational setting the phase average naturally leads to an adiabatic invariant known as the wave action density, cf. for example, [3] for a review of the WKB or SVE method in fluid dynamics. Here we will follow the variational approach of [4, 11] guided by the classical work of [22, 24, 25].

Submesoscale Sea-Surface Motion: Composition of Two Time-Dependent Maps The position and velocity of fluid parcels in motion under gravity on a 2D free surface embedded in $\mathbb{R}^3$ have both horizontal and vertical components. The corresponding flow maps are denoted as the map $\phi_t : \mathbb{R}^2 \to \mathbb{R}^2$ for the horizontal current flow, and as the composite map $\zeta_t \phi_t$ for the vertical elevation of the waves as a function of time and position in $\mathbb{R}^2$. The flow lines of these two components of the flow map of a free surface can be written as

$$r_t = \phi_t r_0 \quad \text{and} \quad z_t = \zeta_t(\phi_t r_0) =: \zeta_t(r_t),$$

where $r_t = (x_t, y_t) \in \mathbb{R}^2$ is the horizontal position along the flow at time t and $\zeta_t(r_t)$ is the vertical elevation at horizontal position r_t at time t, starting at position r_0 at time $t = 0$. Thus, one may say that the initial position of the flow line, r_0, is a Lagrangian coordinate for the horizontal motion, and the horizontal motion is a

Lagrangian coordinate for the vertical motion. That is, the 'footpoint' at time t of the vertical component of the flow map ζ_t is located in the horizontal plane along a curve $\phi_t r_0$ parameterised by time t. Likewise, one can simply say that the wave dynamics is advected, or swept along, by the current dynamics.

Hence, the corresponding horizontal and vertical components of velocity along a stream line r_t in the horizontal plane are defined by,

$$\frac{dr_t}{dt} = \frac{d}{dt}(\phi_t r_0) = \widehat{v}_t(\phi_t r_0) =: \widehat{v}_t(r_t), \quad \text{so} \quad \widehat{v}_t = \frac{d\phi_t}{dt}\phi_t^{-1} \quad \text{and}$$

$$\frac{dz_t}{dt} =: \widehat{w}_t(r_t) = \frac{d}{dt}\big(\zeta_t(\phi_t r_0)\big) = \partial_t \zeta_t(r_t) + \nabla_r \zeta_t(r_t) \cdot \widehat{v}_t(r_t).$$

That is, in the dynamics of free surface flow, the vertical velocity $\widehat{w}(r, t)$ at a given Eulerian point r and time t is related to the wave elevation $\zeta(r, t)$ and horizontal velocity $\widehat{v}(r, t)$ at that point by

$$\widehat{w}(r, t) = \partial_t \zeta(r, t) + \widehat{v}(r, t) \cdot \nabla_r \zeta(r, t).$$

In terms of these fluid variables, one could propose a Hamilton's principle for wave-current interaction of a free surface by following [8] for the variational modelling framework and applying [24, 7] for the potential energy to find[1]

$$0 = \delta S = \delta \int_a^b \ell(\widehat{v}, \zeta, D, \rho)\, dt$$

$$= \delta \int_a^b \int_{\mathcal{D}} \left(\frac{1}{2}\big(|\widehat{v}|^2 + \sigma^2(\partial_t \zeta + \nabla_r \zeta \cdot \widehat{v})^2\big) - \frac{\zeta^2}{2Fr^2}\right) D\rho - p(D-1)\, d^2r\, dt.$$

$$(1)$$

To interpret the variational principle proposed in (1) we rewrite its Lagrangian as a sum of an Eulerian spatial integral and an integral over material mass elements $d^2r_0 = D\rho\, d^2r$ which follow the paths of the horizontal fluid motion, $r(r_0, t) = \phi_t r_0$,

$$0 = \delta S = \delta \int_a^b \int_{\mathcal{D}} \frac{D\rho}{2}|\widehat{v}|^2 - p(D-1)\, d^2r\, dt + \delta \int_a^b \int_{\mathcal{D}_0} \frac{\sigma^2}{2}\dot{\zeta}^2 - \frac{\zeta^2}{2Fr^2}\, d^2r_0\, dt.$$

$$(2)$$

---

[1] In [8] the potential energy was linear in ζ. This linearity neglected the restoring force due to vertical pressure gradient via Archimedes' principle. Adopting the potential energy quadratic in ζ regains this restoring force.

Variations of the first summand in (2) at fixed spatial position (r) yield the Euler fluid equations for 2D divergence free flow with advected buoyancy, $\rho(r, t) = \rho(\phi_t r_0) = \rho_0(r_0)$,

$$\partial_t \widehat{v} + (\widehat{v} \cdot \nabla_r)\widehat{v} = -\frac{1}{\rho}\nabla_r p \quad \text{with} \quad \nabla_r \cdot \widehat{v} = 0. \tag{3}$$

Variations of the second summand in (2) taken at fixed mass element (r_0) yield equations for vertical harmonic oscillations of the elevation of each material mass element

$$\sigma^2 \ddot{\zeta}(r_0, t) = \sigma^2 \frac{d^2\zeta}{dt^2}\bigg|_{r_0} = -\frac{\zeta(r_0, t)}{Fr^2}. \tag{4}$$

The wave-elevation equation in (4) is unrealistic, though, because it implies that fluid mass elements with different labels (r_0) would be oscillating in phase and all with the same frequency, as they follow the flow of the Euler fluid equations (3) for 2D divergence free flow with advected buoyancy. This unrealistic synchronisation and resonance can be removed by including the inertia of each mass element. This can done by including the initial buoyancy of each mass element, as

$$\sigma^2 \ddot{\zeta}(r_0, t) = \sigma^2 \frac{d^2\zeta}{dt^2}\bigg|_{r_0} = -\frac{\rho_{ref}}{\rho_0(r_0)}\frac{\zeta(r_0, t)}{Fr^2}. \tag{5}$$

At this point in our reasoning, we have not yet considered the differences in space and time scales between the fluid flow and the wave activity. In what follows, we will use the simple composition-of-maps idea explained here along with estimates of relative space and time scales to investigate the applicability of this class of models. To improve the applicability of the model comprising (3) and (5) for describing the effects of currents on waves, we will derive a related model in the slowly varying envelope (SVE) approximation. The SVE approximation allows considerations of current and wave dynamics at the same space and time scales.

The comparisons of the simulated solutions of these CoM models with the observations in Figs. 1, 2, 3, and 4 above indicate that these models can indeed produce results that match some aspects of observed features. However, these models are not derived from three dimensional fluid equations. Instead, they are derived from the simple solution Ansatz in Hamilton's principle that the vertical elevation of the sea surface wave activity is carried by divergence-free horizontal fluid motion. The latter assumption is a weakness of the current approach, because it precludes effects of vertical up-welling and down-welling, which are observed to occur along with convergence and divergence of currents [10]. The equations derived here are also not associated with classical surface wave equations such as the nonlinear Schroedinger (NLS) equation, or other celebrated surface wave equations. This departure from the classical water wave literature may be regarded as another weakness of the current approach.

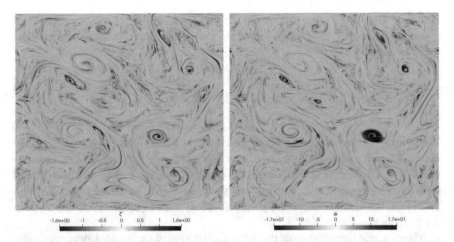

Fig. 4 These 512^2 snapshots of the CoM simulation in the vorticity form (25) shows the elevation ζ in the left panel and the density-weighted vertical velocity $\widetilde{w}$ on the right. The snapshots are taken at the same time and with the same fluid spin-up initial conditions as the snapshots of the simulation of the SVE approximate equations presented in Fig. 3. Overlaying the two figures demonstrates that the resolved features in the ζ distribution in this figure of CoM results are bounded by the SVE wave envelope distribution $|a|^2$ in Fig. 3

Estimating Parameters σ^2 and Fr^2 for Satellite Observations The Lagrangian $\ell(\widehat{v}, \zeta, D, \rho)$ in (1) represents the dimension-free difference of the kinetic and potential energies, augmented by the incompressibility constraint imposed by the Lagrange multiplier p. Two dimension-free parameters (σ^2 and Fr^2) appear in this Hamilton's principle. The coefficient $\sigma^2 = ([H]/[L])^2$ in formula (1) is the square of the vertical-to-horizontal aspect ratio. Typically, for satellite observations of submesoscale dynamics one finds

$$[H] \approx (3 \times 10^{-4} - 3 \times 10^{-3}) \, \text{km} \quad \text{and} \quad [L] \approx (10^{-1} - 10) \, \text{km}, \quad \text{so} \quad \sigma^2 \approx 10^{-3} - 10^{-6} \ll 1$$

for the squared aspect ratio $\sigma^2 \ll 1$ of the height of the waves $[H]$ relative to the breadth $[L]$ of the two-dimensional domain. The squared 'Froude number' Fr^2 in this regime is estimated by the square of the ratio of horizontal and vertical frequency scales at the sea surface,

$$Fr^2 := \left(\frac{[V]/[H]}{N} \right)^2 \approx 1 - 10^4. \tag{6}$$

Here, the horizontal velocity on the sea surface is taken as $[V] = (0.1 - 1)$ m/s, $[H] = (0.3 - 3)$ m. According to [9], the Brunt-Väisälä buoyancy frequency in the sea surface wave regime is given by $N \approx (10^{-3} - 10^{-4})$/s. The ratio of horizontal and vertical frequency scales at the sea surface in (6) is selected for use later in applying the slowly varying envelope (SVE) wave approximation in

Sect. 2.4. Hence, we estimate that the squared product of the 'Froude number' and aspect ratio for satellite observations of the sea surface can reasonably be estimated over the range

$$\sigma^2 Fr^2 := \left(\frac{[V]}{N[L]}\right)^2 \approx 10^{-3} - 10. \tag{7}$$

Modelling the Dynamic Effects of Surface Density Variations As mentioned earlier, the observed oscillations of sea surface waves are by no means simultaneous across the whole domain, although the observations show that they are indeed coordinated spatially with the buoyancy of the fluid. To correct this solution behaviour, the kinetic energy and potential energy need to be de-synchronised from the buoyancy.

The dynamic dependence of the wave kinetic energy on the density is physically required. However, to de-synchronise the wave oscillations we can introduce a constant reference density ρ_{ref} into the wave potential energy, by writing

$$\frac{\zeta^2}{Fr^2} \rightarrow \frac{\rho_{ref}}{\rho} \frac{\zeta^2}{Fr^2} \quad \text{with} \quad \frac{\rho_{ref}}{\rho} \quad \text{of order} \quad O(1). \tag{8}$$

The quantity ρ_{ref} is a constant reference density, and the density ratio $(\rho_{ref}/\rho) = O(1)$.

The density dependence imposed here is important in the dynamics that follows from Hamilton's principle. Substituting the relations in (8) into Hamilton's principle in Eq. (1) leads to the following dimension-free action integral,

$$0 = \delta S = \delta \int_a^b \ell(\widehat{\boldsymbol{v}}, \zeta, D, \rho) \, dt$$

$$= \delta \int_a^b \int_{\mathcal{D}} \left(\frac{1}{2}\left(|\widehat{\boldsymbol{v}}|^2 + \sigma^2(\partial_t \zeta + \nabla_r \zeta \cdot \widehat{\boldsymbol{v}})^2\right) - \frac{\rho_{ref}}{\rho} \frac{\zeta^2}{2Fr^2}\right) D\rho - p(D-1) \, d^2r \, dt. \tag{9}$$

The advected quantities $D(\boldsymbol{r}, t)d^2r$ and $\rho(\boldsymbol{r}, t)$ evolve via push-forward by the horizontal flow map, ϕ_t. For example, $D_t d^2 r_t = \phi_{t*}(D_0 d^2 r_0)$ and $\rho_t = \phi_{t*}\rho_{ref}$ denote, respectively, evolution of the determinant of the Lagrange to Euler map and of the local scalar value of the mass density. Conservation of mass is then expressed as the push-forward relation, $D_t \rho_t d^2 r_t = \phi_{t*}(D_0 \rho_{ref} d^2 r_0)$. The pressure p in (9) acts as a Lagrange multiplier to enforce conservation of area, so that $D_t = 1 = \phi_{t*} D_0$, and the horizontal flow is incompressible, which implies that the horizontal velocity is divergence-free, i.e., $\text{div}_r \widehat{\boldsymbol{v}}(\boldsymbol{r}, t) = 0$. Taking variations of the action integral (9) yields the following set of equations,

$$\delta \widehat{\boldsymbol{v}}: \quad \frac{\delta \ell}{\delta \widehat{\boldsymbol{v}}} = D\rho \big(\widehat{\boldsymbol{v}} \cdot d\boldsymbol{r} + \sigma^2 \widehat{w} \, d\zeta \big) \otimes d^2 r := D\rho \boldsymbol{V} \cdot d\boldsymbol{r} \otimes d^2 r \,,$$

with $\quad \widehat{w} = \partial_t \zeta + \widehat{\boldsymbol{v}} \cdot \nabla_r \zeta \,,$

$$\delta \zeta: \quad \partial_t (\sigma^2 D\rho \widehat{w}) + \operatorname{div}_r (\sigma^2 D\rho \widehat{w}\widehat{\boldsymbol{v}}) - D \frac{\zeta \rho_{ref}}{Fr^2} = 0 \,,$$

$$\delta D: \quad \frac{\delta \ell}{\delta D} = \frac{\rho}{2} \big(|\widehat{\boldsymbol{v}}|^2 + \sigma^2 \widehat{w}^2 \big) - \frac{\rho_{ref} \zeta^2}{2Fr^2} - p =: \rho \widetilde{\varpi} - \widetilde{p} \,,$$

$$\delta \rho: \quad \frac{\delta \ell}{\delta \rho} = \frac{D}{2} \big(|\widehat{\boldsymbol{v}}|^2 + \sigma^2 \widehat{w}^2 \big) =: D\widetilde{\varpi} \,, \quad \widetilde{p} := p + \frac{\rho_{ref} \zeta^2}{2Fr^2} \,,$$

$$\delta p: \quad D - 1 = 0 \quad \Longrightarrow \quad \operatorname{div}_r \widehat{\boldsymbol{v}} = 0 \,.$$

(10)

From their definitions as advected quantities, one also knows that D and ρ satisfy

$$(\partial_t + \mathcal{L}_{\widehat{\boldsymbol{v}}})(D \, d^2 r) = 0 \Longrightarrow \partial_t D + \operatorname{div}_r (D\widehat{\boldsymbol{v}}) = 0 \quad \text{with} \quad D = 1 \,,$$
$$(\partial_t + \mathcal{L}_{\widehat{\boldsymbol{v}}})\rho = 0 \Longrightarrow \partial_t \rho + \widehat{\boldsymbol{v}} \cdot \nabla_r \rho = 0 \,,$$

(11)

where $\mathcal{L}_{\widehat{\boldsymbol{v}}}$ denotes the Lie derivative operation along the horizontal velocity vector field, $\widehat{\boldsymbol{v}}$, which provides coordinate-free brevity in the notation.

Theorem 1 (Kelvin-Noether Circulation Theorem) *Use of the Euler-Poincaré (EP) theorem yields the following Kelvin circulation theorem*

$$\frac{d}{dt} \oint_{c(\widehat{\boldsymbol{v}})} \big(\widehat{\boldsymbol{v}} \cdot d\boldsymbol{r} + \sigma^2 \widehat{w} \, d\zeta \big) = - \oint_{c(\widehat{\boldsymbol{v}})} \frac{1}{\rho} d\widetilde{p} \,.$$

(12)

Proof The Euler-Poincaré (EP) theorem in this case yields

$$(\partial_t + \mathcal{L}_{\widehat{\boldsymbol{v}}}) \frac{\delta \ell}{\delta \widehat{\boldsymbol{v}}} = \frac{\delta \ell}{\delta D} \diamond D + \frac{\delta \ell}{\delta \rho} \diamond \rho := D\nabla_r \frac{\delta \ell}{\delta D} - \frac{\delta \ell}{\delta \rho} \nabla_r \rho \,.$$

(13)

Here the diamond ($\diamond$) operator is defined by

$$\Big\langle \frac{\delta \ell}{\delta a} \diamond a \,, X \Big\rangle_{\mathfrak{X}} =: \Big\langle \frac{\delta \ell}{\delta a} \,, -\pounds_X a \Big\rangle_V \,.$$

(14)

In addition, $X \in \mathfrak{X}$ is a (smooth) vector field defined on $\mathbb{R}^2$ and $a \in V$, a vector space of advected quantities, which are here the scalar function, ρ, and the areal density $D \, d^2 r$. Using the advection relations for D and ρ in (11) and the corresponding variational derivatives in (10) simplifies the EP equation in (14) to

$$(\partial_t + \mathcal{L}_{\widehat{\boldsymbol{v}}}) \Big(\frac{1}{D\rho} \frac{\delta \ell}{\delta \widehat{\boldsymbol{v}}} \Big) = \frac{1}{\rho} \nabla_r \frac{\delta \ell}{\delta D} - \frac{1}{D\rho} \frac{\delta \ell}{\delta \rho} \nabla_r \rho \,.$$

Equation (10) then yields $\quad (\partial_t + \mathcal{L}_{\widehat{\boldsymbol{v}}})\big(\widehat{\boldsymbol{v}} \cdot d\boldsymbol{r} + \sigma^2 \widehat{w} \, d\zeta \big) = -\rho^{-1} d\widetilde{p} + d\widetilde{\varpi} \,.$

(15)

Inserting the last relation into the following standard relation for the time derivative of a loop integral then completes the proof of Eq. (12) appearing in the statement of the theorem,

$$\frac{d}{dt} \oint_{c(\widehat{v})} (\widehat{v} \cdot dr + \sigma^2 \widehat{w} \, d\zeta) = \oint_{c(\widehat{v})} (\partial_t + \mathcal{L}_{\widehat{v}})(\widehat{v} \cdot dr + \sigma^2 \widehat{w} \, d\zeta) = \oint_{c(\widehat{v})} -\rho^{-1} d\widetilde{p} + d\widetilde{\varpi}.$$

(16)

Using the advection relations for D and ρ in (11) again and combining with the variational relations with respect to ζ in (10) simplifies the $\widehat{w}$ and ζ equations, as follows.

$$(\partial_t + \mathcal{L}_{\widehat{v}})\widehat{w} = (\partial_t + \widehat{v} \cdot \nabla_r)\widehat{w} = -\frac{\rho_{ref}}{\sigma^2 Fr^2 \rho} \zeta,$$

$$(\partial_t + \mathcal{L}_{\widehat{v}})\zeta = (\partial_t + \widehat{v} \cdot \nabla_r)\zeta = \widehat{w}.$$

(17)

After deriving these equations, one may finally evaluate the constraint $D = 1$ imposed by the variation in pressure p to obtain further simplifications. □

Corollary 2 (Kelvin-Noether circulation Theorem for the Current) *The Kelvin circulation theorem for the current alone is given by,*

$$\frac{d}{dt} \oint_{c(\widehat{v})} \widehat{v} \cdot dr = -\oint_{c(\widehat{v})} \frac{1}{\rho} dp - d\frac{|\widehat{v}|^2}{2}.$$

(18)

Proof Equation (18) follows by shifting the $\widehat{w} d\zeta$ term in Eq. (38) to the right-hand side, as

$$\frac{d}{dt} \oint_{c(\widehat{v})} \widehat{v} \cdot dr = -\oint_{c(\widehat{v})} \frac{1}{\rho} d\widetilde{p} + \sigma^2 (\partial_t + \mathcal{L}_{\widehat{v}})(\widehat{w} \, d\zeta) - d\widetilde{\varpi}$$

$$= -\oint_{c(\widehat{v})} \frac{1}{\rho} d\widetilde{p} + \sigma^2 ((\partial_t + \widehat{v} \cdot \nabla_r)\widehat{w}) d\zeta + \sigma^2 \widehat{w} d\widehat{w} - d\widetilde{\varpi}$$

$$= -\oint_{c(\widehat{v})} \frac{1}{\rho} d\widetilde{p} - \frac{\rho_{ref}}{Fr^2 \rho} \zeta d\zeta + \sigma^2 \widehat{w} d\widehat{w} - d\widetilde{\varpi}$$

$$= -\oint_{c(\widehat{v})} \frac{1}{\rho} dp - d\frac{|\widehat{v}|^2}{2}$$

$$=: -\oint_{c(\widehat{v})} \frac{1}{\rho} dp - d\frac{|\widehat{v}|^2}{2}.$$

(19)

□

Remark 1 (Separation of Wave and Current Circulation) The decoupling of the Kelvin-Noether circulation theorem into its wave and current components, leading to the reduction of the current flow to the Euler result in Eq. (18), was also observed in [8]. This behaviour is consistent with the Charney-Drazin 'non-acceleration' theorem [6, 23]. Namely, in certain circumstances, wave activity does not create circulation in the mean current. A modification that allows exchange of circulation between wave (vertical) and current (horizontal) components of the flow was proposed in [8]. The instabilities observed around the edges of eddies in the satellite imagery shown in Fig. 1 suggests that a coupling of this sort may exist at high wave number.

Remark 2 It is clear from Eqs. (38)–(18) that generation of circulation of the current by the dynamics in Eq. (15) requires non-zero $\nabla_r \rho \times \nabla_r p$. No current circulation is generated by wave variables in the case of constant buoyancy.

2.3 Thermal Potential Vorticity (TPV) Dynamics on a Free Surface

The momentum map arising from the variations in (10) is given by

$$\frac{1}{D} \frac{\delta \ell}{\delta \widehat{\boldsymbol{v}}} = \rho \widehat{\boldsymbol{v}} \cdot d\boldsymbol{r} + \sigma^2 \rho \widehat{w} d\zeta \,. \tag{20}$$

As expected from the well-known non-acceleration theorem [6, 23], the dynamics of the Euler-Poincaré equations separate (15) gives the dynamics of the fluid and wave components of the momentum one-form (20)

$$
\begin{aligned}
(\partial_t + \mathcal{L}_{\widehat{\boldsymbol{v}}})\left(\rho(\widehat{\boldsymbol{v}} \cdot d\boldsymbol{r})\right) &= -dp + \frac{\rho}{2} d\left(|\widehat{\boldsymbol{v}}|^2\right), \\
(\partial_t + \mathcal{L}_{\widehat{\boldsymbol{v}}})\left(\sigma^2 \rho \widehat{w} d\zeta\right) &= -\frac{\rho_{ref}}{Fr^2} \zeta d\zeta + \sigma^2 \rho \widehat{w} d\widehat{w} \,.
\end{aligned}
\tag{21}
$$

The mass-weighted thermal potential vorticity (TPV) also separates into fluid and wave components $Q = Q_F + Q_W$ with following definitions

$$
\begin{aligned}
Q\, d^2r &= d\left(\rho(\widehat{\boldsymbol{v}} \cdot d\boldsymbol{r} + \sigma^2 \widehat{w} d\zeta)\right) \\
&= d\rho \wedge (\widehat{\boldsymbol{v}} \cdot d\boldsymbol{r} + \sigma^2 \widehat{w} d\zeta) + \rho\left(\widehat{\boldsymbol{z}} \cdot \mathrm{curl}\widehat{\boldsymbol{v}} + \sigma^2 J(\widehat{w}, \zeta)\right) d^2r \\
&= \left(\mathrm{div}(\rho\nabla\psi) + \sigma^2 J(\rho\widehat{w}, \zeta)\right) d^2r \quad \text{when} \quad \widehat{\boldsymbol{v}} = \nabla^{\perp}\psi \quad \text{for} \quad D = 1,
\end{aligned}
$$

with $\quad Q_F := \mathrm{div}(\rho\nabla\psi)\,, \quad Q_W = J\left(\sigma^2 \widetilde{w}, \zeta\right)\,.$

$$\tag{22}$$

where buoyancy weighted vertical velocity is defined as $\widetilde{w} := \rho\widehat{w}$. The dynamics of $Q_F\, d^2r$ and $Q_W\, d^2r$ can be computed from (21) as

$$(\partial_t + \mathcal{L}_{\widehat{v}})(Q_F\, d^2r) = \frac{1}{2}d\rho \wedge d(|\widehat{v}|^2) = \frac{1}{2}J(\rho, |\nabla\psi|^2))\, d^2r\,,$$

$$(\partial_t + \mathcal{L}_{\widehat{v}})(Q_W\, d^2r) = \sigma^2\frac{1}{2}d\rho \wedge d(\widehat{w}^2) = \frac{1}{2}J\left(\rho, \frac{\sigma^2\widetilde{w}^2}{\rho^2}\right)d^2r\,.$$

(23)

From the two relations in (23), one sees that the buoyancy gradient $\nabla\rho$ couples the PV dynamics of the waves (Q_W) and currents (Q_F), each to their corresponding kinetic energy. In the case of constant buoyancy, $d\rho = 0$ in (23); so, the PVs of the waves and currents would be separately advected.

The operator $\mathrm{div}(\rho\nabla)$ is invertible, so long as ρ is a differentiable positive function, which can be ensured by requiring that this condition holds initially. Consequently, the stream function ψ is related to the other fluid variables by

$$\psi := (\mathrm{div}\rho\nabla)^{-1}Q_F\,. \tag{24}$$

The potential vorticity dynamics can then be written in coordinate form as

$$\partial_t Q_F + J(\psi, Q_F) = J\left(\rho, \frac{1}{2}|\nabla_r\psi|^2\right),$$

$$\partial_t Q_W + J(\psi, Q_W) = J\left(\rho, \frac{\sigma^2\widetilde{w}^2}{2\rho^2}\right),$$

$$\text{with} \quad Q_F := \mathrm{div}(\rho\nabla\psi) \quad \text{and} \quad Q_W := J(\sigma^2\widetilde{w}, \zeta)\,,$$

$$\partial_t\rho + J(\psi, \rho) = 0\,,$$

$$\partial_t\zeta + J(\psi, \zeta) = \widehat{w} =: \widetilde{w}/\rho\,,$$

$$\partial_t(\sigma^2\widetilde{w}) + J(\psi, \sigma^2\widetilde{w}) = -\frac{\rho_{ref}\zeta}{Fr^2}\,.$$

(25)

Theorem 3 *The Legendre transform yields the Hamiltonian formulation of our system of wave-current equations (25), which with $\widetilde{w} = \rho\widehat{w}$ may be written in the untangled block-diagonal Poisson form as*

$$\frac{\partial}{\partial t}\begin{bmatrix} Q \\ \rho \\ \sigma^2\widetilde{w} \\ \zeta \end{bmatrix} = \begin{bmatrix} J(Q, \cdot) & J(\rho, \cdot) & 0 & 0 \\ J(\rho, \cdot) & 0 & 0 & 0 \\ 0 & 0 & 0 & -1 \\ 0 & 0 & 1 & 0 \end{bmatrix}\begin{bmatrix} \delta h/\delta Q = \psi \\ \delta h/\delta\rho = \widetilde{\omega} \\ \delta h/\delta(\sigma^2\widetilde{w}) = \widetilde{w}/\rho + J(\zeta, \psi) \\ \delta h/\delta\zeta = -J(\sigma^2\widetilde{w}, \psi) + \frac{\rho_{ref}\zeta}{Fr^2} \end{bmatrix}\,.$$

(26)

The energy Hamiltonian $h(Q, \rho, \widehat{w}, \zeta)$ associated with this system is given by

$$h(Q, \rho, \widetilde{w}, \zeta) = \int \frac{1}{2}\Big(Q - J\big(\sigma^2\widetilde{w}, \zeta\big)\Big)(\mathrm{div}\rho\nabla)^{-1}\Big(Q - J\big(\sigma^2\widetilde{w}, \zeta\big)\Big)$$
$$+ \left(\frac{\sigma^2\widetilde{w}^2}{2\rho^2} + \frac{\rho_{ref}}{\rho}\frac{\zeta^2}{2Fr^2}\right)\rho\, d^2r. \tag{27}$$

Theorem 4 (Casimir Functions) *The Casimir functions, conserved by the relation $\{C_{\Phi,\Psi}, h\} = 0$ with any Hamiltonian $h(\boldsymbol{M}, \boldsymbol{D})$ for the block-diagonal Lie-Poisson bracket in Eq. (26) are given by*

$$C_{\Phi,\Psi} := \int \Phi(\rho) + Q\Psi(\rho)\, d^2r. \tag{28}$$

Proof *The Casimirs $C_{\Phi,\Psi}$ for the direct sum of the Lie-Poisson brackets for Q and ρ and canonical Poisson brackets for $\widetilde{w}$ and ζ follows by direct verification that the $C_{\Phi,\Psi}$ are conserved for any differentiable functions, (Φ, Ψ).* $\qquad\square$

2.4 CoM Equations in the Slowly Varying Envelope (SVE) Approximation

The SVE Solutions Apply to Satellite Observations of Sea Surface Waves From the viewpoint of satellite observations, the vertical motion on the sea surface typically oscillates much more quickly than the rate of change of features in the horizontal motion of the ocean surface currents. In this situation, the standard WKB approximation introduces a solution Ansatz for the slowly varying envelope (SVE) of the rapidly oscillating vertical wave elevation in the standard form [2, 11],

$$\zeta(\boldsymbol{r}, t) = \Re\left(a(\boldsymbol{r}, t)\exp\left(\frac{i\theta(\boldsymbol{r}, t)}{\epsilon}\right)\right) \quad \text{with} \quad \epsilon \ll 1. \tag{29}$$

The SVE solution Ansatz (29) comprises the product of a slowly varying complex amplitude $a(\boldsymbol{r}, t) \in \mathbb{C}$ multiplied by a rapidly oscillating phase $\theta(\boldsymbol{r}, t)/\epsilon \in \mathbb{R}$ with $\epsilon \ll 1$ in which the phase factor $\theta(\boldsymbol{r}, t)$ may also vary slowly as a function of the space and time variables, $(\boldsymbol{r}, t)$.

Following [11], let us substitute the SVE solution Ansatz (29) into Hamilton's principle in (9) and find the condition on the parameter $\epsilon \ll 1$ that will allow higher order wave terms to be neglected. For this, one computes

$$0 = \delta S_{SVE} = \delta \int_a^b \ell_{SVE}(\widehat{\boldsymbol{v}}, D, \rho; a, \theta) \, dt$$

$$= \delta \int_a^b \int_D \frac{1}{2} D\rho |\widehat{\boldsymbol{v}}|^2 - p(D-1) + \frac{\sigma^2}{2} D\rho \left(\left(\frac{d\zeta}{dt} \right)^2 - \frac{\rho_{ref}}{\rho} \frac{\zeta^2}{2\sigma^2 Fr^2} \right) d^2 r \, dt$$

$$= \delta \int_a^b \int_D \frac{1}{2} D\rho |\widehat{\boldsymbol{v}}|^2 - p(D-1)$$

$$+ \frac{\sigma^2}{8} D\rho \left(\left| \frac{da}{dt} \right|^2 + \frac{2}{\epsilon} \frac{d\theta}{dt} \Im \left(a^* \frac{da}{dt} \right) + \frac{|a|^2}{\epsilon^2} \left(\left(\frac{d\theta}{dt} \right)^2 - \frac{\rho_{ref}}{\rho} \frac{\epsilon^2}{\sigma^2 Fr^2} \right) \right) d^2 r \, dt$$

$$\simeq \delta \int_a^b \int_D \frac{1}{2} D\rho |\widehat{\boldsymbol{v}}|^2 - p(D-1)$$

$$+ \frac{\sigma^2 |a|^2}{8\epsilon^2} D\rho \left((\partial_t \theta + \widehat{\boldsymbol{v}} \cdot \nabla_r \theta)^2 - \frac{\rho_{ref}}{\rho} \frac{\epsilon^2}{\sigma^2 Fr^2} \right) d^2 r \, dt + O \left(\frac{\sigma^2}{\epsilon} \right).$$

$$\tag{30}$$

The leading order wave term $O(\epsilon^{-2})$ with $\epsilon \ll 1$ in Hamilton's principle will dominate the solution and the remaining wave terms in the second line of Eq. (31) may be neglected, when[2]

$$\epsilon \ll 1, \quad \frac{\epsilon^2}{\sigma^2 Fr^2} = O(1), \quad \text{and} \quad \sigma^2 Fr^2 \ll 1. \tag{31}$$

According to the estimates in (7) there is a range of physical parameters relevant to satellite observations in which the SVE approximation applies, for $\sigma^2 Fr^2 \ll 1$.

To continue the investigation of the SVE description of wave-current interactions on the sea surface, we take variations of the action integral (31) to find the following set of equations,

$$\delta \widehat{\boldsymbol{v}} : \quad \frac{\delta \ell}{\delta \widehat{\boldsymbol{v}}} = D\rho \left(\widehat{\boldsymbol{v}} \cdot d\boldsymbol{r} + \mathcal{N} d \frac{d\theta}{dt} \right) \otimes d^2 r \quad \text{with} \quad \mathcal{N} := \frac{\sigma^2 |a|^2}{4\epsilon^2},$$

$$\delta |a|^2 : \quad \frac{\delta \ell}{\delta |a|^2} = \frac{\sigma^2}{8 Fr^2} D\rho \left(\left(\frac{d\theta}{dt} \right)^2 - \frac{\rho_{ref}}{\rho} \right) = 0 \quad \text{at} \quad O \left(\frac{\sigma^2}{\epsilon^2} \right)$$

$$\implies \frac{d\theta}{dt} =: -\omega + \widehat{\boldsymbol{v}} \cdot \boldsymbol{k} = \pm \frac{\sqrt{\rho \rho_{ref}}}{\rho} \quad \text{with} \quad \omega(\boldsymbol{r}, t) = -\partial_t \theta \quad \text{and} \quad \boldsymbol{k}(\boldsymbol{r}, t) = \nabla_r \theta,$$

$$\delta \theta : \quad \frac{\delta \ell}{\delta \theta} = 0 \implies \partial_t \mathcal{A} + \text{div}(\mathcal{A} \widehat{\boldsymbol{v}}) = 0, \quad \text{with} \quad \mathcal{A} := D\rho \mathcal{N} \frac{d\theta}{dt} \quad \text{and} \quad \mathcal{N} := \frac{\sigma^2 |a|^2}{4\epsilon^2},$$

$$\delta D : \quad \frac{\delta \ell}{\delta D} = \frac{\rho}{2} |\widehat{\boldsymbol{v}}|^2 - p,$$

$$\delta \rho : \quad \frac{\delta \ell}{\delta \rho} = \frac{D}{2} |\widehat{\boldsymbol{v}}|^2,$$

$$\delta p : \quad D - 1 = 0 \implies \text{div}_r \widehat{\boldsymbol{v}} = 0, \quad \text{Hence,} \quad \partial_t \mathcal{A} + \widehat{\boldsymbol{v}} \cdot \nabla_r \mathcal{A} = 0 \implies \partial_t |a|^2 + \widehat{\boldsymbol{v}} \cdot \nabla_r |a|^2 = 0.$$

$$\tag{32}$$

---

[2] The ratio $\epsilon^2/(\sigma^2 Fr^2) = O(1)$ is required for the rate of change of the phase parameter $\theta(\boldsymbol{r}, t)$ of the SVE wave solution Ansatz (29) to match the time scale of the density $\rho(\boldsymbol{r}, t)$ in Eq. (31).

In the second line of (32) we see that stationarity of the action integral with respect to variations in $|a|^2$ acts as a Lagrange multiplier to impose a constraint which relates the dynamics of the wave phase θ to the buoyancy. This constraint relation involves the Doppler-shifted frequency of the waves, as shown in the third line of (32). In combination with conservation of the wave action density and the divergence free condition on the fluid flow velocity $\widehat{v}$, this constraint relation implies in the last line of (32) that the wave magnitude $|a|^2$ is advected by the fluid flow. Because of the oscillatory nature of the solution Ansatz (29), the sign of the wave phase in $d\theta/dt = \partial_t \theta + \widehat{v} \cdot \nabla_r \theta$ in the second line above is immaterial. Hence, hereafter, we will choose the positive root for $d\theta/dt = \sqrt{\rho \rho_{ref}}/\rho$.

From the conservation of wave action density $\mathcal{A}$ in (32) and the definitions of the advected fluid variables, one finds that $|a|^2$, D and ρ satisfy the following advection relations

$$(\partial_t + \mathcal{L}_{\widehat{v}})(D\,d^2r) = 0 \Longrightarrow \partial_t D + \mathrm{div}_r(D\widehat{v}) = 0 \quad \text{with} \quad D = 1\,,$$

$$(\partial_t + \mathcal{L}_{\widehat{v}})\rho = 0 \Longrightarrow \partial_t \rho + \widehat{v} \cdot \nabla_r \rho = 0\,, \tag{33}$$

$$(\partial_t + \mathcal{L}_{\widehat{v}})|a|^2 = 0 \Longrightarrow \partial_t |a|^2 + \widehat{v} \cdot \nabla_r |a|^2 = 0\,,$$

where $\mathcal{L}_{\widehat{v}}$ denotes the Lie derivative operation along the horizontal velocity vector field, $\widehat{v}$. The Lie derivative notation $\mathcal{L}_{\widehat{v}}$ provides coordinate-free brevity in proving the following Kelvin circulation theorem for thermal wave-current theory.

Theorem 5 (Kelvin-Noether Circulation Theorem) *The variational equations in* (32) *imply the following Kelvin circulation theorem*

$$\frac{d}{dt} \oint_{c(\widehat{v})} \left(\widehat{v} \cdot dr + \mathcal{N} d\frac{d\theta}{dt} \right) = - \oint_{c(\widehat{v})} \frac{1}{\rho} dp\,. \tag{34}$$

Proof The Euler-Poincaré (EP) theorem [16] in this case yields

$$(\partial_t + \mathcal{L}_{\widehat{v}})\frac{\delta\ell}{\delta\widehat{v}} = \frac{\delta\ell}{\delta D} \diamond D + \frac{\delta\ell}{\delta\rho} \diamond \rho := D\nabla_r \frac{\delta\ell}{\delta D} - \frac{\delta\ell}{\delta\rho}\nabla_r\rho\,. \tag{35}$$

Here, the diamond ($\diamond$) operator is defined for a fluid advected quantity f by

$$\left\langle \frac{\delta\ell}{\delta f} \diamond f\,,\, X \right\rangle_{\mathfrak{x}} =: \left\langle \frac{\delta\ell}{\delta f}\,,\, -\pounds_X f \right\rangle_V\,. \tag{36}$$

In (36), $X \in \mathfrak{X}(\mathbb{R}^2)$ is a (smooth) vector field defined on $\mathbb{R}^2$ and $f \in V$ is a vector space of advected quantities. These advected quantities are the scalar function, ρ, and the areal density, $D\,d^2r$.

Upon using the advection relations for D and ρ in (33) and the corresponding variational derivatives in (32), the EP equation in (35) simplifies to

$$(\partial_t + \mathcal{L}_{\widehat{\boldsymbol{v}}})\left(\frac{1}{D\rho}\frac{\delta\ell}{\delta\widehat{\boldsymbol{v}}}\right) = \frac{1}{\rho}\nabla_r\frac{\delta\ell}{\delta D} - \frac{1}{D\rho}\frac{\delta\ell}{\delta\rho}\nabla_r\rho\,.$$

Equation (32) then yields $(\partial_t + \mathcal{L}_{\widehat{\boldsymbol{v}}})\left(\widehat{\boldsymbol{v}}\cdot d\boldsymbol{r} + \mathcal{N}d\frac{d\theta}{dt}\right) = -\rho^{-1}dp + d\left(\frac{1}{2}|\widehat{\boldsymbol{v}}|^2\right).$

$$(37)$$

Inserting the last relation into the following standard relation for the time derivative of a loop integral then completes the proof of Eq. (34) appearing in the statement of the theorem,

$$\frac{d}{dt}\oint_{c(\widehat{v})}\left(\widehat{\boldsymbol{v}}\cdot d\boldsymbol{r} + \mathcal{N}d\frac{d\theta}{dt}\right) = \oint_{c(\widehat{v})}(\partial_t + \mathcal{L}_{\widehat{\boldsymbol{v}}})\left(\widehat{\boldsymbol{v}}\cdot d\boldsymbol{r} + \mathcal{N}d\frac{d\theta}{dt}\right) = \oint_{c(\widehat{v})}-\rho^{-1}dp + d\left(\frac{1}{2}|\widehat{\boldsymbol{v}}|^2\right).$$

$$(38)$$

Note, however, that Eqs. (32) imply the following combination of advected quantities,

$$(\partial_t + \mathcal{L}_{\widehat{v}})\left(\mathcal{N}d\frac{d\theta}{dt}\right) = \frac{\sigma^2}{4Fr^2}(\partial_t + \mathcal{L}_{\widehat{v}})\left(|a|^2d\sqrt{\frac{\rho_{ref}}{\rho}}\right) = 0\,. \qquad (39)$$

Consequently, the wave-momentum 1-form $\mathcal{N}d(\frac{d\theta}{dt})$ is advected by the fluid flow and the Kelvin circulation theorem in Eq. (38) reduces to the standard circulation theorem for the 2D Euler fluid equations.   □

Remark 3 (Separation of Wave and Current Motion in the SVE Approximation) The decoupling of the Kelvin-Noether circulation theorem into its wave and current components for the SVE approximation is inherited from the un-approximated model. When modifications to the un-approximated model which removes this property are added, one would expect the new SVE approximation to lose the non-acceleration result.

Remark 4 Equation (39) implies advection of the 1-form $|a|^2d\rho$, which in turn implies advection of the Jacobian $J(|a|^2, \rho)$. Since the fluid flow is area preserving, $\text{div}\widehat{\boldsymbol{v}} = 0$, the following 2-form will also be advected,

$$(\partial_t + \widehat{\boldsymbol{v}}\cdot\nabla_r)\left(d|a|^2 \wedge d\rho\right) = 0\,. \qquad (40)$$

Thus, the divergence-free flow of $\widehat{\boldsymbol{v}}$ preserves the area element $d|a|^2 \wedge d\rho$. This means that if the gradients $\nabla|a|^2$ and $\nabla\rho$ are not aligned initially, then they will remain so. It also means that equilibrium solutions of (40) will be symplectic manifolds [14].

After deriving these equations, one may finally evaluate the constraint $D = 1$ imposed by the variation in pressure p to obtain further simplifications.

2.5 Thermal Potential Vorticity Dynamics with SVE on a Free Surface

The momentum map arising from the variations of the action in (32) is given by

$$\frac{1}{D}\frac{\delta \ell}{\delta \widehat{\boldsymbol{v}}} = \rho\Big(\widehat{\boldsymbol{v}} \cdot d\boldsymbol{r} + \mathcal{N}d\frac{d\theta}{dt}\Big) \quad \text{with} \quad \mathcal{N} := \frac{\sigma^2 N^2 |a|^2}{4} =: \Gamma |a|^2 \quad \text{and} \quad \frac{d\theta}{dt} = \sqrt{\frac{\rho_{ref}}{\rho}},$$

so $\frac{1}{D}\frac{\delta \ell}{\delta \widehat{\boldsymbol{v}}} = \rho\Big(\widehat{\boldsymbol{v}} \cdot d\boldsymbol{r} + \Gamma |a|^2 d(\sqrt{\rho\rho_{ref}}/\rho)\Big).$

$$(41)$$

According to the Euler-Poincaré equation (37), the dynamics of the fluid and wave components of the 1-form in (41) *separates* into the following equations,

$$(\partial_t + \mathcal{L}_{\widehat{\boldsymbol{v}}})\Big(\rho\big(\widehat{\boldsymbol{v}} \cdot d\boldsymbol{r}\big)\Big) = -dp + \frac{\rho}{2}d\big(|\widehat{\boldsymbol{v}}|^2\big),$$
$$(\partial_t + \mathcal{L}_{\widehat{\boldsymbol{v}}})\big(|a|^2 d\sqrt{\rho\rho_{ref}}\big) = 0.$$

$$(42)$$

This means that the mass-weighted thermal potential vorticity (TPV) dynamics also separates into the following fluid and wave components, $Q = Q_F + Q_W$, given by

$$Q\,d^2r := d\Big(\rho\Big(\widehat{\boldsymbol{v}} \cdot d\boldsymbol{r} + \Gamma |a|^2 d\sqrt{\frac{\rho_{ref}}{\rho}}\Big)\Big)$$
$$= \Big(\mathrm{div}(\rho\nabla\psi) - \Gamma J\big(|a|^2, \sqrt{\rho\rho_{ref}}\big)\Big)d^2r \quad \text{when} \quad \widehat{\boldsymbol{v}} = \nabla^\perp\psi \quad \text{for} \quad D = 1,$$
$$= Q_F\,d^2r + Q_W\,d^2r,$$

with $Q_F := \mathrm{div}(\rho\nabla\psi)$ and $Q_W := \Gamma J\big(\sqrt{\rho\rho_{ref}}, |a|^2\big).$

$$(43)$$

Then, again, the differentials of the separate equations in (42) yield the 'non-acceleration' result,

$$(\partial_t + \mathcal{L}_{\widehat{\boldsymbol{v}}})(Q_F\,d^2r) = \frac{1}{2}d\rho \wedge d|\widehat{\boldsymbol{v}}|^2 = \frac{1}{2}J\big(\rho, |\nabla\psi|^2\big)d^2r,$$
$$(\partial_t + \mathcal{L}_{\widehat{\boldsymbol{v}}})(Q_W\,d^2r) = 0.$$

$$(44)$$

Equivalently, in coordinates one has

$$\partial_t Q_F + \widehat{\boldsymbol{v}} \cdot \nabla Q_F = \frac{1}{2} J\big(\rho, |\nabla \psi|^2\big),$$

$$\partial_t Q_W + \widehat{\boldsymbol{v}} \cdot \nabla Q_W = 0,$$

with $\quad Q_F := \mathrm{div}(\rho \nabla \psi) \quad$ and $\quad Q_W := \Gamma J\big(\sqrt{\rho \rho_{ref}},\ |a|^2\big),$

$$\partial_t \rho + \widehat{\boldsymbol{v}} \cdot \nabla_r \rho = 0 \quad \text{and} \quad \Gamma = \frac{\sigma^2}{4Fr^2} = O(1), \tag{45}$$

$$\partial_t |a|^2 + \widehat{\boldsymbol{v}} \cdot \nabla_r |a|^2 = 0,$$

$$\partial_t \theta + \widehat{\boldsymbol{v}} \cdot \nabla_r \theta = \frac{\sqrt{\rho \rho_{ref}}}{\rho}.$$

The operator $(\mathrm{div}\rho\nabla)$ is invertible, so long as ρ is a differentiable positive function, which can be ensured by requiring that this condition holds initially, since ρ is advected. Consequently, the stream function ψ is related to the other fluid variables by

$$\psi := (\mathrm{div}\rho\nabla)^{-1} Q_F. \tag{46}$$

The dynamics of the equation set (45) explains why the various physical components of the flow coordinate their movements, as seen in satellite observations in Fig. 2. In particular, the motion of buoyancy ρ and squared wave amplitude $|a|^2$ are coordinated with each other through the advection of the momentum 1-form $|a|^2 d\rho$ and the area 2-form $d|a|^2 \wedge d\rho$. Likewise, the motion of the fluid potential vorticity Q_F and the mass density ρ are coordinated with each other through the mass-weighted definition of the stream function in (46). These considerations emphasise again the importance of horizontal buoyancy gradients in sea surface dynamics.

3 Numerical Implementation

Our implementation of the CoM equations (25) and the CoM equations in the SVE approximation (45) used the finite element method (FEM) for the spatial variables. The FEM algorithm we used is based on the algorithm formulated in [15] and is implemented using the Firedrake[3] software. In particular, for (25) we approximated the fluid potential vorticity Q_F, buoyancy ρ, wave elevation ζ and bouyancy weighted wave vertical velocity $\widetilde{w}$ using a first order discrete Galerkin finite element space. Similarly, for (45), we approximated Q_F, ρ, square of the wave amplitude $|a|^2$ and wave phase θ using a first order discrete Galerkin finite element space. The stream function ψ for both models was approximated by using

---

[3] https://firedrakeproject.org/index.html.

a first order continuous Galerkin finite element space. For the time integration, we used the third order strong stability preserving Runge Kutta method [12].

Figures 3 and 4 present snapshots of high resolution runs of the CoM equations and the CoM equations in the SVE approximation. These simulations were run with the following parameters. The domain is $[0, 1]^2$ at a resolution of 512^2. The boundary conditions are periodic in the x direction, and homogeneous Dirichlet for ψ in the y direction. To see the effects of the waves on the currents, the procedure was divided into two stages for both set of equations. The first stage was performed without wave activity for $T_{spin} = 100$ time units starting from the following initial conditions

$$Q_F(x, y, 0) = \sin(8\pi x)\sin(8\pi y) + 0.4\cos(6\pi x)\cos(6\pi y) + 0.3\cos(10\pi x)\cos(4\pi y) +$$

$$0.02\sin(2\pi y) + 0.02\sin(2\pi x),$$

$$\rho(x, y, 0) = 1 + 0.2\sin(2\pi x)\sin(2\pi y) \quad \text{and} \quad \rho_{ref} = 1.$$

$$(47)$$

The purpose of the first stage was to allow the system to spin up to a statistically steady state without any wave activity. The PV and buoyancy variables at the end of the initial spin-up period are denoted as $Q_{spin}(x, y) = Q_F(x, y, T_{spin})$ and $\rho_{spin}(x, y) = \rho(x, y, T_{spin})$. Figures of these variables are shown in Fig. 5. In the second stage, the full simulations including the wave variables were run with the initial conditions for the flow variables being the state achieved at the end of the first stage. To start the second stage for (25), wave variables were introduced with the following initial conditions

$$\zeta(x, y, 0) = \sin(8\pi x)\sin(8\pi y) + 0.4\cos(6\pi x)\cos(6\pi y) + 0.3\cos(10\pi x)\cos(4\pi y) +$$

$$0.02\sin(2\pi y) + 0.02\sin(2\pi x),$$

$$\tilde{w}(x, y, 0) = 0, \quad Q_F(x, y, 0) = Q_{spin}(x, y), \quad \rho(x, y, 0) = \rho_{spin}(x, y),$$

$$\sigma^2 Fr^2 = 10^{-2}.$$

$$(48)$$

To start the second stage for (45), wave variables were introduced with the following initial conditions

$$|a|^2(x, y, 0) = \big(\sin(8\pi x)\sin(8\pi y) + 0.4\cos(6\pi x)\cos(6\pi y) + 0.3\cos(10\pi x)\cos(4\pi y) +$$

$$0.02\sin(2\pi y) + 0.02\sin(2\pi x)\big)^2,$$

$$\theta(x, y, 0) = 0, \quad Q_F(x, y, 0) = Q_{spin}(x, y), \quad \rho(x, y, 0) = \rho_{spin}(x, y).$$

$$(49)$$

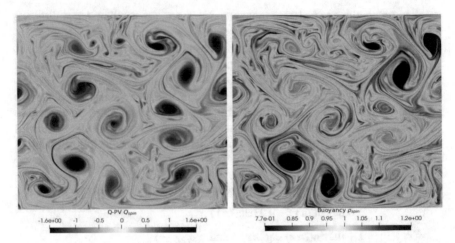

Fig. 5 These figures show the results of the first stage of the simulation in which only fluid motion is present and the wave degrees of freedom are absent. The panels show fluid potential vorticity Q_F (left) and buoyancy ρ (right). The fluid state obtained from the first stage was used as the initial condition for the second stage simulations in which wave variables were included. These distributions of fluid properties show strong spatial coherence. The coordination of wave and fluid properties that emerges in the second stage of the simulations shown in Figs. 3 and 4 arises from the interaction between the wave and current components of the flow which is mediated by the buoyancy gradient

Remark 5 Importantly, the wave phase θ in the second stage was set initially to zero. Thereafter, the wave phase θ increased linearly in time in proportion to the advected quantity $\sqrt{\rho \rho_{ref}}/\rho$ following each flow line, as implied by the last equation in (45).

4   Conclusion and Outlook

This paper models the effects of thermal fronts on the dynamics of the ocean's waves and currents. It introduces and simulates two models of thermal wave-current dynamics on a free surface. The original CoM model is derived from Hamilton's principle via the composition of two maps which represent the horizontal and vertical motion respectively. The second, a slowly varying envelope (SVE) model, is introduced via the standard WKB approximation which takes advantage of large separation of the space-time scales between the slow horizontal currents and fast vertical oscillations. In particular, the second model introduces the WKB solution Ansatz into Hamilton's principle, whereupon the time integral averages over the phases of the rapid oscillations that are out of resonance with the slowly varying envelope. Model runs of both models are presented in which the buoyancy mediates the dynamics of the currents and waves, as seen in Figs. 3 and 4. These simulations also validate the use of the WKB approximation for two reasons. First, the resolved small scale wave features of the original CoM model lie primarily within the envelope defined by the SVE approximate model. This means that the dynamics of the spatial features of the SVE approximate model are consistent with those of the original CoM model, although the resolved space and time scales differ. Secondly, requiring that $\epsilon^2/(Fr^2\sigma^2) = O(1)$ ensures that the time scale for the wave envelope dynamics matches that of the fluid motion.

Nonetheless, the two models introduced here merit further study in several directions. For example, it remains to: (1) quantify the correlations observed visually; (2) determine their rate of formation; and (3) parameterise the model for comparison and analysis of the satellite data on which their derivations were based. Furthermore, the models discussed here involve only variables that are evaluated on the free surface and therefore they neglect bathymetry. A scientific challenge persists in understanding regions of the ocean where bathymetry has profound effects on the observable surface dynamics, such as in the Lofoten vortex [21]. This is a multiscale issue that might be addressed by including mesoscale modulations of the sub-mesoscale models derived here. One candidate for providing the mesoscale modulations would the thermal quasi-geostrophic (TQG) model in which bathymetry has recently been included [15].

The currents are modelled here by the two dimensional incompressible Euler equations, as seen in Eqs. (2) and (3). Incompressibility is a reasonable assumption in some regions of the ocean, for example when the quasigeostrophic approximation is valid. There are regions in the upper ocean where other equations are more suitable for modelling currents, and the development and investigation of such two dimensional models is an open problem which warrants further consideration.

As mentioned in Remark 1, the wave component of the model presented here does not create circulation in the currents. The instabilities present in satellite simulations indicate that additional modelling is needed to fully capture this effect. Future work will investigate approaches for modelling these instabilities.

Many other questions remain about wave-current interaction. The full extent of submesoscale ocean dynamics is by no means adequately described by existing models. For example, we have little understanding of the formation and dynamics of various sea-surface phenomena, including the so-called 'spirals on the sea' [18]. Other questions are emerging because the ocean has absorbed in excess of 90% of the heat present in the earth system as a result of human activity during the post-industrial era [19]. The absorption of heat from the warming atmosphere is ongoing and it is forecast to become more dramatic. This absorption has resulted in 'marine heat waves', which are predicted to increase in frequency and severity. These changes to the upper ocean, where most of this heat is stored, could have a profound effect on the dynamical landscape of our oceans. These effects may, in turn, influence our weather and climate systems. Over the millennia, the ocean has approached statistical equilibrium under its current forcing conditions. Using modelling terminology, one says the ocean is well 'spun-up'. However, the continued warming of the ocean is likely to influence the number and intensity of thermal fronts. One hopes that mathematical models will provide a useful framework for estimating some of the potential impacts of these thermal fronts on atmospheric effects, as well.

Acknowledgments We are grateful to our friends and colleagues who have generously offered their time, thoughts, and encouragement in the course of this work during the time of COVID-19. Thanks to A. Arnold, B. Chapron, D. Crisan, E. Luesink, A. Mashayek, and J. C. McWilliams for their thoughtful comments and discussions. Particular thanks to B. Chapron, for extensive discussions of satellite oceanography and for providing the satellite data in Figs. 1 and 2. We also thank B. Fox-Kemper for constructive discussions of modelling approaches in physical oceanography. These discussions helped us clarify the distinction between the present CoM modelling approach and the classical balance equation approach. The authors are grateful for partial support, as follows. DH for European Research Council (ERC) Synergy grant STUOD - DLV-856408; RH for the EPSRC scholarship (Grant No. EP/R513052/1); and OS for the EPSRC Centre for Doctoral Training in the Mathematics of Planet Earth (Grant No. EP/L016613/1).

References

1. Aulicino, G., Cotroneo, Y., Ruggiero, P.D., Buono, A., Corcione, V., Nunziata, F. and Fusco, G., 2022. Remote Sensing Applications in Satellite Oceanography. In Measurement for the Sea (pp. 181–209). Springer, Cham.
2. Bretherton, F. and Garret, C., 1968. Wave trains in inhomogeneous moving media. Proc. R. Soc. Lond. A 302, 529–554.
3. Bühler, O., 2014. *Waves and mean flows*. Cambridge University Press.
4. Burby, J.W. and Ruiz, D.E., 2020. Variational nonlinear WKB in the Eulerian frame. Journal of Mathematical Physics, 61(5), p. 053101.
5. Chapron, B., Kudryavtsev, V.N., Collard, F., Rascle, N., Kubryakov, A.A. and Stanichny, S.V., 2020. Studies of Sub-Mesoscale Variability of the Ocean Upper Layer Based on Satellite Observations Data. Physical Oceanography, 27(6), pp. 619–630. https://archimer.ifremer.fr/doc/00682/79420/82002.pdf
6. Charney, J. G. and P. G. Drazin, 1961: Propagation of planetary-scale disturbances from the lower into the upper atmosphere. J. Geophys. Res., 66, 83–110.
7. Craig, W., 2016. On the Hamiltonian for water waves. arXiv preprint arXiv:1612.08971.
8. Crisan, D., Holm, D.D. and Street, O.D., 2021. Wave-current interaction on a free surface. Stud Appl Math. 147:1277–1338. https://doi.org/10.1111/sapm.12425
9. Dong, J., Fox-Kemper, B., Zhang, H. and Dong, C., 2020. The scale of submesoscale baroclinic instability globally. Journal of Physical Oceanography, 50(9), pp. 2649–2667. https://doi.org/10.1175/JPO-D-20-0043.1
10. Fox-Kemper, B., Johnson, L. and Qiao, F., 2022. Ocean near-surface layers. In Ocean Mixing (pp. 65–94). Elsevier.
11. Gjaja, I. and Holm, D.D., 1996. Self-consistent wave-mean flow interaction dynamics and its Hamiltonian formulation for a rotating stratified incompressible fluid, *Physica D*, **98** 343–378. https://doi.org/10.1016/0167-2789(96)00104-2
12. Gottlieb, S. On High Order Strong Stability Preserving Runge–Kutta and Multi Step Time Discretizations. J Sci Comput 25, 105–128 (2005). https://doi.org/10.1007/s10915-004-4635-5
13. Gula, J., Taylor, J., Shcherbina, A. and Mahadevan, A., 2022. Submesoscale processes and mixing. In *Ocean Mixing* (pp. 181–214). Elsevier.
14. Holm, D.D., 2011. *Geometric mechanics-Part I: Dynamics and symmetry*. World Scientific Publishing Company.

15. Holm, D.D., Luesink, E and Pan, W, 2021. Stochastic mesoscale circulation dynamics in the thermal ocean. Physics of Fluids, 33, 046603. https://doi.org/10.1063/5.0040026
16. Holm, D.D., Marsden, J.E. and Ratiu, T.S., 1998. The Euler–Poincar´ equations and semidirect products with applications to continuum theories. Advances in Mathematics, 137(1), pp.1–81. https://doi.org/10.1006/aima.1998.1721
17. Morrow, R., Fu, L.-L., Ardhuin, F., Benkiran, M., Chapron, B., Cosme, E., d'Ovidio, F., Farrar, J.T., Gille, S.T. [et al.], 2019. Global Observations of Fine-Scale Ocean Surface Topography with the Surface Water and Ocean Topography (SWOT) Mission. Frontiers in Marine Science, 6, 232. https://doi:10.3389/fmars.2019.00232
18. Munk, W., Armi, L., Fischer, K. and Zachariasen, F., 2000. Spirals on the sea. Proceedings of the Royal Society of London. Series A: Mathematical, Physical and Engineering Sciences, 456(1997), pp. 1217–1280.
19. H.-O. Pörtner, D.C. Roberts, V. Masson-Delmotte, P. Zhai, M. Tignor, E. Poloczanska, K. Mintenbeck, A. Alegría, M. Nicolai, A. Okem, J. Petzold, B. Rama, N.M. Weyer (eds.), 2019. IPCC Special Report on the Ocean and Cryosphere in a Changing Climate. In press.
20. Rascle, N., Molemaker, J., Marié, L., Nouguier, F., Chapron, B., Lund, B. and Mouche, A., 2017. Intense deformation field at oceanic front inferred from directional sea surface roughness observations. Geophysical Research Letters, 44(11), pp. 5599–5608. https://doi.org/10.1002/2017GL073473
21. Volkov, D. L., Kubryakov, A. A., and Lumpkin, R., Formation and variability of the Lofoten basin vortex in a high-resolution ocean model, Deep Sea Res., Part I 105, 142–157 (2015). https://doi.org/10.1016/j.dsr.2015.09.001, Google Scholar
22. Voronovich, A. G. (1976), Propagation of internal and surface gravity waves in the approximation of geometrical optics, Izv. Atmos. Ocean. Phys., 12 p. 850–857.
23. White, A.A., 1986. Finite amplitude, steady Rossby waves and mean flows: Analytical illustrations of the Charney-Drazin non-acceleration theorem. Quarterly Journal of the Royal Meteorological Society, 112(473), pp.749–773. https://doi.org/10.1002/qj.49711247311
24. Whitham, G.B., 1967. Variational Methods and Applications to Water Waves, Proc. Roy. Soc. London. Series A, Mathematical and Physical Sciences, Vol. 299, No. 1456, A Discussion on Nonlinear Theory of Wave Propagation in Dispersive Systems (Jun. 13, 1967), pp. 6–25 https://www.jstor.org/stable/2415780
25. Whitham, G.B., 2011. Linear and nonlinear waves (Vol. 42). John Wiley & Sons.
26. Yurovskaya, M., Rascle, N., Kudryavtsev, V., Chapron, B., Marié, L. and Molemaker, J., 2018. Wave spectrum retrieval from airborne sunglitter images. Remote sensing of Environment, 217, pp.61–71. https://doi.org/10.1016/j.rse.2018.07.026

Variational Stochastic Parameterisations and Their Applications to Primitive Equation Models

Ruiao Hu and Stuart Patching

Abstract We present a numerical investigation into the stochastic parameterisations of the Primitive Equations (PE) using the Stochastic Advection by Lie Transport (SALT) and Stochastic Forcing by Lie Transport (SFLT) frameworks. These frameworks were chosen due to their structure-preserving introduction of stochasticity, which decomposes the transport velocity and fluid momentum into their drift and stochastic parts, respectively. In this paper, we develop a new calibration methodology to implement the momentum decomposition of SFLT and compare with the Lagrangian path methodology implemented for SALT. The resulting stochastic Primitive Equations are then integrated numerically using a modification of the FESOM2 code. For certain choices of the stochastic parameters, we show that SALT causes an increase in the eddy kinetic energy field and an improvement in the spatial spectrum. SFLT also shows improvements in these areas, though to a lesser extent. SALT does, however, have the drawback of an excessive downwards diffusion of temperature.

Keywords Primitive equations · Geometric mechanics · FESOM2 · Stochastic parameterisation

1 Introduction

Uncertainty can be present in ocean models due to a number of factors including, but not limited to: small-scale processes not resolved by the grid; observation error; model error; numerical error and unrealistic viscosities imposed to ensure numerical stability. Several stochastic parameterisation techniques [PZ14, Ber05, Mem14, Hol15, HH21] have been proposed recently as ways of representing uncertainty in ocean models. Because these parameterisations are probabilistic, it is possible to

R. Hu · S. Patching (✉)
Department of Mathematics, Imperial College London, London, UK
e-mail: ruiao.hu15@imperial.ac.uk; s.patching17@imperial.ac.uk

© The Author(s) 2023
B. Chapron et al. (eds.), *Stochastic Transport in Upper Ocean Dynamics*,
Mathematics of Planet Earth 10, https://doi.org/10.1007/978-3-031-18988-3_9

generate ensemble forecasts [CCH+19, CCH+20, Cot20, UJPD21] with associated means and variances, which can then be applied to data assimilation. This work will focus on two frameworks which introduce stochasticity in a way that preserves certain fundamental and desirable properties of fluid flows. These frameworks are: Stochastic Advection by Lie Transport (SALT) [Hol15] and Stochastic Forcing by Lie Transport (SFLT) [HH21]. Both SALT and SFLT are derived from variational principles, from which we may observe the geometric structure of the fluid equations and the conservation laws which are inherited.

The key assumption of SALT is the decomposition of transport velocity into a slow mean part and a fast, rapidly fluctuating part around the mean. In the limit of high fluctuation frequency, one can use homogenisation theory to transform the rapidly-fluctuating component to a sum of stochastic vector fields [CGH17]. Thus, the modification from the deterministic flow is the addition of stochastic vector fields to the transport velocity. This stochastic modification has been shown [Hol15] to preserve the Kelvin circulation theorem and the advection equation for potential vorticity. In the case where buoyancy obeys an advection relation, the potential vorticity is conserved along particle paths. However, SALT violates energy conservation since stochastic Hamiltonians are introduced into the variational principle. The application of the SALT in quasi-geostrophic (QG) models and the 2D Euler equations has been investigated before in [CCH+20, CCH+19, Cot20]. However, these models are too simplistic to be used in operational ocean simulations, and the majority of ocean codes (e.g. MOM5 [GBB+00], ICON [Kor17], MITgcm [MAH+97], FESOM2 [DSWJ16]) solve the Primitive Equations (PE). For this reason, if SALT is to be employed for use in practical applications, it must be adapted for use in PE. This introduces additional features to the model as compared the QG or 2D Euler: in PE there are advected quantities such as temperature and salinity, which in the SALT framework are advected by the stochastic velocity. There is, moreover, a subtlety in the pressure arising from the imposition of a semi-martingale Lagrange multiplier in the incompressibility condition of the variational principle [SC21].

An alternative stochasatic parameterisation is the more recent SFLT framework [HH21]. Derived via a Lagrange-d'Alembert principle, SFLT allows the addition of arbitrary stochastic forcings to the evolution equations of the momentum and of the advected quantities. This modification differs from SALT, as stochasticity is added in the variational principle *after* taking variations of the Hamiltonian for the deterministic system . By considering the Lie-Poisson bracket of the system, we choose the forcing to be of a particular form that preserves, on every realisation of the noise, the original (deterministic) Hamiltonian. For PE, the Hamiltonian is given in Eq. (2). However, the addition of energy preserving forces will modify the Kelvin circulation theorem. In the current work, we will consider the case where the stochastic forcing is in the energy conserving form and applied to the momentum equation. As in the SALT case, stochastic pressure terms will appear in the momentum equation due to the imposition of semi-martingale Lagrange multiplier in the incompressibility constraint. Prior to the present work, SFLT has not been implemented into numerical models.

The rest of the paper is structured as follows. In Sect. 2, we derive PE with both SALT and SFLT from a variational principle and we show the conservation properties from the resulting equations. In Sect. 3, we consider calibration procedures to calculate the stochastic parameters of SALT and SFLT. In particular, we use the Lagrangian paths method of [CCH+20] but also consider a simpler technique, that of Eulerian differences, which we propose is more appropriate for use in SFLT. In Sect. 4, we present numerical results of applying SALT and SFLT to FESOM2 [DSWJ16] (see Sect. 5), demonstrating the different effects of these stochastic frameworks and the sensitivity to the choice of parameters.

2 Stochastic Primitive Equations

2.1 Variational Principles for Stochastic Primitive Equations

Variational principles may be used to derive systems of fluid equations [HMR98, HSS09] which obey conservation laws such as the Kelvin circulation theorem. To derive the Primitive Equations from a variational principle, the appropriate Lagrangian is [HSS09]:

$$l(\mathbf{u}, D, T, S) = \int \left(\frac{1}{2} |\mathbf{u}|^2 + \mathbf{u} \cdot \mathbf{R} - V(T, S, z) \right) D d^3 x, \tag{1}$$

where $\mathbf{u} = (u, v)$ is the horizontal velocity vector field, $\mathbf{R}$ is the Coriolis potential, which satisfies curl $\mathbf{R} = f(y)\hat{\mathbf{z}}$ with $f(y) = 2\Omega \cos y$ and $\Omega = 2\pi/\text{day}$ is the rotational frequency of the earth. T and S are the temperature and salinity respectively; these are tracers advected by the fluid. D is the Jacobian of the flow map g_t that maps a fluid particle at initial position $\mathbf{x}_0$ to its position $\mathbf{x}_t = g_t \mathbf{x}_0$ at time t. V is the potential energy, which has explicit dependence on T and S, as well as on the vertical coordinate z. The three-dimensional velocity shall be denoted $\mathbf{v} = (\mathbf{u}, w)$.

In order to obtain the correct hydrostatic balance condition the potential energy should obey $\frac{\partial V}{\partial z}(T, S, z) = g(1+b)$ where the partial derivative is taken with respect to z at constant T, S. b is the buoyancy, given by the equation of state $b = b(T, S, z)$.

It is convenient here to use the Clebsch version of the variational principle [CH09] in Hamiltonian form. The Hamiltonian is given by Legendre transformation as $h(\mathbf{m}^h, D, T, S) := \langle \mathbf{u}, \frac{\delta l}{\delta \mathbf{u}} \rangle - l(\mathbf{u}, D, T, S)$ where $\mathbf{m}^h := \frac{\delta l}{\delta \mathbf{u}} = D(\mathbf{u} + \mathbf{R})$ is the horizontal momentum. We have also defined the inner product $\langle p, q \rangle = \int p \cdot q d^3 x$. We shall use the same angle-bracket notation for all such pairings, when p and q are dual variables, e.g. vector field and 1-form density; or a scalar and a density. The Hamiltonian can be written explicitly as:

$$h(\mathbf{m}^h, D, T, S) = \int \left(\frac{1}{2} \left| \frac{\mathbf{m}^h}{D} - \mathbf{R} \right|^2 + V(T, S, z) \right) D d^3 x . \qquad (2)$$

In the Clebsch variational principle when SALT or SFLT are present, the (3-dimensional) transport velocity $d\boldsymbol{\chi}$ is defined to be a stochastic process. The form of $d\boldsymbol{\chi}$ is defined using Lagrange-multiplier constraints to impose the transport equations $(d + \mathcal{L}_{d\boldsymbol{\chi}}) a = 0$, where $a \in \{D, T, S\}$ [see SW68]. Here we remark that for clarity, we denote by an italic d the spatial differential and a straight red d for the stochastic time-increment. $\mathcal{L}_{d\boldsymbol{\chi}}$ denotes the Lie derivative, which is a differential operator with a form that depends on the object on which it acts. We remark here that there is a slight abuse of notation and we shall write D as a shorthand for $D d^3 x$ so that this is a density 3-form and the Lie derivative is given by $\mathcal{L}_{d\boldsymbol{\chi}} D = \nabla \cdot (d\boldsymbol{\chi} D)$. T and S are scalars, so we have $\mathcal{L}_{d\boldsymbol{\chi}} T := d\boldsymbol{\chi} \cdot \nabla T$ and similarly for S. In order to obtain the incompressibility of the transport velocity $d\boldsymbol{\chi}$, we include an additional constraint to set $D = 1$ where the Lagrange multiplier will be interpreted as the pressure. Since the Hamiltonian h only depends on the horizontal momentum $\mathbf{m}^h$, we need to include an extra constraint so that the vertical component of the momentum is set to zero; this will give us hydrostatic balance.

The defining feature of SALT is that the transport velocity is the sum of the drift velocity and a number of stochastic corrections to the drift:

$$d\boldsymbol{\chi}(\mathbf{x}, t) := \mathbf{v}(\mathbf{x}, t)dt + \sum_i \boldsymbol{\xi}_i(\mathbf{x}, t) \circ dW_t^i , \qquad (3)$$

where $\boldsymbol{\xi}_i(\mathbf{x}, t)$ are arbitrary vector fields. We remark here that Eq. (3) is a stochastic process at fixed Eulerian points $\mathbf{x}$ and we do not solve for this process explicitly. $d\boldsymbol{\chi}$ is distinct from the particle trajectories $\mathbf{x}_t$, which evolve in time according to $d\mathbf{x}_t = \mathbf{v}(\mathbf{x}_t, t)dt + \sum_i \boldsymbol{\xi}_i(\mathbf{x}_t, t) \circ dW_t^i$ and will be used during calibration procedures in Sect. 3. We can impose the form of the transport velocity specified in Eq. (3) by including in the action some additional stochastic Hamiltonians $\sum_i h_i(\mathbf{m}^h) \circ dW_t^i$ where the horizontal component of the parameters is given by $\boldsymbol{\xi}_i^h(\mathbf{x}, t) = \frac{\delta h_i}{\delta \mathbf{m}^h}$. The three-dimensional momentum is denoted $\mathbf{m} = (\mathbf{m}^h, m_3)$. We note that in principle $\boldsymbol{\xi}_i$ may depend on time; however, we shall henceforth assume for simplicity that $\boldsymbol{\xi}_i = \boldsymbol{\xi}_i(\mathbf{x})$ is a function of space only. When h_i are independent of $\mathbf{m}^h$, we have the relation $d\boldsymbol{\chi}(\mathbf{x}, t) := \mathbf{v}(\mathbf{x}, t)dt$, so that $d\boldsymbol{\chi}$ reduces to the original deterministic transport.

SFLT is included [HH21] via a Lagrange-d'Alembert term $\langle \delta d\boldsymbol{\chi}, \mathbf{F} \rangle$ added to the variation of the action δS. Since this is added after variations of the action are taken, the forcing $\mathbf{F}$ can in principle be arbitrary. Overall, the variational principle takes the following form:

$$0 = \delta S = \delta \int \langle d\boldsymbol{\chi}, \mathbf{m} \rangle - h(\mathbf{m}^{(h)}, D, T, S)dt - \langle d\zeta, m_3 \rangle - \langle dP, D - 1 \rangle$$

$$+ \langle \alpha, (d + \mathcal{L}_{d\boldsymbol{\chi}}) D \rangle + \langle \beta, (d + \mathcal{L}_{d\boldsymbol{\chi}}) T \rangle + \langle \gamma, (d + \mathcal{L}_{d\boldsymbol{\chi}}) S \rangle$$

$$- \underbrace{\sum_{i=1}^{N_\xi} h_i \left(\mathbf{m}^{(h)}, \boldsymbol{\xi}_i^{(h)} \right) \circ dW_t^i}_{\text{SALT}} - \underbrace{\int \langle \delta d\boldsymbol{\chi}, \mathbf{F} \rangle}_{\text{SFLT}} . \tag{4}$$

The first two lines of Eq. (4) are what would be included in the unmodified variational principle. $d\zeta$ is a Lagrange multiplier, enforcing $m_3 = 0$ and after taking variations can be interpreted as the vertical component of the stochastic transport velocity. Indeed, we may expand $d\zeta = w dt + \sum_i \xi_i^{(z)} \circ dW_t^i$; note that here $d\zeta$ is varied and so the third component of $\boldsymbol{\xi}_i$ is treated as a variable in the action, whereas the horizontal components are treated as fixed parameters. The final term on the top line enforces incompressibility, and the Lagrange multiplier dP must be stochastic since a semi-martingale Lagrange multiplier is required to enforce a condition on the semi-martingale D [see SC21]. On the second line the quantities α, β, γ are Lagrange multipliers enforcing the fact that D, T, S are advected quantities. The final line contains the modifications required to include SALT or SFLT; we shall not in practice use both SALT and SFLT together, but for compactness of the presentation we include them together here. The first modification, giving SALT, consists of a sum of N_ξ Hamiltonians multiplied by Stratonovich noise. The second, additional term is a Lagrange-d'Alembert term which introduces a shift $\mathbf{F}$ in the momentum. We remark that by including further Lagrange-d'Alembert terms such as $\langle \delta\alpha, dF_D \rangle$ or $\langle \delta\beta, dF_T \rangle$ etc. we may add arbitrary forcings to the right-hand side of the equations for the advected tracers. However, we do not consider this here.

The equations resulting from the variational principle $\delta S = 0$ are:

$$\delta \mathbf{m}^h: \qquad d\boldsymbol{\chi}^{(h)} = \frac{\delta h}{\delta \mathbf{m}^h} dt + \sum_i \frac{\delta h_i}{\delta \mathbf{m}^h} \circ dW_t^i ; \tag{5a}$$

$$\delta m_3: \qquad d\chi^{(z)} = d\zeta ; \tag{5b}$$

$$\delta d\boldsymbol{\chi}: \qquad \mathbf{m} = \alpha \diamond D + \beta \diamond T + \gamma \diamond S + \mathbf{F} ; \tag{5c}$$

$$\delta\alpha: \qquad (d + \mathcal{L}_{d\boldsymbol{\chi}}) D = 0 ; \tag{5d}$$

$$\delta\beta: \qquad (d + \mathcal{L}_{d\boldsymbol{\chi}}) T = 0 ; \tag{5e}$$

$$\delta\gamma: \qquad (d + \mathcal{L}_{d\boldsymbol{\chi}}) S = 0 ; \tag{5f}$$

$$\delta D: \qquad (d + \mathcal{L}_{d\boldsymbol{\chi}}) \alpha = -\left(dP + \frac{\delta h}{\delta D} dt \right) ; \tag{5g}$$

$$\delta T: \qquad (d + \mathcal{L}_{d\boldsymbol{\chi}}) \beta = -\frac{\delta h}{\delta T} dt ; \tag{5h}$$

$$\delta S: \qquad \left(\mathrm{d} + \mathcal{L}_{\mathrm{d}\boldsymbol{\chi}}\right)\gamma = -\frac{\delta h}{\delta S}\mathrm{d}t\,; \tag{5i}$$

$$\delta \mathrm{d}P: \qquad D = 1\,; \tag{5j}$$

$$\delta \mathrm{d}\zeta: \qquad m_3 = 0\,; \tag{5k}$$

The diamond in Eq. (5c) is a binary operator acting on two variables that are dual with respect to the inner product $\langle\cdot,\cdot\rangle$ (e.g. or scalar and density) and giving a 1-form density. Explicitly, for two dual variables p, q and an arbitrary vector field X : the diamond is defined by the relation $\langle p \diamond q, X\rangle = -\langle p, \mathcal{L}_X q\rangle$. We can compute these explicitly as follows:

$$\frac{\delta h}{\delta D} \diamond D = D\nabla\frac{\delta h}{\delta D}\,, \qquad \frac{\delta h}{\delta T} \diamond T = -\frac{\delta h}{\delta T}\nabla T\,, \qquad \frac{\delta h}{\delta S} \diamond S = -\frac{\delta h}{\delta S}\nabla S\,. \tag{6}$$

We note that the form of $\mathrm{d}\boldsymbol{\chi}$ as given in Eq. (3) is not an input to the variational principle, but a consequence of it. Indeed, we obtain Eq. (3) by defining $\mathbf{v} := \left(\frac{\delta h}{\delta \mathbf{m}^h}, w\right)$ and $\boldsymbol{\xi}_i := \left(\frac{\delta h}{\delta \mathbf{m}^h}, \xi_i^{(z)}\right)$. The horizontal velocity is therefore $\mathbf{u} = \frac{\delta h}{\delta \mathbf{m}^h} = \frac{\mathbf{m}^h}{D} - \mathbf{R}$. The fact that $D = 1$, combined with Eq. (5d) gives the incompressibility condition $\nabla \cdot \mathrm{d}\boldsymbol{\chi} = \nabla^{(h)} \cdot \mathrm{d}\boldsymbol{\chi}^{(h)} + \frac{\partial}{\partial z}\mathrm{d}\zeta = 0$. By Doob-Meyer decomposition [Doo53, Mey62, Mey63], we can split the incompressibility condition into its drift part and stochastic oscillations. Thus we are able to compute $w, \xi_i^{(z)}$ in terms of $\mathbf{u}$ and $\boldsymbol{\xi}^{(h)}$ respectively:

$$\nabla^{(h)} \cdot \mathbf{u} + \frac{\partial w}{\partial z} = 0\,, \qquad \nabla^{(h)} \cdot \boldsymbol{\xi}_i^{(h)} + \frac{\partial \xi_i^{(z)}}{\partial z} = 0\,. \tag{7}$$

Boundary conditions at $z = 0$ allow us to integrate Eq. (7) in the vertical direction. To obtain the momentum equation we apply $\left(\mathrm{d} + \mathcal{L}_{\mathrm{d}\boldsymbol{\chi}}\right)$ to both sides of Eq. (5c) and use the fact that the Lie derivative obeys a Leibniz rule with respect to the diamond operator. After some re-arranging, we obtain:

$$\left(\mathrm{d} + \mathcal{L}_{\mathrm{d}\boldsymbol{\chi}}\right)\left(\frac{\mathbf{m}^h - \mathbf{F}}{D} \cdot \mathrm{d}\mathbf{x}\right) = -\mathrm{d}\left(\left(\frac{\delta h}{\delta D} - V\right)\mathrm{d}t + \mathrm{d}P\right) + \frac{\partial V}{\partial z}\mathrm{d}z\mathrm{d}t\,. \tag{8}$$

We shall show in Sect. 2.2 that the SFLT terms will conserve energy if we require that the momentum shift $\mathbf{F}$ takes a particular form, which is that it satisfies $\left(\mathrm{d} + \mathcal{L}_{\mathrm{d}\boldsymbol{\chi}}\right)\mathbf{F} = \mathcal{L}_{\mathbf{v}}\mathrm{d}\Phi$, for some stochastic process $\mathrm{d}\Phi$. In this work, we shall assume further that $\mathrm{d}\Phi$ has the form $\mathrm{d}\Phi = \sum_I \boldsymbol{\phi}_I \circ \mathrm{d}B_t^I$ for some spatially dependent parameters $\boldsymbol{\phi}_I$ and with B_t^I being a set of independent Brownian motions. Because the momentum $\mathbf{m} = (\mathbf{m}^h, 0)$ has only horizontal components, we shall assume that $\boldsymbol{\phi}_I$ also have only horizontal components. Moreover, we can expand the pressure in terms of its drift component and Brownian increments: $\mathrm{d}P =$

$pdt + \sum_i p_i \circ dW_t^i + \sum_I p_I \circ dB_t^I$. Thus, writing $\mathbf{m} = \mathbf{u} + \mathbf{R}$ and expanding $d\chi$ in terms of $\mathbf{v}$ and $\boldsymbol{\xi}_i$, we find that Eq. (8) becomes:

$$
d\mathbf{u} + \left[\nabla \cdot (\mathbf{v}\mathbf{u}) + f\hat{\mathbf{z}} \times \mathbf{v} + \nabla p + g(1 + b)\hat{\mathbf{z}} \right] dt
$$

$$
+ \sum_i \left[\nabla \cdot (\boldsymbol{\xi}_i \mathbf{u}) + f\hat{\mathbf{z}} \times \boldsymbol{\xi}_i + \nabla \boldsymbol{\xi}_i \cdot \mathbf{u} + \nabla (p_i + \boldsymbol{\xi}_i \cdot \mathbf{R}) \right] \circ dW_t^i \tag{9}
$$

$$
- \sum_I \left[\nabla \cdot (\mathbf{v}\boldsymbol{\phi}_I) - \nabla \boldsymbol{\phi}_I \cdot \mathbf{v} - \nabla (p_I - \mathbf{v} \cdot \boldsymbol{\phi}_I) \right] \circ dB_t^I = 0.
$$

The first line of Eq. (9) contains the terms of the deterministic momentum equation, the second line contains the SALT terms and the final line contains the SFLT contributions. Equation (9) is a three-dimensional equation, but the third component is the (diagnostic) hydrostatic balance condition rather than a prognostic evolution equation for w. In the cases of SALT and SFLT hydrostatic balance includes additional constraints on the stochastic parts of the pressure dP:

$$
\frac{\partial p}{\partial z} = -g(1 + b), \qquad \frac{\partial p_i'}{\partial z} = -\frac{\partial \boldsymbol{\xi}_i}{\partial z} \cdot \mathbf{u}, \qquad \frac{\partial p_I'}{\partial z} = -\frac{\partial \boldsymbol{\phi}_I}{\partial z} \cdot \mathbf{v}, \tag{10}
$$

where we have the definitions of the shifted stochastic pressure terms $p_i' := p_i + \boldsymbol{\xi}_i \cdot \mathbf{R}$ and $p_I' := p_I - \mathbf{v} \cdot \boldsymbol{\phi}_I$. We solve Eq. (10) by imposing the following surface pressure boundary conditions:

$$
p|_{z=0} = g\eta, \qquad p_i'|_{z=0} = \psi_i, \qquad p_I'|_{z=0} = \psi_I, \tag{11}
$$

where η is the free surface height. The boundary condition on p is that used in the linear free surface approximation, which is employed in FESOM2 [DSWJ16]. ψ_i and ψ_I are functions only of the horizontal direction and are arbitrary. They may be used to introduce some stochastic atmospheric forcing at the ocean surface, but we do not consider this in the present work. For simplicity we shall set $\psi_i = \psi_I = 0$ for all i, I. Solving Eq. (10) with the boundary conditions in Eq. (11) gives us the following:

$$
p = g(\eta - z) + g \int_z^0 b\, dz', \tag{12a}
$$

$$
p_i' = \psi_i + \int_z^0 \frac{\partial \boldsymbol{\xi}_i}{\partial z} \cdot \mathbf{u}\, dz', \tag{12b}
$$

$$
p_I' = \psi_I + \int_z^0 \frac{\partial \boldsymbol{\phi}_I}{\partial z} \cdot \mathbf{v}\, dz'. \tag{12c}
$$

A more exact condition on the deterministic pressure would be $p|_{z=\eta} = 0$. Using this gives almost the same result for p except that the upper limit of the integral will instead be η.

The equation for the evolution of the free surface height η is obtained by integrating the incompressibility condition and using appropriate surface boundary conditions. For the linear free surface approximation we take $w|_{z=0}dt = d\eta$; at the bottom boundary $z = -H(x, y)$ we have $d\mathbf{\chi}|_{z=-H} \cdot \nabla(z + H) = 0$. Thus, integrating the incompressibility condition in the vertical direction from $z = -H$ to $z = 0$ we find, in the linear free surface case:

$$d\eta + \nabla \cdot \int_{-H}^{0} \mathbf{u}dt \, dz = 0. \tag{13}$$

Again, the more exact boundary condition would be $d\mathbf{\chi}|_{z=\eta} \cdot \nabla(z - \eta) = d\eta$ ad in this case Eq. (13) is modified by $\mathbf{u}dt \to \mathbf{u}dt + \sum_i \mathbf{\xi}_i \circ dW_t^i$ and the upper limit of the integral will be η rather than 0. However, for our numerical simulations we use the linear free surface.

From Eqs. (5e) and (5f) we have the advection equations:

$$dT + \mathbf{v} \cdot \nabla T dt + \sum_i \mathbf{\xi}_i \cdot \nabla T \circ dW_t^i = 0, \tag{14}$$

$$dS + \mathbf{v} \cdot \nabla S dt + \sum_i \mathbf{\xi}_i \cdot \nabla S \circ dW_t^i = 0, \tag{15}$$

for temperature and salinity respectively. The horizontal component of the momentum equation (9), along with the solutions Eq. (10) for pressure (with the equation of state $b = b(T, S, z)$), the incompressibility conditions Eq. (7), the tracer advection equations (14) and (15) and the linear free surface equation (13) give us a complete set of fluid equations, the Primitive Equations with SALT and SFLT.

2.2 Conservation Laws

The key benefit of the SALT and SFLT frameworks is that they retain some of the fundamental conservation properties possessed by the deterministic equations. By writing the Primitive Equations in the geometric form given in Eqs. (5d)–(5f) and (8), we may demonstrate the effect of the stochastic frameworks on these conservation laws. First, we consider energy conservation. The total energy is equal to the Hamiltonian, as given in Eq. (2). For convenience of notation, we define $\tilde{h}(\mathbf{m}, D, T, S, w) = h(\mathbf{m}^h, D, T, S) + \langle m_3, w \rangle$. h and $\tilde{h}$ are equal on solutions of the equations, but we have $\frac{\delta \tilde{h}}{\delta \mathbf{m}} = \mathbf{v}$. By direct calculation, the time evolution of the energy is given by:

$$dh = \sum_i \left[\left\langle \frac{\delta \tilde{h}}{\delta \mathbf{m}}, \sum_i \mathcal{L}_{\xi_i} \mathbf{m}^h \right\rangle + \left\langle g(1+b), \xi_i^{(z)} \right\rangle \right] \circ dW_t^i$$

$$- \left\langle \frac{\delta \tilde{h}}{\delta \mathbf{m}}, (d + \mathcal{L}_{d\chi}) \mathbf{F} \right\rangle . \tag{16}$$

Thus, the energy conservation property is violated by the stochastic terms. The two terms on the right-hand side of the pairing in Eq. (16) come from SALT and SFLT respectively. However, as shown in [HH21], the energy deviation from SFLT can be nullified by choosing $(d + \mathcal{L}_{d\chi}) \mathbf{F} = \sum_I \mathcal{L}_{\mathbf{v}} \boldsymbol{\phi}_I \circ dB_t^I$ for some parameters $\boldsymbol{\phi}_I(\mathbf{x})$. Indeed, by the anti-symmetry of the vector field commutator:

$$\left\langle \frac{\delta \tilde{h}}{\delta \mathbf{m}}, \sum_I \mathcal{L}_{\mathbf{v}} \boldsymbol{\phi}_I \circ dB_t^I \right\rangle = \left\langle \left[\frac{\delta \tilde{h}}{\delta \mathbf{m}}, \mathbf{v} \right], \sum_I \boldsymbol{\phi}_I \circ dB_t^I \right\rangle = 0, \tag{17}$$

where the square bracket $[\cdot]$, denotes the commutator of vector fields. Thus, energy conservation is broken by SALT but preserved by a class of stochastic forcing in SFLT. In the remainder of the paper, we shall assume the stochasticity introduced by SFLT are in the energy preserving form.

The next conservation law we consider is the Kelvin circulation theorem. The evolution of the circulation corresponding to Eq. (8) is given by:

$$d \oint_{C(t)} \frac{\mathbf{m}^h}{D} \cdot d\mathbf{x} = -g \oint_{C(t)} b(T, S, z) dz \, dt + \sum_I \oint_{C(t)} (\text{curl} \, \boldsymbol{\phi}_I \times \mathbf{v}) \cdot d\mathbf{x} \circ dB_t^I , \tag{18}$$

where $C(t)$ is a closed loop moving with the transport velocity $d\chi$. We see that SALT affects the circulation theorem only by modifying the advection of the loop; thus the circulation theorem for SALT is the same as in the deterministic case, but with the circulation considered around a stochastically-transported loop. Therefore, circulation is generated only by buoyancy gradients being misaligned with the vertical direction. In SFLT, on the other hand, there are additional forces introduced, which generate the circulation of fluid momentum.

The evolution of potential vorticity associated with Eq. (8) can be expressed as

$$(d + d\chi \cdot \nabla) q = \frac{1}{D} \boldsymbol{\omega} \cdot \nabla \left(\frac{\partial b}{\partial z} d\chi^{(z)} \right) + \frac{1}{D} \nabla b \cdot \sum_I [\nabla \cdot (\mathbf{v} \omega_I) - \omega_I \cdot \nabla \mathbf{v}] \circ dB_t^I , \tag{19}$$

where $\boldsymbol{\omega} := \text{curl} \left(\mathbf{m}^h / D \right)$ is the relative vorticity, $\omega_I = \text{curl} \, \boldsymbol{\phi}_I$ is the stochastic vorticity generated by SFLT and $q := \frac{1}{D} \boldsymbol{\omega} \cdot \nabla b$ is the potential vorticity. Similar to the Kelvin circulation theorem, SALT introduces stochasticity in the transport velocity $d\chi$, while SFLT introduces stochastic forces that act on the advection of fluid potential vorticity. If we assume that the buoyancy has no explicit dependence on the vertical coordinate, i.e. $\frac{\partial b}{\partial z} = 0$, then q is purely advected by the flow in the absence of SFLT.

3 Calibration of the Stochastic Parameters

3.1 Lagrangian Paths

In order to calibrate the parameters ξ_i used in SALT we propose to use the method of Lagrangian paths introduced in [CCH+19, CCH+20].

First, we perform a fine-grid model run, which we shall take to be the 'truth'. Resulting from this run we get an output velocity $\mathbf{v}(\mathbf{x}, t)$ saved at times $t \in \{t_1, \ldots, t_{N-1}, t_N\}$, where the time interval between subsequent sample times, $t_{i+1} - t_i$, is greater than the velocity decorrelation time, defined to be the smallest τ at which the auto-correlation function $C(\tau)$ is less than e^{-1}. Suppose the fine-grid resolution is M times that of the coarse grid, in which case the coarse-grid time step is given by $\Delta t_c := M \Delta t_f$, where Δt_f is the time step for the fine-grid model run. In order to compute Lagrangian paths we also save $\mathbf{v}(\mathbf{x}, t)$ at $t \in \{t_i, t_i + \Delta t_f, \ldots, t_i + (M-1)\Delta t_f\}$ for each $i = 1, \ldots, N$.

To obtain the corresponding coarse-grid velocity $\bar{\mathbf{v}}(\bar{\mathbf{x}}, t)$ from $\mathbf{v}(\mathbf{x}, t)$, we apply a coarse-graining operator to $\mathbf{v}(\mathbf{x}, t)$, which consists of a local average over fine-grid points, to obtain a velocity $\bar{\mathbf{v}}(\bar{\mathbf{x}}, t)$ defined on the coarse grid. Considering a distribution of tracer particles whose initial positions $\mathbf{x}_0^r$ are the (three-dimensional) coordinates of the coarse-grid nodes (enumerated by r), we compute Lagrangian paths on the fine grid and coarse grid respectively:

$$\mathbf{x}_f^r\left(t_i + M\Delta t_f\right) := \mathbf{x}_0^r + \sum_{m=0}^{M-1} \mathbf{v}\left(\mathbf{x}_f^r\left(t_i + m\Delta t_f\right), t_i + m\Delta t_f\right)\Delta t_f, \tag{20a}$$

$$\mathbf{x}_c^r\left(t_i + M\Delta t_f\right) := \mathbf{x}_0^r + \bar{\mathbf{v}}\left(\mathbf{x}_c^r(t_i), t_i\right)\Delta t_c, \tag{20b}$$

where $\mathbf{x}_f^f\left(t_i + M\Delta t_f\right)$ and $\mathbf{x}_c^r\left(t_i + M\Delta t_f\right)$ are the Lagrangian paths computed as integral curves of $\mathbf{v}_f$ and $\bar{\mathbf{v}}$ respectively; the integral is carried out over one coarse-grid time-step, which is equivalent to M fine-grid time steps. We can then define the difference $\Delta\mathbf{x}_{r,i} = \Delta\mathbf{x}(t_i, \mathbf{x}_0^r) := \mathbf{x}_f^r\left(t_r + M\Delta t_f\right) - \mathbf{x}^c\left(t_r + M\Delta t_f\right)$ and apply the method of [HJS07] to compute the Empirical Orthogonal Functions (EOFs). To summarise, we subtract off the time mean to define $\Delta\mathbf{x}_{r,i}' := \Delta\mathbf{x}_{r,i} - \frac{1}{N}\sum_{i=0}^{N-1}\Delta\mathbf{x}_{r,i}$. In the x-direction we then have a matrix with components $\Delta x_{r,i}'$. From this we construct the matrix $\Lambda^{(x)}$ which has components $\Lambda_{rs}^{(x)} = \frac{1}{N}\sum_{i=0}^{N-1}\Delta x_{r,i}'\Delta x_{s,i}'$. The EOFs in the x-direction are then defined to be the eigenvectors of the matrix $\Lambda^{(x)}$

which we denote as $a_i^{(x)}$, for $i = 1 \dots N$. They are normalised in the sense that $\sum_{\mathbf{x}} a_i^{(x)}(\mathbf{x}) a_j^{(x)}(\mathbf{x}) = \delta_{ij}$, where the sum is over all grid points. We apply the same process to the y-component $\Delta y'_{r,i}$ to obtain N eigenvectors in the y-direction, which we denote $a_i^{(y)}$. We do not compute the eigenvectors for the z-direction since these will be obtained from the incompressibility condition.

We remark that the method we have used here, in which we compute the EOFs of each component of $\Delta \mathbf{x}$ separately, is different from the method found in other sources [e.g. HLB96], in which the components are computed together and we obtain a set of two-component eigenvectors $\mathbf{a}_i$ immediately, with one eigenvalue λ_i corresponding to each of these EOFs. However, this method was attempted for SALT runs in the current set-up and the results of model runs were less successful. For this reason we have chosen to compute the components separately.

Thus, in our case we have N eigenvectors in each of the horizontal directions and these will have associated eigenvalues $\lambda_i^{(x)}$ and $\lambda_i^{(y)}$. We define the horizontal components of $\boldsymbol{\xi}_i$ by a re-scaling of these eigenvectors. The magnitude of the eigenvalue $\lambda_i^{(x)}$ gives an indication of how much of the variance is captured by the corresponding eigenvector. Therefore, we choose to scale the parameters so that $\left\langle \boldsymbol{\xi}_i^{(h)}, \boldsymbol{\xi}_i^{(h)} \right\rangle \propto \lambda_i$. Moreover, in order to ensure that the different methods for computing $\boldsymbol{\xi}_i$ may be compared fairly, we require that the L_2-norm of the sum be the same for each method. Thus we impose the following:

$$\frac{1}{V_{tot}} \sum_{i=1}^{N_\xi} \left\langle \boldsymbol{\xi}_i^{(h)}, \boldsymbol{\xi}_i^{(h)} \right\rangle = \gamma^2 \tag{21}$$

where γ is a constant with units $ms^{-1/2}$, which we shall choose later; V_{tot} is the total volume of the domain. and $N_\xi \leq N$ is the number of EOFs we choose to keep for our model runs. The total integral, denoted by angle brackets, is defined by $\langle \mathbf{a}, \mathbf{b} \rangle := \sum_{\mathbf{x}} \mathbf{a}(\mathbf{x}) \cdot \mathbf{b}(\mathbf{x}) V(\mathbf{x})$. We can achieve the required properties by choosing the following scaling:

$$\xi_i^{(x)}(\mathbf{x}) = \gamma \sqrt{\frac{\lambda_i^{(x)}}{\lambda_{tot}} \cdot \frac{V_{tot}}{V(\mathbf{x})}} a_i^{(x)}(\mathbf{x}) \tag{22}$$

where $V(\mathbf{x})$ is the volume of the grid cell located at $\mathbf{x}$ and we have defined $\lambda_{tot} := \sum_{i=1}^{N_\xi} (\lambda_i^{(x)} + \lambda_i^{(y)})$. After computing the horizontal components in this way, $\boldsymbol{\xi}_i^{(h)}$ are then smoothed to zero near the boundaries in order to enforce the impermeability condition at the boundary, $\boldsymbol{\xi}_i^{(h)} \cdot \mathbf{n} = 0$, where $\mathbf{n}$ is the normal to the boundary and $\boldsymbol{\xi}_i^{(h)} = (\xi_i^{(x)}, \xi_i^{(y)})$ is the horizontal part of $\boldsymbol{\xi}_i = (\xi_i^{(x)}, \xi_i^{(y)}, \xi_i^{(z)})$.

For the z-component we use the incompressibility condition Eq. (7) along with the impermeability condition $\boldsymbol{\xi}_i \cdot \nabla (z + H) = 0$ at the lower boundary $z = -H$ to obtain:

$$\xi_i^{(z)} = -\nabla^{(h)} \cdot \int_{-H}^{z} \boldsymbol{\xi}_i^{(h)} dz, \tag{23}$$

where $\nabla^{(h)} = (\frac{\partial}{\partial x}, \frac{\partial}{\partial y})$ is the horizontal gradient. This method for computing the vertical component of $\boldsymbol{\xi}_i$ is applicable to any system of fluid equations with an incompressibility condition. We could, alternatively, compute all three components of $\boldsymbol{\xi}_i$ as EOFs of the three components of $\Delta \mathbf{x}$. However, the resulting three-component vector $\boldsymbol{\xi}_i$ will not be guaranteed to be divergence-free. We would then need to subtract off the divergent part $\boldsymbol{\xi}_i \to \boldsymbol{\xi}_i' = \boldsymbol{\xi}_i - \nabla \Delta^{-1} (\nabla \cdot \boldsymbol{\xi}_i)$ where Δ^{-1} is the inverse Laplacian. However, computing the divergent part of the vector $\boldsymbol{\xi}_i$ is computationally expensive; moreover, the components of $\boldsymbol{\xi}_i$ computed in this way will not be guaranteed to be orthogonal with respect to $\langle \cdot, \cdot \rangle$. Thus in this paper we consider only the $\boldsymbol{\xi}_i$ for which the vertical components are computed from integrating the incompressibility condition.

3.2 Eulerian Differences

To calibrate the parameters $\boldsymbol{\phi}_I$ used in SFLT we propose an alternative method by using differences in fixed Eulerian coordinates. Consider the deterministic momentum equation given by:

$$(d + \mathcal{L}_{\mathbf{v}dt}) \left(\mathbf{m}^h \right) = -\left(p + \frac{\delta h}{\delta D} \right) \diamond D dt - \frac{\delta h}{\delta T} \diamond T dt - \frac{\delta h}{\delta S} \diamond S dt, \tag{24}$$

and the SFLT equation:

$$(d + \mathcal{L}_{\bar{\mathbf{v}}dt}) \left(\bar{\mathbf{m}}^h \right) - \sum_I \mathcal{L}_{\bar{\mathbf{v}}} (d\Phi) = -\left(p + \frac{\delta h}{\delta \bar{D}} \right) \diamond \bar{D} dt - \frac{\delta h}{\delta \bar{T}} \diamond \bar{T} dt - \frac{\delta h}{\delta \bar{S}} \diamond \bar{S} dt, \tag{25}$$

where the notation $(\bar{\cdot})$ are used on the variables of the SFLT equations to emphasise the difference between deterministic and stochastic variables. The goal of the stochastic parameterisation is to decompose the "true" fluid flow to a slow drift component and a rapid fluctuating component whose amplitude can be estimated from data. In the example of estimating the momentum fluctuation $d\Phi$ of $\bar{\mathbf{m}}^h$, we denote the slow drift component as $\bar{\mathbf{m}}^h$ and we seek the solution to the minimisation problem

$$\min_{d\Phi} \mathbb{E} \left[\left| d\mathbf{m}^h - d\bar{\mathbf{m}}^h \right|^2 \right]. \tag{26}$$

Assuming D, T and S do not have rapidly fluctuating components, the minimisation problem becomes

$$\min_{d\Phi} \mathbb{E}\left[\left|\mathcal{L}_{\mathbf{v}}(\mathbf{m}^h dt) - \mathcal{L}_{\bar{\mathbf{v}}}(\bar{\mathbf{m}}^h dt - d\Phi)\right|^2\right]. \qquad (27)$$

We see that this minimisation problem can be solved by taking $d\Phi = \left(\mathbf{m}^h - \bar{\mathbf{m}}^h\right) dt = (\mathbf{u} - \bar{\mathbf{u}}) dt$. Therefore, we define the differences

$$\Delta\mathbf{x}_{r,I} := \Delta\mathbf{x}(t_I, \mathbf{x}_0^r) = \left[\mathbf{u}(t_I, \mathbf{x}_0^r) - \bar{\mathbf{u}}(t_I, \mathbf{x}_0^r)\right]\Delta t_c \qquad (28)$$

for $I = 1, \ldots, N$. We then assume the expansion $d\Phi = \sum_{I=1}^{N} \phi_I \circ dB_t^I$. As before, we subtract the time-mean to obtain $\Delta\mathbf{x}'_{r,I} = \Delta\mathbf{x}_{r,I} - \frac{1}{N}\sum_{I=0}^{N-1} \mathbf{x}_{r,I}$ and then compute the EOFs exactly as we did in Sect. 3.1 to get our parameters ϕ_I.

In both methods we initially compute horizontal components of the stochastic parameters using EOFs, but for SALT there is the additional step of integrating the incompressibility condition to obtain the vertical component. The vertical component is not needed for ϕ_I since it is a part of the decomposition of the fluid momentum $\mathbf{m}$, the vertical part of which vanishes in the Primitive Equations. In fully three-dimensional models in which the vertical component of the momentum is non-zero, the Eulerian differences of the momenta will be a three-dimensional object and one can compute all three components of the parameters ϕ_I using EOFs.

We can also consider using Eulerian differences as an option for ξ_i in SALT. This effectively means approximating the fine-grid Lagrangian path by taking only one time-step in the coarse grid: $\mathbf{x}_f \approx \mathbf{v}_f(\mathbf{x}_0, t)\Delta t_c$. We can expect that this will be a reasonable approximation for small M, but for larger M the Lagrangian paths method will diverge from the Eulerian differences. In our numerical investigations in SALT we shall consider ξ_i computed from both the Lagrangian paths method and Eulerian differences method. For SFLT we also consider ϕ_I computed from Lagrangian paths (for completeness) as well as those computed by Eulerian differences as described above.

4 Results

We solve the Primitive Equations using the FESOM2 code on a rectangular domain $[0, 40°] \times [30°, 60°] \times [0, -H]$, where $H = 1600\text{m}$ is the depth of the domain and the bathymetry is flat. Impermeability conditions are imposed at all boundaries. The model is spun up for three years from zero initial velocity and an initial temperature profile given by $T(z) = T_0 + \frac{\lambda}{\alpha\rho_0}\left((1 - \beta)\tanh\left(\frac{z}{z_0}\right) + \beta\frac{z}{H}\right)$, which is based on the test case described in [RDH+12, SDB+16]. We take $T_0 = 25°C$, $\beta = 0.05$, $\lambda = 5\text{kgm}^{-3}$, $z_0 = 300\text{m}$, $\rho_0 = 1030\text{kgm}^{-3}$, and $\alpha = 0.00025\text{K}^{-1}$. For simplicity, salinity is kept constant and we use a linear equation of state which depends only on temperature: $b = -\alpha(T - 10°C)$. The flow is driven by a wind forcing in the upper layer given by $\tau(x, y) = \frac{-\tau_0 \Delta z_0}{\rho_0} \cos\left(\frac{\pi y}{15°}\right) \hat{\mathbf{x}}$, where $\Delta z_0 = 10$m is the thickness of the upper layer; $\tau_0 = 0.2$ms^{-2} is the wind strength. The vertical discretisation consists of 23 layers, with layer thicknesses increasing with depth. For the horizontal discretisation we take a fine grid of spacing $1/4°$ and a coarse grid of spacing $1/2°$. At the latitudes we are considering, $1/4°$ corresponds to an eddy-permitting model, while $1/2°$ may be considered non-eddy resolving [see Hal13]. We run the deterministic model on the fine grid and the coarse grid, and carry out the SALT and SFLT runs on the coarse grid. All coarse-grid runs are begun from the same initial condition, being the final time snapshot after the three-year spin-up period; the fine-grid run is begun from the end of the three-year spin-up on the fine grid. We save data in each case at intervals of 15 days, over a time period of 10 years, for a total of 240 snapshots. From the fine grid data we have the 'truth' velocity $\mathbf{v}_f$. To this we apply a coarse-graining $\bar{\mathbf{v}}$; we then follow the procedures outlined in Sect. 3 to compute $\boldsymbol{\xi}_i^{(h)}$ and $\boldsymbol{\phi}_I$. However, there is no canonical choice for how the coarse-graining should be done. We consider a filter defined by an equally-weighted nine-point average over nearest neighbours, and we denote this filter $\mathcal{F}$; this filter, applied once, has a width equal to the spacing on the coarse grid, i.e. $1/2°$. The coarse-graining will then be done by applying this filter N_{filt} times successively, then projecting onto the coarse grid. Thus, the smoothing filter applied N_{filt} times will be denoted $\mathcal{F}^{N_{filt}}$; this has a width $N_{filt}/2$ degrees with a stronger weighting for points closer to the centre of the filter. We consider the cases $N_{filt} = 1, 4, 32$.

From the deterministic model run, we have velocities saved at 240 time snapshots, so we can use these to compute 240 EOFs. We do this for both the Lagrangian paths method and the Eulerian differences method, for each of the three choices of N_{filt}; this gives a total of six sets of parameters. In our model runs we shall choose to keep $N_\xi = N_\phi = 32$ of these parameters for each run. In Fig. 1 we plot the square-root of the sum of the squares of these parameters (before re-scaling by γ) as a field in space. From Fig. 1 it appears the differences between Lagrangian paths or Eulerian differences are minimal. We remark that here the time-steps on the fine and coarse grids differ only by a factor of 2; it is expected that if a bigger difference in resolution is used, then more steps will be needed in computing the Lagrangian paths and therefore the corresponding parameters will differ more substantially. The number of times we apply the smoothing operator, however, has a much greater effect and we see significantly different fields with $N_{filt} = 32$ than we do with $N_{filt} = 4$ or $N_{filt} = 1$. Indeed, it appears from Fig. 1 that the weaker filter causes the parameters to be more strongly concentrated around the western boundary, whereas for the stronger filter the parameters are spread more across the domain.

The cumulative spectra of the EOFs are shown in Fig. 2. These spectra show us how many EOFs are needed to capture a given percentage of the total variability; or conversely, how much variance is captured by a given number of EOFs. We show in each case how much variability is captured by using 32 EOFs. In all cases the Lagrangian paths method gives a slightly higher variability captured, though

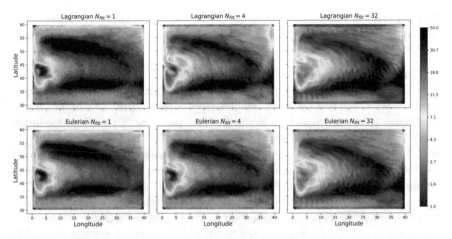

Fig. 1 $\frac{1}{\gamma}\left(\frac{1}{N_\xi}\sum_{i=1}^{N_\xi}\boldsymbol{\xi}_i^{(h)}\cdot\boldsymbol{\xi}_i^{(h)}\right)^{1/2}$ in the upper fluid layer for different methods of computing $\boldsymbol{\xi}_i^{(h)}$. Top row: $\boldsymbol{\xi}_i^{(h)}$ computed from Lagrangian paths for different strengths of smoothing filter. Bottom row: $\boldsymbol{\xi}_i^{(h)}$ computed from Eulerian differences for different strengths of smoothing filter

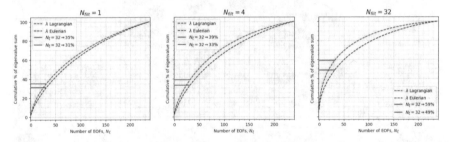

Fig. 2 Eigenvalue spectra of zonal $\boldsymbol{\xi}_i$, plotted for three different values for N_{filt}. On each panel is shown the spectrum for the EOFs calculated by Lagrangian trajectories and Eulerian differences. The horizontal lines show what the percentage of the total variance is captured by choosing $N_\xi = 32$ EOFs

the difference is small, especially for the smaller values of N_{filt}. A much bigger variability is captured, however, in the $N_{filt} = 32$ case when compared with the $N_{filt} = 1$ case.

We implemented SALT and SFLT into FESOM2 (see Appendix section) and ran the model with each choice of parameters and with the appropriate re-scaling as detailed above. For all SALT runs we use $N_\xi = 32$ with the scaling $\gamma = 2 \times 10^{-3}ms^{-1/2}$. For SFLT we also take $N_\phi = 32$ but scale the parameters with $\gamma = 10^2 ms^{-1/2}$. This re-scaling is chosen empirically taking γ to be the largest value possible that will not result in model blow-up. It appears that the magnitude of parameters that we are able to use for SFLT is much higher. This is possibly due to the fact that SFLT does not involve any direct modification of the tracer equation.

SALT, on the other hand, includes an advection of the temperature by the stochastic transport velocity; using higher values for this velocity may destabilise the tracer equation and cause model blow-up.

The results of these runs are shown in Figs. 3, 4, 5. Figure 3 shows the eddy kinetic energy (EKE), defined by $E = \frac{1}{2}|\mathbf{u} - \langle \mathbf{u} \rangle|^2$, where $\langle \mathbf{u} \rangle$ is the time-averaged velocity. We notice that the eddy kinetic energy is significantly less in the coarse-grid deterministic run than it is in the fine-grid run. This is probably due to the fact that small scales are less present in the coarse-grid flow, and in the coarse-grid model the viscosity used is greater and so kinetic energy is dissipated at a faster rate. However, when we include SALT there is, for most choices of ξ_i, a notable increase in EKE across the domain, particularly around the western boundary. The exception is in the cases in which the coarse-grained velocities $\bar{\mathbf{v}}$ used to calculate ξ_i are defined with only one application of the smoothing operator, as shown in panels (c) and (d) in Fig. 3. This could be because, from Fig. 2, the inclusion of 32 ξ_i captures a smaller amount of the total variability; it may also be that the effect of

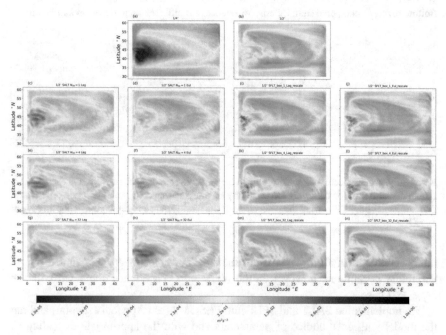

Fig. 3 Time-average of eddy kinetic energy at depth 16 m below the surface. Panel (**a**) is from the high-resolution (1/4°) deterministic model, while (**b**) is from the low-resolution (1/2°) deterministic model. Panels (**c**), (**g**), (**k**) are the results of model runs at 1/2° with SALT, where ξ_i are computed using Lagrangian differences using a coarse velocity defined by applying the smoothing filter 1, 4 and 32 times respectively. Panels (**d**), (**h**), (**l**) are also SALT runs but ξ_i are computed from Eulerian differences rather than Lagrangian trajectories. Panels (**e**), (**i**), (**m**) are SFLT runs with ϕ_I computed from Lagrangian trajectories, while (**f**), (**j**), (**n**) have ϕ_I computed from Eulerian differences

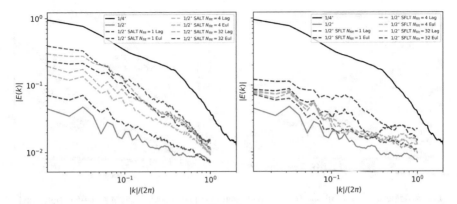

Fig. 4 Spectra of eddy kinetic energy for SALT (left panel) with ξ_i calculated from Lagrangian paths and from Eulerian differences; and for SFLT (right panel) with ϕ_I calculated from Lagrangian paths and from Eulerian differences. Also included in each plot are the spectra for the deterministic runs on the fine and coarse grids. Spectra are calculated in the x-direction at fixed $y = 45\frac{1}{6}^\circ$ by $\left|\hat{E}(k)\right| := \frac{1}{t_{max}} \int_0^{t_{max}} \left|\int E(x,t)e^{ikx}dx\right| dt$. Here $t_{max} = 10$years and $t = 0$ corresponds to the beginning of the model run, after spin-up

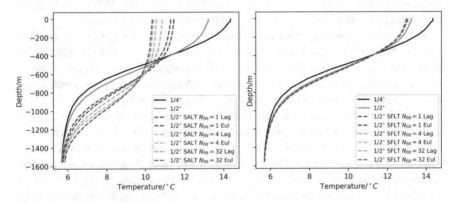

Fig. 5 Vertical profiles of temperature horizontally-averaged across the domain after 10 years of model time. The left-hand panel shows the results from the SALT runs, alongside the deterministic runs. The right-hand panel shows the results from the SFLT runs, alongside the deterministic runs

the ξ_is is more spread out across the domain, as shown in Fig. 1, which overall has a greater impact than having them more highly concentrated in one region. For SFLT there is only a modest improvement in the EKE field, and the effect is similar for all choices of the parameters. In all cases there appears to be little difference between the Eulerian differences method and the Lagrangian paths method when the same N_{filt} is used.

We can also consider the spatial spectra, as shown in Fig. 4. There we see that the $1/4^\circ$ run contains higher EKE at all scales than the low-resolution run. Every SALT run succeeds in increasing the energy at almost all scales and in shifting the

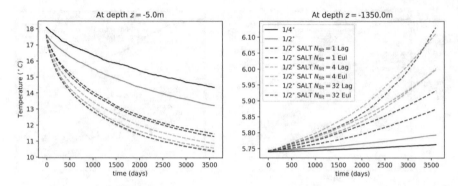

Fig. 6 Time series of spatially-averaged temperature fields for SALT runs at $z = -5$ m (left panel) and $z = -1350$ m (right panel)

spectrum towards that of the $1/4°$ run. The most significant improvements are seen in the run with Eulerian parameters computed with $N_{filt} = 32$; in contrast, there is only a small change from the deterministic run when the $N_{filt} = 1$ Eulerian parameters are used. For SFLT the improvement is again less noticeable, with all choices of parameters only giving a slight increase in EKE at all scales.

Since we are working with the Primitive Equations, the buoyancy can have a large effect on the fluid flow. We therefore consider the temperature, which determines buoyancy directly via the linear equation of state. Figure 5 shows vertical temperature profiles at the end of the ten-year run. In the coarse-grid model there is a slightly lower average temperature in the upper layers of the fluid, and slightly higher temperatures in the lower layers. However, with SALT included there is, for some choices of parameters, a significant reduction in temperature in the upper layers, while at lower depths the temperature increases relative to the deterministic model. Considering the time series of spatially averaged temperature at $z = -5$ m and $z = -1350$ m in Fig. 6, we see the downwards diffusion effects are persistent in time. In the deterministic case we see that the coarse-grid model has a stronger downwards diffusion of temperature than the fine-grid run. The inclusion of SALT also accelerates this downward-diffusion effect. It therefore appears that the calibrated stochastic terms we have included in the temperature equation with SALT cause a downwards-diffusion effect. Indeed, an additional SALT run (not shown), in which the stochastic terms were not included in the temperature advection, did not display this downwards diffusion behaviour. Thus, further investigation will be required in order to determine how to avoid the excessive downwards diffusion in the tracer equation while maintaining a positive effect on the EKE field. SFLT has very little effect on the temperature field when compared to that of the low-resolution model. This is expected however since there are no direct stochastic effects in the temperature equation. Comparing with SALT runs where the temperature downwards-diffusion effect is present against the SFLT runs, we believe that the temperature is the dominant force for the evolution of velocity, at least at the resolutions we have considered here. Then, the limited effects on EKE by the SFLT framework are explained as it does not affect the driving temperature fields directly. It remains part of future work to consider the case where SFLT is added to the temperature field.

5 Summary and Discussion

This work lays the groundwork for the application of two relatively new stochastic parameterisation frameworks to the Primitive Equations. The first, SALT, has hitherto only been applied to simple idealised ocean models such as QG and 2D Euler. The second, SFLT, had not been investigated numerically prior to the present work. We have demonstrated some of the desirable theoretical properties of the stochastic Primitive Equations with the noise added in these ways. Notably, the preservation of a circulation theorem for SALT and energy conservation for SFLT. We have proposed to calculate the parameters ξ_i governing SALT and ϕ_I governing SFLT by two different methods: Lagrangian paths and Eulerian differences. We find that there are no significant differences between the two methods, either in the parameters themselves or in the results of model runs. In this case it is preferable to use the Eulerian differences method, as the parameters in this case are computationally less expensive to compute. However, we have used a set-up in which the fine-grid resolution is only is only 2 times the coarse-grid resolution. However, using a larger ratio of grid resolutions would mean more time-steps are needed in the Lagrangian paths and so may give different EOFs that differ more significantly than what we have observed here. We do observe, however, that there are large sensitivities to the choice of smoothing used in defining the coarse-grained velocity, from which the parameters are calculated. In the SALT case, the model runs using parameters calculated with a strong smoothing filter show a significant improvement in the eddy kinetic energy field at all depths, as well as in the eddy kinetic energy spectrum. In the SFLT case, the improvement in EKE field and EKE spectrum are more modest compared to the improvement by SALT due to the lack of direct stochastic effects to the driving temperature fields. Considering the temperature profile, however, we observe that SALT causes significant additional downward diffusion when compared with the deterministic model. It remains an open problem to devise a method to avoid this effect. The answer may lie in a different method for configuring the parameters ξ_i or it may be the case that this is a property intrinsic to SALT. In either case, further study is needed in this direction.

The stochastic parameterisation frameworks considered in this paper distils all uncertainties of the ocean models into the stochastic parameters ξ_i and ϕ_I. However, the effects of these stochastic parameterisations could be limited by the model, both physically and numerically. Examples of the limiting factors for the Primitive Equations are the forcing from the temperature field and artificial viscosity imposed for numerical stability. The interplay between numerical effects such as artificial viscosity and stochastic parameterisation is particularly interesting for future work. This is due to different numerical viscosity are imposed at different mesh resolutions to numerical stability which influences the calibration process. Thus, we expect there are limits to the effects of SALT and SFLT for low-resolution simulations where viscosity are dominant. In high-resolution simulations, we expect to see further effects of stochasticity as the influence of viscosity diminishes. After all, the problem of stochastic parameterisations are not just model-dependent, it also dependent on the numerical method solving it.

Acknowledgments We are grateful to our friends and colleagues who have generously offered their time, thoughts, and encouragement in the course of this work during the time of COVID-19. Thanks to P. Berloff, C. Cotter, S. Danilov, D. Holm, S. Juricke, E. Luesink, W. Pan and all members of the Geometric Mechanics group at Imperial College for their thoughtful comments and discussions. We acknowledge the Alfred Wegener Institute for the use of their computing facilities. The authors are grateful for partial support, as follows. RH for the EPSRC scholarship (Grant No. EP/R513052/1); and SP for the EPSRC Centre for Doctoral Training in the Mathematics of Planet Earth (Grant No. EP/L016613/1). Finally, we thank the anonymous reviewer for giving useful feedback on our manuscript.

Appendix: Numerical Implementation

In order to apply SALT and SFLT to FESOM2 we adapt the time-stepping scheme to include the appropriate stochastic terms. Details of the original (deterministic) time-stepping are given in [DSWJ16]. We modify the scheme from FESOM2 to a two-step Heun-type method [BBT04]; we choose this because of the use of Stratonovich integrals, to which the Heun method converges. The first step in the method is to compute the modified pressure:

$$
\hat{p}_h^n = \rho_0 g \int_{-H}^{z} b(T^{n+1/2}) dz' + \sum_i \int_{-H}^{z} \frac{\partial \boldsymbol{\xi}_i}{\partial z} \cdot \mathbf{u}^n dz' \frac{\Delta W_{n+1}^i}{\Delta t}
$$

$$
+ \sum_I \int_{-H}^{z} \frac{\partial \boldsymbol{\phi}_I}{\partial z} \cdot \mathbf{u}^n dz' \frac{\Delta B_{n+1}^I}{\Delta t}
\tag{29}
$$

where ΔW_{n+1}^i and ΔB_{n+1}^I are independent, normally-distributed random variables with mean 0 and variance Δt. For the sake of conciseness we shall assume that the buoyancy depends only on temperature T, and that salinity is kept constant; however, extending the method to include additional tracers should be straightforward. The advective, diffusive and pressure parts of the momentum right-hand-side are then computed:

$$
\Delta \hat{\mathbf{u}}^{n+1} = \hat{\mathbf{R}}^{n+1/2} - \nabla \left(\hat{p}_h^n + \eta^n \right) \Delta t + \mathbf{D} \left(\mathbf{u}^n, \Delta \hat{\mathbf{u}}^{n+1} \right)
\tag{30}
$$

where $\hat{\mathbf{R}}^{n+1/2}$ is an Adams-Bashforth interpolation of the advective and Coriolis terms. In fact we have $\hat{\mathbf{R}}^{n+1/2} = \left(\frac{3}{2} + \epsilon\right)\hat{\mathbf{R}}^n - \left(\frac{1}{2} + \epsilon\right)\hat{\mathbf{R}}^{n-1}$, where $\hat{\mathbf{R}}^n = \mathbf{R}\left[\mathbf{v}^n \Delta t, \mathbf{u}^n\right] + \sum_i \left(\mathbf{R}\left[\boldsymbol{\xi}_i, \mathbf{u}^n\right] - \nabla^{(h)}\boldsymbol{\xi}_i \cdot \mathbf{u}^n\right) \Delta W_{n+1}^i - \sum_I \mathbf{R}\left[\mathbf{v}^n, \boldsymbol{\phi}_I \Delta B_{n+1}^I\right]$ and $\mathbf{R}\left[\mathbf{v}, \mathbf{u}\right] := -\nabla \cdot (\mathbf{v}\mathbf{u}) - \mathbf{f} \times \mathbf{v}$. $\mathbf{D}$ includes the horizontal and vertical diffusion terms, as well as the external wind forcing.

The change in free surface height $\Delta\hat{\eta}^{n+1}$ is computed implicitly:

$$\left(1 - g\Delta^2\nabla \cdot \int_{-H}^0 \nabla\,(\cdot)\,dz\right)\Delta\hat{\eta}^{n+1} = -\nabla \cdot \int_{-H}^0 \nabla \cdot \left(\mathbf{u}^n + \Delta\hat{\mathbf{u}}^{n+1}\right)dz\Delta t$$

(31)

Once this has been solved we can finally compute the stepped-forward horizontal velocity:

$$\hat{\mathbf{u}}^{n+1} = \mathbf{u}^n + \Delta\hat{\mathbf{u}}^{n+1} - g\Delta t\nabla\Delta\hat{\eta}$$

(32)

Then we solve for the total layer thickness $\bar{h}$, which in the continuous case is the same as the free surface height η; in the discrete case, however, they are different and we compute:

$$\hat{\bar{h}}^{n+3/2} = \bar{h}^{n+1/2} - \nabla \cdot \int_{-H}^0 \hat{\mathbf{u}}^{n+1}dz\Delta t$$

In our present set-up we then set the free-surface height as a linear interpolation of the total layer heights:

$$\hat{\eta}^{n+1} = \theta\hat{\bar{h}}^{n+3/2} + (1 - \theta)\hat{\bar{h}}^{n+1/2}$$

(33)

where $\theta \in [0, 1]$ is an arbitrary parameter, which we set equal to 1.

Since we have the horizontal velocity we may compute the vertical velocity:

$$\hat{w}^{n+1} = -\nabla \cdot \int_{-H}^z \hat{\mathbf{u}}^{n+1}dz'$$

(34)

The newly-computed three-dimensional velocity, along with the stochastic SALT velocity, is then used to advect the tracer:

$$\hat{T}^{n+3/2} = T^{n+1/2} - R_T\left[T^{n+1/2}, T^{n-1/2}, \hat{\mathbf{v}}^{n+1}\Delta t + \boldsymbol{\xi}_i\Delta W_{n+3/2}\right] + K\left[T^{n+1/2}\right]$$

(35)

where R_T denotes the advection scheme and K is the diffusion. From these steps we compute intermediate values $\hat{X}^{n+1} := \left(\hat{\mathbf{u}}^{n+1}, \hat{\eta}^{n+1}, \hat{\bar{h}}^{n+3/2}, \hat{T}^{3/2}\right)$ from values at the previous two time steps: $\mathbf{u}^n, \mathbf{u}^{n-1}, \bar{h}^{n+1/2}, T^{n+1/2}, T^{n-1/2}$. We may write this schematically as:

$$\hat{X}^{n+1} = X^n + \mathcal{F}\left[X^n, X^{n-1}\right] \qquad (36)$$

where $\mathcal{F}$ is an operator representing the computations outlined above. For the corrector step we follow the same steps as above, to compute $\mathcal{F}\left[\hat{X}^{n+1}, X^n\right]$ and we have the overall evolution given by:

$$X^{n+1} = X^n + \frac{1}{2}\left[\mathcal{F}\left[X^n, X^{n-1}\right] + \mathcal{F}\left[\hat{X}^{n+1}, X^n\right]\right] \qquad (37)$$

This method differs from the usual Heun method because the right-hand side depends on the previous two time-steps, rather than just the previous one. It remains to prove that adding the stochasticity with this method does converge to the required Stratonovich integrals.

Bibliography

[BBT04] K. Burrage, P. M. Burrage, and T. Tian. Numerical methods for strong solutions of stochastic differential equations: an overview. *Proceedings of the Royal Society of London. Series A: Mathematical, Physical and Engineering Sciences*, 460(2041):373–402, 2004.

[Ber05] Pavel S. Berloff. Random-forcing model of the mesoscale oceanic eddies. *Journal of Fluid Mechanics*, 529:71–95, 2005.

[CCH+19] Colin Cotter, Dan Crisan, Darryl D. Holm, Wei Pan, and Igor Shevchenko. Numerically Modeling Stochastic Lie Transport in Fluid Dynamics. *Multiscale Modeling & Simulation*, 17(1):192–232, 2019.

[CCH+20] Colin Cotter, Dan Crisan, Darryl D. Holm, Wei Pan, and Igor Shevchenko. Modelling uncertainty using stochastic transport noise in a 2-layer quasi-geostrophic model, 2020.

[CGH17] C. J. Cotter, G. A. Gottwald, and D. D. Holm. Stochastic partial differential fluid equations as a diffusive limit of deterministic Lagrangian multi-time dynamics. *Proceedings of the Royal Society A: Mathematical, Physical and Engineering Sciences*, 473(2205):20170388, 2017.

[CH09] C. J. Cotter and D. D. Holm. Continuous and Discrete Clebsch Variational Principles. *Foundations of Computational Mathematics*, 9(2):221–242, 2009.

[Cot20] Cotter, Colin and Crisan, Dan and Holm, Darryl and Pan, Wei and Shevchenko, Igor. Data Assimilation for a Quasi-Geostrophic Model with Circulation-Preserving Stochastic Transport Noise. *Journal of Statistical Physics*, 179(5):1186–1221, 2020.

[Doo53] Joseph Leo Doob. *Stochastic processes*, volume 10. New York Wiley, 1953.

[DSWJ16] S. Danilov, D. Sidorenko, Q. Wang, and T. Jung. The Finite-volumE Sea ice-Ocean Model (FESOM2). *Geoscientific Model Development Discussions*, pages 1–44, 2016.

[GBB+00] Stephen M. Griffies, Claus Böning, Frank O. Bryan, Eric P. Chassignet, Rüdiger Gerdes, Hiroyasu Hasumi, Anthony Hirst, Anne-Marie Treguier, and David Webb. Developments in ocean climate modelling. *Ocean Modelling*, 2(3):123–192, 2000.

[Hal13] Robert Hallberg. Using a resolution function to regulate parameterizations of oceanic mesoscale eddy effects. *Ocean Modelling*, 72:92–103, 2013.

[HH21] Darryl D. Holm and Ruiao Hu. Stochastic effects of waves on currents in the ocean mixed layer. *Journal of Mathematical Physics*, 62(7):073102, 2021.

[HJS07] A. Hannachi, I. T. Jolliffe, and D. B. Stephenson. Empirical orthogonal functions and related techniques in atmospheric science: A review. *International Journal of Climatology*, 27(9):1119–1152, 2007.

[HLB96] Philip Holmes, John L. Lumley, and Gal Berkooz. *Turbulence, Coherent Structures, Dynamical Systems and Symmetry*. Cambridge Monographs on Mechanics. Cambridge University Press, 1996.

[HMR98] Darryl D Holm, Jerrold E Marsden, and Tudor S Ratiu. The Euler-Poincaré Equations and Semidirect Products with Applications to Continuum Theories. *Advances in Mathematics*, 137(1):1–81, 1998.

[Hol15] Darryl D. Holm. Variational principles for stochastic fluid dynamics. *Proceedings of the Royal Society A: Mathematical, Physical and Engineering Sciences*, 471(2176):20140963, 2015.

[HSS09] Darryl D Holm, Tanya Schmah, and Cristina Stoica. *Geometric mechanics and symmetry: from finite to infinite dimensions*, volume 12. Oxford University Press, 2009.

[Kor17] Peter Korn. Formulation of an unstructured grid model for global ocean dynamics. *Journal of Computational Physics*, 339:525–552, 2017.

[MAH+97] John Marshall, Alistair Adcroft, Chris Hill, Lev Perelman, and Curt Heisey. A finite-volume, incompressible Navier Stokes model for studies of the ocean on parallel computers. *Journal of Geophysical Research: Oceans*, 102(C3):5753–5766, 1997.

[Mem14] Etienne Mémin. Fluid flow dynamics under location uncertainty. *Geophysical & Astrophysical Fluid Dynamics*, 108(2):119–146, 2014.

[Mey62] Paul-André Meyer. A decomposition theorem for supermartingales. *Illinois Journal of Mathematics*, 6(2):193–205, 1962.

[Mey63] Paul-André Meyer. Decomposition of supermartingales: the uniqueness theorem. *Illinois Journal of Mathematics*, 7(1):1–17, 1963.

[PZ14] PierGianLuca Porta Mana and Laure Zanna. Toward a stochastic parameterization of ocean mesoscale eddies. *Ocean Modelling*, 79:1–20, 2014.

[RDH+12] T Ringler, D Danilov, R Hallberg, A Adcroft, P Berloff, and P Gent. A test case for the assessment of Simulating Ocean Mesoscale Activity (SOMA). *Unpublished manuscript.*, 2012.

[SC21] O. D. Street and D. Crisan. Semi-martingale driven variational principles. *Proceedings of the Royal Society A: Mathematical, Physical and Engineering Sciences*, 477(2247):20200957, 2021.

[SDB+16] Dmitry V Sein, Sergey Danilov, Arne Biastoch, Jonathan V Durgadoo, Dmitry Sidorenko, Sven Harig, and Qiang Wang. Designing variable ocean model resolution based on the observed ocean variability. *Journal of Advances in Modeling Earth Systems*, 8(2):904–916, 2016.

[SW68] R. L. Seliger and Gerald Beresford Whitham. Variational principles in continuum mechanics. *Proceedings of the Royal Society of London. Series A. Mathematical and Physical Sciences*, 305(1480):1–25, 1968.

[UJPD21] Takaya Uchida, Quentin Jamet, Andrew Poje, and William K Dewar. An ensemble-based eddy and spectral analysis, with application to the Gulf Stream. *Journal of Advances in Modeling Earth Systems*, page e2021MS002692, 2021.

A Pathwise Parameterisation for Stochastic Transport

Oana Lang and Wei Pan

Abstract In this work we set the stage for a new probabilistic pathwise approach to effectively calibrate a general class of stochastic nonlinear fluid dynamics models. We focus on a 2D Euler SALT equation, showing that the driving stochastic parameter can be calibrated in an optimal way to match a set of given data. Moreover, we show that this model is robust with respect to the stochastic parameters.

1 Introduction

A fundamental challenge in observational sciences, such as weather forecasting and climate change predictions, is the modelling of uncertainty due, for example, to unknown or neglected physical effects, and incomplete information in both the data and the formulation of the theoretical models for prediction. Various dynamical parameterisation approaches have been proposed to tackle this challenge, see e.g. [6], [4], [11], [5], [1]. Of particular interest are the recently developed Data Driven models, that accommodate uncertainty by predicting both the expected future measurement values and their uncertainties, based on input from measurements and statistical analysis of the initial data. To effectively incorporate uncertainty in the data driven approach, such predictions are made in a probabilistic sense. Additionally, a data assimilation procedure is used to take into account the time integrated information obtained from the data being observed along the solution path during the forecast interval as "in flight corrections".

In the geoscience community, *data assimilation* (DA) refers to a set of methodologies designed to efficiently combine past knowledge of a geophysical system (in the form of a *numerical model*) with new information about that system (in the form of *observations*). DA is a central component of Numerical Weather Prediction where it is used to improve forecasting by adjusting the model parameters and reducing the

O. Lang (✉) · W. Pan
Imperial College London, London, UK
e-mail: o.lang15@imperial.ac.uk; wei.pan@imperial.ac.uk

© The Author(s) 2023
B. Chapron et al. (eds.), *Stochastic Transport in Upper Ocean Dynamics*,
Mathematics of Planet Earth 10, https://doi.org/10.1007/978-3-031-18988-3_10

uncertainties. To achieve this, a stochastic feedback loop between the model and the observation may be introduced: the assimilation of more data during the prediction interval will then decrease the uncertainty of the forecasts based on the initial data, by selecting the more likely paths as more observational data is collected. This is the basis of the so-called *ensemble data assimilation* which uses a set of model trajectories that are intermittently updated according to data.

A key step for ensuring the successful application of the combined stochastic parameterisation and data assimilation procedure, is the "correct" calibration of stochastic model parameters. For Stochastic Advection by Lie Transport (SALT) and Location Uncertainty (LU) models, current numerical methods for calibration, see [4], [1], [5], [12], have largely been inspired by the physical interpretation of the models derivations. More specifically on the assumption that the flow map is decoupled into a slow scale mean part and a fast scale fluctuating part. In the references mentioned before, it was shown that these methods are effective and led to successful combination of data driven models and state of the art data assimilation techniques.

In this work, we wish to investigate the feasibility and viability of probabilistic pathwise approach for calibration. Our general aim is to explore such ideas for a wide class of nonlinear stochastic transport models. This will be very useful in data assimilation problems, as in real world applications the signal is usually observed through discrete observations, but no results of this type for SALT or LU models have been obtained before. Currently, Lagrangian particle trajectories are simulated starting from each point on both the physical grid and its refined version, then the differences between the particle positions are used to calibrate the noise. This is computationally expensive and not fully justified from a theoretical perspective. In the same spirit as [3] but with a more complicated noise term and without any smoothing effects of a Laplacian, we propose an approach which uses high-frequency in time and low-frequency in space observations of a single path of the solution, to rigorously infer properties of the stochastic parameters. The knowledge of the noise is crucial for determining the behaviour of the solution and for assessing to what degree the solution of the coarse resolution SPDE deviates from the solution of the fine resolution PDE in the model reduction procedure, so an optimal calibration of the noise parameters is relevant from both a theoretical and an applied perspective.

In this work we look at stochastic calibration for the two-dimensional incompressible Euler equation in vorticity form. This stochastic equation models the local rotation of a fluid flow in the presence of spatial uncertainties and it has been derived from fundamental principles in [6]. This equation is a key ingredient in modelling phenomena in oceanography and in order to ensure that it efficiently encodes the small-scale variability in the upper part of the ocean, one needs to specify the stochastic parameters based on real observations. One of the main issues in parameter estimation using real data is the fact that the model parameters do not map to observations in a unique way (*model identifiability* problem, see e.g. [2]). For this reason, we believe that a probabilistic approach is much more suitable.

The 2D Euler equation in the form derived in [6] and studied in [4], [5] and [8] is given by:

$$d\omega_t + u_t \cdot \nabla \omega_t dt + \sum_{i=1}^{\infty} \xi_i \cdot \nabla \omega_t \circ dW_t = 0 \tag{1}$$

where $u = (u^1, u^2)$ is the fluid velocity, $\omega = curl\ u = \partial_2 u_1 - \partial_1 u_2$ is the vorticity, $(\xi_i)_i$ are divergence-free time-independent vector fields such that

$$\sum_{i=1}^{\infty} \|\xi_i\|_{k+1,\infty}^2 < \infty \tag{2}$$

and $(W^i)_{i \in \mathbb{N}}$ is a sequence of independent Brownian motions. Global well-posedness for Eq. (1) has been proven in [8] and the numerical and data assimilation perspective has been studied in [4] and [5]. In [8] the authors have shown that Eq. (1) admits a unique pathwise solution which belongs to the Sobolev space $\mathcal{W}^{k,2}(\mathbb{T}^2)$ ($k \geq 2$) when $\omega_0 \in \mathcal{W}^{k,2}(\mathbb{T}^2)$ and which can be extended to $L^\infty(\mathbb{T}^2)$ when $\omega_0 \in L^\infty(\mathbb{T}^2)$.

In this paper we consider the following SPDE on the two-dimensional torus $\mathbb{T}^2 = \mathbb{R}^2/\mathbb{Z}^2$, driven by a 1-dimensional Brownian motion W:

$$d\omega_t + u_t \cdot \nabla \omega_t dt + \xi \cdot \nabla \omega_t \circ dW_t = 0 \tag{3}$$

where u and ω are as above and $\circ$ denotes Stratonovich integration. We impose the following condition on the stochastic parameter ξ, in the same spirit as (2):

$$\|\xi\|_{k+1,\infty}^2 < \infty \tag{4}$$

with $k > 4$. This condition ensures that for any $f \in \mathcal{W}^{2,2}(\mathbb{T}^2) \cap \mathcal{W}^{2,\infty}(\mathbb{T}^2)$,

$$\|\xi \cdot \nabla f\|_2^2 \leq C \|f\|_{1,2}^2 \qquad \|\xi \cdot \nabla(\xi \cdot \nabla f)\|_2^2 \leq C \|f\|_{2,2}^2 \tag{5}$$

$$\|\xi \cdot \nabla f\|_\infty^2 \leq C \|f\|_{1,\infty}^2 \qquad \|\xi \cdot \nabla(\xi \cdot \nabla f)\|_\infty^2 \leq C \|f\|_{2,\infty}^2. \tag{6}$$

Remark 1 We can view the stochastic part as a space-time noise (ξ, W) where the spatial component is given by ξ and the time component is a standard Brownian motion. This perspective is many times useful in numerical applications where $(\xi \circ dW_t) \cdot \nabla$ is implemented as a random operator applied to the solution ω.

The problem of parameter estimation, known also as *statistical inference*, is technically challenging for such (infinite-dimensional) SPDEs driven by transport noise, as most methods used in the literature benefit from a diagonalizable structure of the underlying space-covariance matrices. This structure is specific for additive noise and therefore it does not apply in our case. Also, most results are obtained for stochastic variations of the heat equation, which contain a smoothing Laplace operator (see for instance [3]). Our model does not contain a Laplacian a priori, and therefore we cannot exploit the properties of a heat kernel. These makes the analysis much harder.

Contributions of the Paper

In this work, we focus on Eq. (3) from two perspectives:

- First, we show that the driving stochastic parameter ξ can be calibrated in an optimal way to match a set of high-frequency in time given data. This is done using a forced and damped version of the equation and a parametric form of the stream function and the corresponding stochastic parameter which is implemented using an orthonormal basis. Our technique can be explicitly applied to calibrate the 2D Euler model using *real* oceanic data and we intend to do this in coming work.

- Second, we show that the original 2D Euler model is *robust* with respect to the stochastic parameters ξ in the sense that if we consider two couples (ω^1, ξ^1) and (ω^2, ξ^2) which solve Eq. (3), then the L^2 distance between ω^1 and ω^2 can be controlled using the initial conditions and the difference between ξ^1 and ξ^2 only (see Sect. 4). This is important in applications as it shows that if we consider approximate values for ξ, the corresponding model solution remains close to the true solution.

Structure of the Paper

In Sect. 2 below we present the problem formulation. In Sect. 3 we introduce the methodology. In Sect. 4 we prove the robustness of the original model and in Sect. 5 we present the numerical results.

2 Problem Formulation

Let $(\Omega, \mathcal{F}, (\mathcal{F}_t)_{t \geq 0}, \mathbb{P})$ be a filtered probability space and W a one-dimensional Brownian motion adapted to the complete and right-continuous filtration $(\mathcal{F}_t)_{t \geq 0}$.

Let $h : \mathbb{R} \to \mathbb{R}$ be a smooth function representing some observation map. We assume we have available a finite sequence of *high frequency* in time snapshots of observed vorticity fields, that are denoted by $h(\omega^*)_{t_i}(x) := h(\omega^*_{t_i})(x)$, $i = 1, \ldots, N$, and are adapted to $(\mathcal{F}_t)_{t \geq 0}$. We take the view that the $h(\omega^*)_{t_i}$'s are the given observation *data*. We further assume that $\omega^*_{t_i} \in \mathcal{W}^{k,2}(\mathbb{T}^2)$, $k > 4$.

Writing ω_ξ to denote solutions to the model (3) for a given vector field ξ, the generic problem we are interested in is to find a ξ so that solutions to (3) matches the data as best as possible, i.e.

$$\arg \min_{\xi} \|\omega^* - \omega_\xi\| \tag{7}$$

for some suitable norm.[1]

The dimension of the observations currently coincides with the number of sources of noise, that is we have a *determined* system. However, in practice this is not always a realistic assumption and in future work we will look at *underdetermined* or *overcomplete* systems i.e. when the number of noise sources is larger than the dimension of the observation operator.

In general, the infinite dimensional optimisation problem (7) may be too hard to solve in practice. We thus make concrete the form of ξ. Let $(\mathfrak{e}_j)_{j \in \mathbb{N}}$ be an orthonormal basis in $L^2(\mathbb{T}^2)$. We assume the following parametric form for the stream function of ξ, which is henceforth denoted by ζ,

$$\zeta(x) = \sum_{j=1}^{\infty} \alpha_j \mathfrak{e}_j, \tag{8}$$

where α_j are reals. Then

$$\xi(x) = \nabla^\perp \zeta(x) = \sum_{j=1}^{\infty} \alpha_j \nabla^\perp \mathfrak{e}_j(x) \tag{9}$$

and the optimisation problem (7) then reduces to finding the coefficients α_j.

3 Methodology

For a stochastic process X_t defined on a filtered probability space, its *quadratic variation* is defined by

$$[X]_t := \lim_{\max_j \Delta t_j \to 0} \sum_{i=1}^{n} |X_{t_i} - X_{t_{i-1}}|^2, \tag{10}$$

where $t_0 = 0 < t_1 < \cdots < t_n = t$ is a partition of the interval $[0, t]$, $\Delta t_i := |t_i - t_{i-1}|$, and the convergence is in the sense of probability (see e.g. [7]).

From (3) and (9) we have

$$\omega_t(x) = \omega_0(x) - \int_0^t B_s(x; \omega) \, ds - \int_0^t \sum_j \alpha_j \nabla^\perp \mathfrak{e}_j(x) \cdot \nabla \omega_t(x) \circ dW_t \tag{11}$$

in which for notation simplicity, we have introduced $B_s(x; \omega) := \mathbf{u}_s(x) \cdot \nabla \omega_s(x)$.

---

[1] By the assumed regularity of h, any solution to (7) is also a solution to $\arg \min_{\xi} \|h(\omega^*) - h(\omega_\xi)\|$.

Using Itô's lemma, and following standard results on the quadratic variation of semimartingales, it is straightforward to show that

$$[h(\omega)]_t = \sum_{i,j=1}^{\infty} \alpha_i \alpha_j \int_0^t \langle h'(\omega_s), \nabla^{\perp} \mathbf{e}_i \cdot \nabla \omega_s \rangle \langle h'(\omega_s), \nabla^{\perp} \mathbf{e}_j \cdot \nabla \omega_s \rangle \, ds. \quad (12)$$

Due to global existence and uniqueness of solutions to (3), $[h(\omega)]_t$ exists globally $\mathbb{P}$-almost surely. Thus the right hand side of (23) can be arbitrarily well approximated by its truncation for all t i.e. for a given $\epsilon > 0$, there exists M_ϵ such that

$$\left| [h(\omega)]_t - \sum_{i,j=1}^{M_\epsilon} \alpha_i \alpha_j \int_0^t \langle h'(\omega_s), \nabla^{\perp} \mathbf{e}_i \cdot \nabla \omega_s \rangle \langle h'(\omega_s), \nabla^{\perp} \mathbf{e}_j \cdot \nabla \omega_s \rangle \, ds \right| < \epsilon.$$
$$(13)$$

Additionally, from the computational perspective, for any fixed M_ϵ, the linear map

$$\mathbf{A}_{ij} := \int_0^t \langle h'(\omega_s), \nabla^{\perp} \mathbf{e}_i \cdot \nabla \omega_s \rangle \langle h'(\omega_s), \nabla^{\perp} \mathbf{e}_j \cdot \nabla \omega_s \rangle \, ds \quad (14)$$

that defines the truncated quadratic form is symmetric and positive definite,[2] and thus can be diagonalised by a unitary linear map. Doing so, we obtain the following linear problem

$$[h(\omega)]_t = \sum_{j=1}^{M_\epsilon} \tilde{\alpha}_j^2 \lambda_j + \epsilon', \quad (15)$$

where ϵ' denotes the truncation error of (23), λ_j are the eigenvalues of the associated linear map, and $\tilde{\alpha}_j$'s are the original α values which get rescaled by the unitary matrix from the diagonalisation.

We can estimate $[h(\omega)]_t$ using the high frequency in time data $h(\omega^*)$ and (10), assuming the discrete sum converges fast enough,

$$[h(\omega)]_t \approx \widehat{[h(\omega)]}_{t,N} := \sum_{i=1}^{N} |h(\omega^*)_{t_i} - h(\omega^*)_{t_{i-1}}|^2. \quad (16)$$

The estimate $\widehat{[h(\omega)]}_{t,N}$ could then be used in (15) to get an estimate for the $\tilde{\alpha}$. One could then recover the original α's by applying the unitary linear map that's associated with the diagonalisation of $\mathbf{A}_{ij}$.

---

[2] Since $[h(\omega)]_t$ is strictly positive.

Example 1 Let h be the identity map. Let $\mathfrak{e}_\kappa = e^{i\kappa \cdot x}$ be the Fourier basis. Then we have

$$\widehat{[\omega]}_{t,N} = \sum_{\substack{i,j=1 \\ \text{with } \kappa_i, \kappa_j \in \mathbb{Z}^2}}^{\infty} \alpha_i \alpha_j \int_0^t \mathfrak{e}_{\kappa_i} \mathfrak{e}_{\kappa_j} (\kappa_i^\perp \cdot \nabla \omega_s) \, (\kappa_j^\perp \cdot \nabla \omega_s) \, ds. \tag{17}$$

In Sect. 5 we test numerically Eq. (17) for an idealised example, and show we can adequately recover the basis coefficients using our methodology.

Example 2 In this example, we assume the data are the kinetic energy of the flow,

$$E_t := \frac{1}{2} \int_{\mathbb{T}^2} |\mathbf{u}_t|^2 dx. \tag{18}$$

Thus the data are "indirect" information about the vorticity. Note that the energy data is feasible for SALT models as energy is not a conserved quantity of SALT.

Below, we avoid calculating the pressure term of the Euler system by utilising the Biot-Savart operator K that links the velocity field to the vorticity field in Eq. (3). For further discussions on this topic see [9] or [10]. We have

$$\mathbf{u}(\mathbf{x}) = (K \star \omega)(\mathbf{x}) = \int_{\mathbb{T}^2} K(\mathbf{x} - \mathbf{y})\omega(\mathbf{y})d\mathbf{y} \tag{19}$$

where

$$K(\mathbf{x}) = \sum_{\kappa \in \mathbb{Z}^2 \setminus \{0\}} \frac{i\kappa^\perp}{\|\kappa\|^2} e^{i\kappa \cdot \mathbf{x}}. \tag{20}$$

It is known that, for any $k \geq 0$, there exists a constant $C_{k,2}$, that is independent of $\mathbf{u}$, and such that

$$\|\mathbf{u}\|_{k+1,2} \leq C_{k,2} \|\omega\|_{k,2}.$$

If $\psi : \mathbb{T}^2 \times [0, \infty) \to \mathbb{R}$ is a solution for $\Delta \psi = -\omega$ then $\mathbf{u} = \nabla^\perp \psi$ solves $\omega = \text{curl } \mathbf{u}$, so $\mathbf{u} = -\nabla^\perp \Delta^{-1} \omega$. The reconstruction of $\mathbf{u}$ from ω is ensured by the incompressibility condition $\nabla \cdot \mathbf{u} = 0$ and a periodic, distributional solution of $\Delta \psi = -\omega$ is given by

$$\psi(\mathbf{x}) = (G \star \omega)(\mathbf{x})$$

where G is the Green's function of the operator $-\Delta$ on $\mathbb{T}^2$

$$G(\mathbf{x}) = \sum_{\kappa \in \mathbb{Z}^2 \setminus \{0\}} \frac{e^{i\kappa \cdot \mathbf{x}}}{\|\kappa\|^2}$$

and $\kappa = (\kappa_1, \kappa_2)$, $\kappa^\perp = (\kappa_2, -\kappa_1)$.

Combining (11) with the Biot-Savart law (19) we obtain

$$\mathbf{u}_t(\mathbf{x}) = \mathbf{u}_0(\mathbf{x}) - \int_0^t \int_{\mathbb{T}^2} K(\mathbf{x} - \mathbf{y}) B_s(\mathbf{y}; \omega) d\mathbf{y} ds - \int_0^t \int_{\mathbb{T}^2} K(\mathbf{x} - \mathbf{y}) \xi(\mathbf{y}) \cdot \nabla \omega_s(\mathbf{y}) d\mathbf{y} \circ dW_s \tag{21}$$

Using Itô's lemma, we obtain

$$E_t - E_0 = -\int_0^t \langle \mathbf{u}_s, K \star (B_s - \frac{1}{2}\xi \cdot \nabla(\xi \cdot \nabla \omega_s)) \rangle ds - \int_0^t \langle \mathbf{u}_s, K \star (\xi \cdot \nabla \omega_s) \rangle dW_s \tag{22}$$

where $\langle \cdot, \cdot \rangle$ is the standard $L^2(\mathbb{T}^2)$ pairing. Thus

$$[E]_t = \sum_{i,j=1}^\infty \alpha_i \alpha_j \int_0^t \langle \mathbf{u}_s, K \star (\nabla^\perp \mathbf{e}_j \cdot \nabla \omega_s) \rangle \langle \mathbf{u}_s, K \star (\nabla^\perp \mathbf{e}_i \cdot \nabla \omega_s) \rangle ds. \tag{23}$$

4 Robustness

Theorem 2 *Let ω^1, ω^2 be two solutions of the 2D Euler equation (3) and ξ^1, ξ^2 the corresponding stochastic parameters for each of these two solutions. More precisely, (ω^ℓ, ξ^ℓ) for $\ell = 1, 2$ solves*

$$d\omega_t^\ell + u_t^\ell \cdot \nabla \omega_t^\ell dt + \xi^\ell \cdot \nabla \omega_t^\ell dW_t = \frac{1}{2}\xi^\ell \cdot \nabla \left(\xi^\ell \cdot \nabla \omega_t^\ell \right). \tag{24}$$

Then for any $p \geq 2$ there exist some constants[3] $C = C(p, T), C_{1,p}, C_{2,p}$, such that

$$\mathbb{E}\left[\sup_{t \in [0,T]} e^{-\gamma(t)} \|\omega_t^1 - \omega_t^2\|_2^{2p} \right] \leq C_{p,T} \left(\|\omega_0^1 - \omega_0^2\|_2^{2p} + \|\xi^1 - \xi^2\|_2^{2p} + \|\xi^1 - \xi^2\|_{1,2}^{2p} \right) \tag{25}$$

where

---

[3] In this theorem all constants generically denoted by $C, C_{p,T}, C_{1,p}, C_{2,p}, \tilde{C}$ may differ from line to line and from term to term.

$$\gamma(t) := C_{1,p} \int_0^t \|\omega_r^1\|_{k,2}^2 dr + C_{2,p} t$$

and $k > 4$.

Proof of Theorem 2 Let $\bar{\omega} := \omega^1 - \omega^2, \bar{u} = u^1 - u^2, \bar{\xi} = \xi^1 - \xi^2$. Then $\bar{\omega}$ satisfies

$$d\bar{\omega}_t + (\bar{u}_t \cdot \nabla \omega_t^1 + u_t^2 \cdot \nabla \bar{\omega}_t) dt + \left(\xi^1 \cdot \nabla \omega_t^1 - \xi^2 \cdot \nabla \omega_t^2 \right) dW_t$$

$$= \frac{1}{2} \left(\xi^1 \cdot \nabla(\xi^1 \cdot \nabla \omega_t^1) - \xi^2 \cdot \nabla(\xi^2 \cdot \nabla \omega_t^2) \right) dt.$$

By the Itô formula:

$$d\|\bar{\omega}_t\|_2^2 = -2\langle \bar{\omega}_t, \xi^1 \cdot \nabla \omega_t^1 - \xi^2 \cdot \nabla \omega_t^2 \rangle dW_t - 2\langle \bar{\omega}_t, \bar{u}_t \cdot \nabla \omega_t^1 + u_t^2 \cdot \nabla \bar{\omega}_t \rangle dt$$

$$+ \left(\langle \bar{\omega}_t, \xi^1 \cdot \nabla(\xi^1 \cdot \nabla \omega_t^1) - \xi^2 \cdot \nabla(\xi^2 \cdot \nabla \omega_t^2) \rangle \right. \tag{26}$$

$$\left. + \langle \xi^1 \cdot \nabla \omega_t^1 - \xi^2 \cdot \nabla \omega_t^2, \xi^1 \cdot \nabla \omega_t^1 - \xi^2 \cdot \nabla \omega_t^2 \rangle \right) dt.$$

We make the following notations

$$m_t := \|\bar{\omega}_t\|_2^2 \qquad Z := \|\bar{\xi}\|_2^2 \qquad L := \|\bar{\xi}\|_{1,2}^2$$

$$A_t := -2\langle \bar{\omega}_t, \bar{u}_t \cdot \nabla \omega_t^1 + u_t^2 \cdot \nabla \bar{\omega}_t \rangle + \langle \bar{\omega}_t, \xi^1 \cdot \nabla(\xi^1 \cdot \nabla \omega_t^1) - \xi^2 \cdot \nabla(\xi^2 \cdot \nabla \omega_t^2) \rangle$$

$$+ \langle \xi^1 \cdot \nabla \omega_t^1 - \xi^2 \cdot \nabla \omega_t^2, \xi^1 \cdot \nabla \omega_t^1 - \xi^2 \cdot \nabla \omega_t^2 \rangle$$

$$D_t := \int_0^t \langle \bar{\omega}_s, \xi^1 \cdot \nabla \omega_s^1 - \xi^2 \cdot \nabla \omega_s^2 \rangle dW_s$$

$$\phi(t) := C\|\omega_t^1\|_{k,2}^2 + C$$

$$\psi(t) := (C\|\omega_t^1\|_{k,2}^2 + C)Z + C\|\omega_t^1\|_{k,2}^2 L$$

$$\tilde{Z} := C\|\omega_t^1\|_{k,2}^2 Z$$

Then we can write (26) as

$$dm_t = A_t dt - 2dD_t$$

We want to estimate each of the terms which appear in (26). The difference of the nonlinear terms is analysed explicitly in [8] pp. 9:

$$\langle \bar{\omega}_t, \bar{u}_t \cdot \nabla \omega_t^1 \rangle \leq \|\bar{\omega}_t\|_2 \|\bar{u}_t\|_4 \|\nabla \omega_t^1\|_4 \leq C\|\bar{\omega}_t\|_2^2 \|\omega_t^1\|_{k,2} = C\|\omega_t^1\|_{k,2}^2 m_t$$

We used here that $\|\nabla\omega_t^1\|_4 \leq C\|\omega_t^1\|_{k,2}$ and $\|\bar{u}_t\|_4 \leq C\|\bar{u}_t\|_{1,2} \leq C\|\bar{\omega}_t\|_2$. Also, since u^2 is divergence-free, $\langle\bar{\omega}_t, u_t^2 \cdot \nabla\bar{\omega}_t\rangle = -\frac{1}{2}\int_{\mathbb{T}^2}(\nabla \cdot u_t^2)(\bar{\omega}_t)^2 dx = 0$. We estimate the difference terms which include ξ^1 and ξ^2 in Lemma 3 below. Note here that the term $\langle\bar{\omega}_t, \xi^2 \cdot \nabla\left(\xi^2 \cdot \nabla\bar{\omega}_t\right)\rangle$ is negative. Using these estimates and Lemma 3 below we have that

$$A_t dt \leq \psi(t)dt + \phi(t)m_t dt.$$

Then

$$d\left(e^{-\int_0^t \phi(s)ds}m_t\right) = e^{-\int_0^t \phi(s)ds}(dm_t - \phi(t)m_t dt)$$

$$\leq e^{-\int_0^t \phi(s)ds}(\psi(t)dt - 2dD_t).$$

After raising everything to the power $p \geq 2$,[4] taking the supremum over $t \in [0, T]$ and then the expectation, we obtain

$$\mathbb{E}\left[\sup_{t\in[0,T]}\left(e^{-\int_0^t \phi(s)ds}m_t\right)^p\right] \leq C_p m_0^p + C_p\mathbb{E}\left[\sup_{t\in[0,T]}\left|\int_0^t e^{-\int_0^s \phi(r)dr}\psi(s)ds\right|^p\right]$$

$$+ C_p\mathbb{E}\left[\sup_{t\in[0,T]}\left|\int_0^t e^{-\int_0^s \phi(r)dr}dD_s\right|^p\right]$$

$$(27)$$

For the stochastic integral we use the Burkholder-Davis-Gundy inequality: for arbitrary $p \geq 2$ and a martingale M_t there exists a constant C_p such that[5]

$$\mathbb{E}\left[\sup_{t\in[0,T]}|M_t|^p\right] \leq C_p\mathbb{E}\left[[M]_T^{p/2}\right]$$

where $[M]_t$ is the quadratic variation of the martingale M_t. In our case

---

[4] We use here and below that $|a + b|^p \leq 2^{p-1}(|a|^p + |b|^p)$, $p \geq 2$.

[5] In this proof C, C_p are generic constants which may differ from line to line and from term to term.

$$M_t := \int_0^t e^{-\int_0^s \phi(r)dr} \, dD_s$$

and then

$$[M]_t = \int_0^t e^{-2\int_0^s \phi(r)dr} \, d[D]_s = \int_0^t e^{-2\int_0^s \phi(r)dr} |\langle \bar{\omega}_s, \xi^1 \cdot \nabla \omega_s^1 - \xi^2 \cdot \nabla \omega_s^2 \rangle|^2 ds.$$

Therefore[6]

$$\mathbb{E}\left[[M]_T^{p/2}\right] = \mathbb{E}\left[\left(\int_0^T e^{-2\int_0^s \phi(r)dr} |\langle \bar{\omega}_s, \xi^1 \cdot \nabla \omega_s^1 - \xi^2 \cdot \nabla \omega_s^2 \rangle|^2 ds\right)^{p/2}\right]$$

$$\leq C_{p,T}\mathbb{E}\left[\int_0^T e^{-p\int_0^s \phi(r)dr} |\langle \bar{\omega}_s, \xi^1 \cdot \nabla \omega_s^1 - \xi^2 \cdot \nabla \omega_s^2 \rangle|^p ds\right]$$

$$\leq C_{p,T} \int_0^T \mathbb{E}\left[\sup_{r\in[0,s]} e^{-p\int_0^r \phi(q)dq} \left(m_r^p + \tilde{Z}^p\right)\right] ds.$$

Using these estimates in (27) we obtain

$$\mathbb{E}\left[\sup_{t\in[0,T]}\left(e^{-\int_0^t \phi(s)ds} m_t\right)^p\right] \leq C_p m_0^p + C_{p,T}\mathbb{E}\left[\sup_{t\in[0,T]}\int_0^t e^{-p\int_0^s \phi(r)dr} \psi(s)^p ds\right]$$

$$+ C_{p,T} \int_0^T \mathbb{E}\left[\sup_{r\in[0,s]} e^{-p\int_0^r \phi(q)dq} \left(m_r^p + \tilde{Z}^p\right)\right] ds$$

(28)

For the second term on the right hand side of (28) we use that, since Z is deterministic and by [8] the 2D Euler equation (3) has a unique global solution in $\mathcal{W}^{k,2}(\mathbb{T}^2)$ for $k \geq 2$, there exist $\tilde{C}_p^1, \tilde{C}_p^2$ such that for all $t \in [0, T]$

---

[6] We use here the control obtained for Q in Lemma 3. More precisely: since $Q \leq Cm_t + \tilde{Z}$ then $Q^p \leq C_p(m_t^p + \tilde{Z}^p)$.

$$\mathbb{E}\left[\sup_{s\in[0,t]}\psi(s)^p\right] = \mathbb{E}\left[\sup_{s\in[0,t]}\left((C\|\omega_s^1\|_{k,2}^2 + C)Z + C\|\omega_s^1\|_{k,2}^2 L\right)^p\right]$$

$$\leq Z^p\mathbb{E}\left[\sup_{s\in[0,t]}(C\|\omega_s^1\|_{k,2}^2 + C)^p\right] + L^p\mathbb{E}\left[\sup_{s\in[0,t]}(C\|\omega_s^1\|_{k,2}^2)^p\right]$$

$$\leq \tilde{C}_p^1 Z^p + \tilde{C}_p^2 L^p.$$

The same argument is used to control $\displaystyle\int_0^T \mathbb{E}\left[\sup_{r\in[0,s]} e^{-p\int_0^r \phi(q)dq}\tilde{Z}^p\right] ds$ in the

third term of (28). Then

$$\mathbb{E}\left[\sup_{t\in[0,T]}\left(e^{-\int_0^t \phi(s)ds} m_t\right)^p\right] \leq C_{p,T}^1(m_0^p + Z^p + L^p)$$

$$+ C_{p,T}^2\int_0^T \mathbb{E}\left[\sup_{r\in[0,s]}\left(e^{-\int_0^r \phi(q)dq} m_r\right)^p\right] ds.$$

Then by Gronwall lemma

$$\mathbb{E}\left[\sup_{t\in[0,T]}\left(e^{-\int_0^t \phi(s)ds} m_t\right)^p\right] \leq e^{\int_0^T C_{p,T}^2 ds}\left(m_0^p + \int_0^T C_{p,T}^1(m_0^p + Z^p + L^p)ds\right)$$

$$\leq e^{C(T)}\left(m_0^p + T(C_{p,T}^1(m_0^p + Z^p + L^p))\right).$$

So we finally obtain that

$$\mathbb{E}\left[\sup_{t\in[0,T]} e^{-\gamma(t)}\|\omega_t^1 - \omega_t^2\|_2^{2p}\right]$$

$$\leq C_{p,T}\left(\|\omega_0^1 - \omega_0^2\|_2^{2p} + \|\xi^1 - \xi^2\|_2^{2p} + \|\xi^1 - \xi^2\|_{1,2}^{2p}\right), \quad p \geq 2$$

where

$$\gamma(t) := p\int_0^t \phi(r)dr.$$

$\square$

Lemma 3 *Let (ω_t^1, ξ^1) and (ω_t^2, ξ^2) be two solutions of the 2D Euler equation with $\bar{\omega}_t := \omega_t^1 - \omega_t^2$ and $\bar{\xi} := \xi^1 - \xi^2$. Then there exist constants C[7] such that the following estimates hold:*

$$Q := |\langle \bar{\omega}_t, \xi^1 \cdot \nabla \omega_t^1 - \xi^2 \cdot \nabla \omega_t^2 \rangle| \leq C \|\bar{\omega}_t\|_2^2 + C \|\omega_t^1\|_{k,2}^2 \|\bar{\xi}\|_2^2.$$

$$A := \langle \xi^1 \cdot \nabla \omega_t^1 - \xi^2 \cdot \nabla \omega_t^2, \xi^1 \cdot \nabla \omega_t^1 - \xi^2 \cdot \nabla \omega_t^2 \rangle \leq C \|\bar{\omega}_t\|_2^2 + C \|\omega_t^1\|_{k,2}^2 \|\bar{\xi}\|_2^2$$

$$|B| \leq (C \|\omega_t^1\|_{k,2}^2 + C) \|\bar{\omega}_t\|_2^2 + C \|\bar{\xi}\|_2^2 + C \|\omega_t^1\|_{k,2}^2 \|\bar{\xi}\|_{1,2}^2$$

where

$$B := \langle \bar{\omega}_t, \xi^1 \cdot \nabla \left(\xi^1 \cdot \nabla \omega_t^1 \right) - \xi^2 \cdot \nabla \left(\xi^2 \cdot \nabla \omega_t^2 \right) \rangle.$$

and $k > 4$.

Proof For the difference terms which include ξ^1 and ξ^2 we use that

$$\xi^1 \cdot \nabla \omega_t^1 - \xi^2 \cdot \nabla \omega_t^2 = \bar{\xi} \cdot \nabla \omega_t^1 + \xi^2 \cdot \nabla \bar{\omega}_t.$$

We have

$$Q = |\langle \bar{\omega}_t, \xi^1 \cdot \nabla \omega_t^1 - \xi^2 \cdot \nabla \omega_t^2 \rangle|$$

$$\leq |\langle \omega_t^1 - \omega_t^2, (\xi^1 - \xi^2) \cdot \nabla \omega_t^1 \rangle| + |\langle \omega_t^1 - \omega_t^2, \xi^2 \cdot \nabla (\omega_t^1 - \omega_t^2) \rangle|$$

$$\leq \frac{1}{2} \|\omega_t^1 - \omega_t^2\|_2^2 + \frac{1}{2} \|\nabla \omega_t^1\|_\infty^2 \|\xi^1 - \xi^2\|_2^2$$

$$\leq \frac{1}{2} \|\bar{\omega}_t\|_2^2 + \frac{C}{2} \|\omega_t^1\|_{k,2}^2 \|\bar{\xi}\|_2^2$$

with $k \geq 3$, since the second scalar product is zero due to the fact that $\nabla \cdot \xi^2 = 0$. Also

$$A = \langle \xi^1 \cdot \nabla \omega_t^1 - \xi^2 \cdot \nabla \omega_t^2, \xi^1 \cdot \nabla \omega_t^1 - \xi^2 \cdot \nabla \omega_t^2 \rangle = \|\xi^1 \cdot \nabla \omega_t^1 - \xi^2 \cdot \nabla \omega_t^2\|_2^2$$

$$\leq \|(\xi^1 - \xi^2) \cdot \nabla \omega_t^1\|_2^2 + \|\xi^2 \cdot \nabla (\omega_t^1 - \omega_t^2)\|_2^2$$

$$\leq \|\xi^1 - \xi^2\|_2^2 \|\nabla \omega_t^1\|_\infty^2 + C \|\omega_t^1 - \omega_t^2\|_2^2$$

$$\leq C \|\omega_t^1\|_{k,2}^2 \|\bar{\xi}\|_2^2 + C \|\bar{\omega}_t\|_2^2$$

where $k \geq 3$. For the higher order term we have

---

[7] C differs from line to line and from term to term depending on the Sobolev embedding we use.

$$B = \langle \omega_t^1 - \omega_t^2, \xi^1 \cdot \nabla \left(\xi^1 \cdot \nabla \omega_t^1 \right) - \xi^2 \cdot \nabla \left(\xi^2 \cdot \nabla \omega_t^2 \right) \rangle$$

$$= \langle \omega_t^1 - \omega_t^2, (\xi^1 - \xi^2) \cdot \nabla (\xi^1 \cdot \nabla \omega_t^1) \rangle$$

$$+ \langle \omega_t^1 - \omega_t^2, \xi^2 \cdot \nabla \left((\xi^1 - \xi^2) \cdot \nabla \omega_t^1 \right) \rangle$$

$$+ \langle \omega_t^1 - \omega_t^2, \xi^2 \cdot \nabla \left(\xi^2 \cdot \nabla (\omega_t^1 - \omega_t^2) \right) \rangle$$

$$=: a + b + c.$$

Note that c is negative:

$$\langle \omega_t^1 - \omega_t^2, \xi^2 \cdot \nabla \left(\xi^2 \cdot \nabla (\omega_t^1 - \omega_t^2) \right) \rangle = -\langle \xi^2 \cdot \nabla (\omega_t^1 - \omega_t^2), \xi^2 \cdot \nabla (\omega_t^1 - \omega_t^2) \rangle$$

$$= -\| \xi^2 \cdot \nabla (\omega_t^1 - \omega_t^2) \|_2^2$$

$$\leq 0$$

so $|B| \leq |a| + |b|$. We estimate $|a|$ as follows:

$$|a| = |\langle \omega_t^1 - \omega_t^2, (\xi^1 - \xi^2) \cdot \nabla (\xi^1 \cdot \nabla \omega_t^1) \rangle| \leq \frac{1}{2} \| \nabla (\xi^1 \cdot \nabla \omega_t^1) \|_\infty^2 \| \omega_t^1 - \omega_t^2 \|_2^2 + \frac{1}{2} \| \xi^1 - \xi^2 \|_2^2$$

$$\leq \frac{C}{2} \| \omega_t^1 \|_{2,\infty}^2 \| \omega_t^1 - \omega_t^2 \|_2^2 + \frac{1}{2} \| \xi^1 - \xi^2 \|_2^2$$

$$\leq \frac{C}{2} \| \omega_t^1 \|_{k,2}^2 \| \bar{\omega}_t \|_2^2 + \frac{1}{2} \| \bar{\xi} \|_2^2$$

with $k > 4$. Likewise, we estimate $|b|$:

$$|b| = |\langle \omega_t^1 - \omega_t^2, \xi^2 \cdot \nabla \left((\xi^1 - \xi^2) \cdot \nabla \omega_t^1 \right) \rangle| \leq \frac{1}{2} \| \omega_t^1 - \omega_t^2 \|_2^2 + \frac{1}{2} \| \xi^2 \cdot \nabla \left((\xi^1 - \xi^2) \cdot \nabla \omega_t^1 \right) \|_2^2$$

$$=: \frac{1}{2} \| \omega_t^1 - \omega_t^2 \|_2^2 + \frac{1}{2} \mathcal{K}.$$

Now

$$\mathcal{K} \leq \| \xi^2 \cdot \nabla (\xi^1 - \xi^2) \cdot \nabla \omega_t^1 \|_2^2 + \| \xi^2 \cdot (\xi^1 - \xi^2) \cdot \nabla (\nabla \omega_t^1) \|_2^2 := \mathcal{K}_1 + \mathcal{K}_2$$

where

$$\mathcal{K}_1 \leq \| \xi^2 \cdot \nabla \omega_t^1 \|_\infty^2 \| \nabla (\xi^1 - \xi^2) \|_2^2$$

$$\leq C \| \omega_t^1 \|_{1,\infty}^2 \| \xi^1 - \xi^2 \|_{1,2}^2$$

$$\leq C \| \omega_t^1 \|_{k,2}^2 \| \xi^1 - \xi^2 \|_{1,2}^2$$

and

$$\mathcal{K}_2 \leq C \|\xi^2 \cdot \nabla(\nabla \omega_t^1)\|_4^2 \|\xi^1 - \xi^2\|_4^2$$

$$\leq C \|\xi^2 \cdot \nabla(\nabla \omega_t^1)\|_{1,2}^2 \|\xi^1 - \xi^2\|_{1,2}^2$$

$$\leq C \|\omega_t^1\|_{k,2}^2 \|\xi^1 - \xi^2\|_{1,2}^2$$

for $k > 4$. Then

$$\mathcal{K} \leq 2C \|\omega_t^1\|_{k,2}^2 \|\bar{\xi}\|_{1,2}^2$$

and therefore

$$|b| \leq \frac{1}{2} \|\bar{\omega}_t\|_2^2 + C \|\omega_t^1\|_{k,2}^2 \|\bar{\xi}\|_{1,2}^2$$

which gives

$$|B| \leq \left(\frac{C}{2} \|\omega_t^1\|_{k,2}^2 + \frac{1}{2} \right) \|\bar{\omega}_t\|_2^2 + \frac{1}{2} \|\bar{\xi}\|_2^2 + C \|\omega_t^1\|_{k,2}^2 \|\bar{\xi}\|_{1,2}^2.$$

$\square$

5 Numerical Results

In this section, we show the results we obtained for Example 1 in Sect. 3. We implemented the main equation (3) with added forcing and damping, on a unit square domain with doubly periodic boundary conditions,

$$d\omega_t + \mathbf{u}_t \cdot \nabla \omega_t dt + \xi \cdot \nabla \omega_t \circ dW_t = (Q - r\omega_t)dt \tag{29}$$

where we chose $r = 0.001$ and $Q(x) = 0.01(\cos(8\pi y) + \sin(8\pi x))$. Note that, since the added forcing term is of bounded variation, (17) is unchanged for (29).

We considered a ξ whose parametric form with respect to the Fourier basis consists of only one α. The stream function of our chosen ξ is given by

$$\zeta(x, y) = \alpha \left(\cos(k_1 2\pi x) \cos(k_2 2\pi y) - \sin(k_1 2\pi x) \sin(k_2 2\pi y) \right). \tag{30}$$

Note that

$$\zeta = \frac{\alpha}{2} (e^{i2\pi k \cdot x} + e^{-i2\pi k \cdot x}), \tag{31}$$

and

$$\xi = i\alpha\pi (e^{i2\pi k \cdot x} - e^{-i2\pi k \cdot x})k^{\perp}. \tag{32}$$

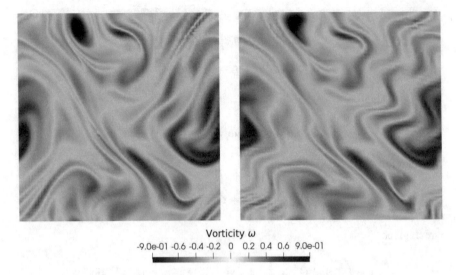

Vorticity ω

-9.0e-01 -0.6 -0.4 -0.2 0 0.2 0.4 0.6 9.0e-01

Fig. 1 Snapshots of the numerical solution $\omega(t, x)$ to (29) at times $t = 0$ (left), and $t = 1$ (right)

To discretise (29), we followed the methods documented in [4]—a mixed Finite Element method was used for the spatial derivatives, and an explicit strong stability preserving Runge-Kutta scheme of order 3 was used for the time derivative. We added the forcing and damping terms to help with maintaining the statistical homogeneity of the numerical solution, once it has reached a spun-up state from some set initial state. Our choice for the set initial state was

$$\omega(0, x, y) = \sin(8\pi x)\sin(8\pi y) + 0.4\cos(6\pi x)\cos(6\pi y)$$
$$+ 0.3\cos(10\pi x)\cos(4\pi y) + 0.02\sin(2\pi y) + 0.02\sin(2\pi x).$$
$$(33)$$

Spatially, we chose the grid size 64×64 cells. We first spun-up the system until it reached a statistical equilibrium state. This statistical equilibrium state was then set as the initial condition for our experiment. Figure 1 shows a snapshot of the obtained initial condition. Over the spin-up phase, we used $\alpha = 0.000001$ and $k^\mathsf{T} = (2, 4)$.

The time horizon for the experiment data was chosen to be the unit interval, i.e. we generated data $\omega^*(t_i, x)$ for $0 = t_0 < t_1 < \cdots < t_N = 1$. See Fig. 1 for snapshots of $\omega^*(0, \mathbf{x})$ and $\omega^*(1, \mathbf{x})$. When generating the data, we used the larger value of $\alpha = 0.001$. This was to avoid any possible numerical issues[8] when we attempted to recover α from data.

Assuming we know in-advance the exact Fourier wavenumber k, the linear system for estimation reduces to

---

[8] When α is small, α^2 is close to machine precision.

$$[\widehat{\omega}]_{t,N}(\mathbf{x}) := \sum_{i=1}^{N} (\omega_{t_i}(\mathbf{x}) - \omega_{t_{i-1}}(\mathbf{x}))^2 \approx \alpha^2 4\pi^2 \, B(t,k,\mathbf{x}) \mathfrak{e}'_k(\mathbf{x}) \tag{34}$$

where

$$B(t,k,\mathbf{x}) := \int_0^t (k^\perp \cdot \nabla \omega_s(\mathbf{x}))^2 ds \tag{35}$$

and

$$\mathfrak{e}'_k(x,y) := (\cos(k_1 2\pi x)\sin(k_2 2\pi y) + \sin(k_1 2\pi x)\cos(k_2 2\pi y))^2 . \tag{36}$$

Thus our estimate for α is given by

$$\widehat{\alpha}_N^2 = \frac{1}{4\pi^2} \frac{\int_{\mathbb{T}^2} [\widehat{\omega}]_{t,N}(\mathbf{x}) dx}{\int_{\mathbb{T}^2} B(t,k,\mathbf{x}) \mathfrak{e}'_k(\mathbf{x}) dx} . \tag{37}$$

Remark 4 In (37), we applied spatial averaging to stabilise estimation.

Remark 5 The assumption that we know k in advance is of course too strong from the applications viewpoint. The aim of this experiment is to test the strength of the pathwise approach under the assumption of "perfect knowledge". If we cannot accurately recover α in this case, then getting a good estimate for α using the pathwise approach may be too difficult or impractical in more realistic scenarios.

Figure 2 shows snapshots of $[\widehat{\omega}]_{t,N}(\mathbf{x})$ and $B(t,k,\mathbf{x}) \mathfrak{e}'_k(\mathbf{x})$. We applied (37) for different values of N. In each case, the time integral that constitutes $B(t,k,\mathbf{x})$ was approximated using a simple trapezoidal rule, for which the same N number of data snapshots were used. Figure 3 shows the results for the relative error

$$\mathrm{err}_N = \frac{|\alpha - \widehat{\alpha}_N|}{\alpha} \tag{38}$$

for the different values of N. The results show that, in the worst case of $N = 2500$, the relative error was no greater than 0.89. This translates to an absolute error of range of 0.001 ± 0.00089. The best case was when all 200,000 data samples were used to estimate α, the relative error in that case was 0.00135. This suggests convergence and stabilisation of the sum for $[\widehat{\omega}]_t$.

For future work, we aim to test the pathwise approach for cases in which we do not know the exact selection of basis elements for ξ. Further, we wish to extend and test these ideas on coarse grained PDE data and compare with the results that were obtained in [4] using previously developed calibration methods.

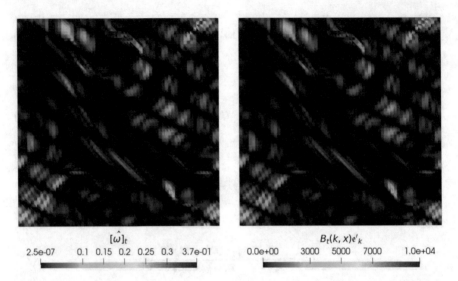

Fig. 2 Shown on the left is a snapshot of the estimate $[\widehat{\omega}]_t$, which was computed using $N = 200{,}000$ data samples. Shown on the right is a snapshot of the basis element $B_t(k, x) (\cos(k_1 2\pi x) \sin(k_2 2\pi y) + \sin(k_1 2\pi x) \cos(k_2 2\pi y))^2$, which was approximated using the same N number of data samples

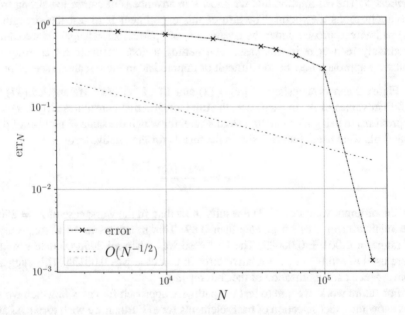

Fig. 3 The plot (in log log scale) shows the relative error err_N defined in (38) as a function of N. err_N was computed for $N = 2500, 5000, 10{,}000, 20{,}000, 40{,}000, 50{,}000, 66{,}667, 100{,}000, 200{,}000$

Acknowledgments The authors would like to thank Prof Dan Crisan for the many helpful suggestions and constructive ideas he shared with them during the preparation of this work. They also thank Prof Darryl Holm, Prof Bertrand Chapron, Prof Etienne Mémin, and the whole STUOD team for many inspiring discussions they had during the STUOD meetings.

Funding
Both authors were partially supported by the European Research Council (ERC) under the European Union's Horizon 2020 Research and Innovation Programme (ERC, Grant Agreement No 856408).

Appendix

Lemma 6 (Gronwall Lemma) *Let* $\beta : [0, T] \rightarrow [0, \infty)$ *be a non-negative absolutely continuous function that satisfies for a.e. t*

$$d\beta(t) \leq \phi(t)\beta(t)dt + \psi(t)dt$$

where ϕ, ψ *are non-negative integrable functions on* $[0, T]$. *Then*

$$\beta(t) \leq e^{\int_0^t \phi(s)ds} \left(\beta(0) + \int_0^t \psi(s)ds \right)$$

for all $t \in [0, T]$.

References

1. Brecht, R., Li, L., Bauer, W., Mémin, E. (2021), Rotating shallow water flow under location uncertainty with a structure-preserving discretization, Journal of Advances in Modeling Earth Systems, 13, e2021MS002492, https://doi.org/10.1029/2021MS002492
2. Browning AP, Warne DJ,Burrage K, Baker RE, Simpson MJ, Identifiability analysis for stochastic differential equation models in systems biology. J. R. Soc. Interface 17: 20200652. (2020) https://doi.org/10.1098/rsif.2020.0652
3. Chong, C., High-frequency analysis of parabolic stochastic PDES, The Annals of Statistics 2020, Vol. 48, No. 2, 1143–1167.
4. Cotter, C., Crisan, D., Holm, D., Pan, W., Shevchenko, I., 2019. Numerically modelling stochastic Lie transport in fluid dynamics. Multiscale Model. Simul. 17, 192–232.
5. Cotter, C., Crisan, D., Holm, D., Pan, W., Shevchenko, I., 2020a. A particle filter for stochastic advection by Lie transport (SALT): A case study for the damped and forced incompressible 2D Euler equation. SIAM/ASA J. Uncertain. Quantif. 8, 1446–1492.
6. Holm, D., Variational principles for stochastic fluid dynamics, Proc. R.Soc.A 471:20140963, 2015.
7. Karatzas, I., Shreve, S.E., Brownian Motion and Stochastic Calculus, Springer, 1998, ISBN 0-387-97655-8.

8. Lang, O., Crisan, D., Well-posedness for a stochastic 2D Euler equation with transport noise, Stoch PDE: Anal Comp (Jan 2022), https://doi.org/10.1007/s40072-021-00233-7.
9. Majda, A., Bertozzi, A., Vorticity and Incompressible Flow, Cambridge University Press, 2007, ISBN 0 521 63948 4 paperback.
10. Marchioro, C., Pulvirenti, M., Mathematical Theory of Incompressible Nonviscous Fluids, Springer-Verlag, 1994, 978-0387940441.
11. Mémin, E., Fluid dynamics under location uncertainty, Journal Geophysical & Astrophysical Fluid Dynamics, Volume 108 (2014).
12. Valentin Resseguier, Long Li, Gabriel Jouan, Pierre Dérian, Etienne Mémin, and Bertrand Chapron, New trends in ensemble forecast strategy: uncertainty quantification for coarse-grid computational fluid dynamics, Archives of Computational Methods in Engineering, 28(1):215–261, 2021.

Stochastic Parameterization with Dynamic Mode Decomposition

Long Li, Etienne Mémin, and Gilles Tissot

Abstract A physical stochastic parameterization is adopted in this work to account for the effects of the unresolved small-scale on the large-scale flow dynamics. This random model is based on a stochastic transport principle, which ensures a strong energy conservation. The dynamic mode decomposition (DMD) is performed on high-resolution data to learn a basis of the unresolved velocity field, on which the stochastic transport velocity is expressed. Time-harmonic property of DMD modes allows us to perform a clean separation between time-differentiable and time-decorrelated components. Such random scheme is assessed on a quasi-geostrophic (QG) model.

Keywords Stochastic parameterization · Dynamical system · Data-driven

1 Introduction

The modelling under location uncertainty (LU) setting has shown to provide consistent physical representations of fluid dynamics [10, 12]. This representation introduces a random component to describe the unresolved flow components. This enables to consider less dissipative systems than the classical large-scale counterparts. Nevertheless, the ability of such a model to represent faithfully the uncertainties associated to the actual unresolved small scales highly depends on the definition of the random component and on its evolution along time. Unsurprisingly, stationarity/time-varying and homogeneity/inhomogeneity characteristics have strong influences on the results [1, 2]. Another important aspect concerns the ability to include in the noise representation a stationary drift component associated to the temporal mean of the high-resolution fluctuations. As shown in this paper such stationary drift can be elegantly introduced in the noise through Girsanov theorem.

L. Li (✉) · E. Mémin · G. Tissot
Inria Rennes - Bretagne Atlantique, Campus de Beaulieu, Rennes, France
e-mail: long.li@inria.fr

© The Author(s) 2023   179
B. Chapron et al. (eds.), *Stochastic Transport in Upper Ocean Dynamics*,
Mathematics of Planet Earth 10, https://doi.org/10.1007/978-3-031-18988-3_11

Yet, large-scale persistent components associated to the high resolution fluctuations are not strictly stationary and slowly varying quasi-periodic components might be important to include. To that purpose we devise a noise generation scheme relying on the dynamic mode decomposition [13]. Such a decomposition or other related techniques aiming to provide a spectral representation of the Koopman operator [11] will allow us to represent the noise as a superposition of random and deterministic harmonics oscillators. The first ones are attached to the fast components whereas the latter represent the slow fluctuations components. As demonstrated in Sect. 4, this strategy brings us a very efficient technique for ocean double-gyres configuration.

2 Modelling Under Location Uncertainty

In this section, we briefly review the LU setting and the associated random QG model that will be used for the numerical evaluations.

2.1 Stochastic Flow

The evolution of Lagrangian particle trajectory (X_t) under LU is described by the following stochastic differential equation (SDE):

$$dX_t(x) = v\big(X_t(x), t\big)\, dt + \sigma\big(X_t(x), t\big)\, dB_t, \qquad X_0(x) = x \in \mathcal{D}, \qquad (1)$$

where v denotes the time-smooth resolved velocity that is both spatially and temporally correlated, σdB_t stands for the fast oscillating unresolved flow component (also called *noise* in the following) that is only correlated in space, and $\mathcal{D} \subset \mathbb{C}^d$ ($d = 2$ or 3) is a bounded spatial domain.

We now give the mathematical definitions of the noise. In the following, let us fix a finite time $T < \infty$ and the Hilbert space $H = (L^2(\mathcal{D}))^d$ with the inner product $\langle f, g \rangle_H = \int_{\mathcal{D}} (f^\dagger g)(x)\, dx$ and the norm $\|f\|_H = \langle f, f \rangle_H^{1/2}$, where $\bullet^\dagger$ stands for transpose-conjugate operation. Then, $\{B_t\}_{0 \leq t \leq T}$ is an H-valued cylindrical Brownian motion (see definition in [4]) on a filtered probability space $(\Omega, \mathcal{F}, \{\mathcal{F}_t\}_{0 \leq t \leq T}, \mathbb{P})$, with the covariance operator $\mathrm{diag}(\mathbf{I}_d)$ (where $\mathbf{I}_d$ is an d-dimensional vector of identity operators). For each $(\omega, t) \in \Omega \times [0, T]$ constraining, $\sigma(\cdot, t)[\bullet]$ to be a (random) Hilbert-Schmidt integral operator on H with a bounded matrix kernel $\breve{\sigma} = (\breve{\sigma}_{ij})_{i,j=1,\ldots,d}$ such that

$$\sigma(x, t) f = \int_{\mathcal{D}} \breve{\sigma}(x, y, t) f(y)\, dy, \qquad f \in H, \qquad x \in \mathcal{D}. \qquad (2a)$$

Its adjoint operator $\sigma^*(\cdot, t)[\bullet]$ satisfying $\langle \sigma(\cdot, t)f, g \rangle_H = \langle f, \sigma^*(\cdot, t)g \rangle_H$ reads:

$$\sigma^*(x,t)\,g = \int_{\mathcal{D}} \breve{\sigma}^\dagger(x,y,t)g(y)\,dy, \quad g \in H, \quad x \in \mathcal{D}. \tag{2b}$$

The composite operator $\sigma(\cdot,t)\sigma^*(\cdot,t)[\bullet]$ is trace class on H and admits eigenfunctions $\xi_n(\cdot,t)$ with eigenvalues $\lambda_n(t)$ satisfying $\sum_{n\in\mathbb{N}} \lambda_n(t) < +\infty$. The noise can then be equally defined by the spectral decomposition:

$$\sigma(x,t)\,dB_t = \sum_{n\in\mathbb{N}} \lambda_n^{1/2}(t)\xi_n(x,t)\,d\beta_n(t), \tag{3}$$

where β_n are independent standard Brownian motions. In addition, we assume that the operator-space-valued process $\{\sigma(\cdot,t)[\bullet]\}_{0\leq t\leq T}$ is stochastically integrable, i.e. $\mathbb{P}\big[\int_0^T \sum_{n\in\mathbb{N}} \lambda_n(t)\,dt < +\infty\big] = 1$. From [4], the stochastic integral $\{\int_0^t \sigma(\cdot,s)\,dB_s\}_{0\leq t\leq T}$ is a continuous square integrable H-valued martingale, hence a centered Gaussian process, $\mathbb{E}_{\mathbb{P}}[\int_0^t \sigma(\cdot,s)\,dB_s] = 0$, of bounded variance, $\mathbb{E}_{\mathbb{P}}\big[\|\int_0^t \sigma(\cdot,s)\,dB_s\|_H^2\big] < +\infty$. Moreover, the joint quadratic variation process of the noise, evaluated at the same point $x \in \mathcal{D}$, is given by

$$\Big\langle \int_0^{\cdot} \sigma(x,s)\,dB_s, \int_0^{\cdot} \sigma(x,s)\,dB_s \Big\rangle_t = \int_0^t a(x,s)\,ds \tag{4a}$$

$$a(x,t) = \int_{\mathcal{D}} \breve{\sigma}(x,y,t)\breve{\sigma}^\dagger(y,x,t)\,dy = \sum_{n\in\mathbb{N}} \lambda_n(t)\big(\xi_n\xi_n^\dagger\big)(x,t). \tag{4b}$$

We remark that real-valued noise can be achieved by adding the constraint that both eigenfunctions, eigenvalues and the standard Brownian motions in (3) are organised in complex-conjugated pairs. In that case, its joint quadratic variation process is real-valued as well.

The previous formulations consist of only a zero-mean and temporally uncorrelated noise. However, this might not be enough and including a mean or time-correlated component of the unresolved velocity field could be of crucial importance to obtain a relevant model. For instance, the eddy parametrization proposed by [15] is decomposed into a deterministic mean term and a stochastic term of zero-mean. For the double-gyre circulation configuration, the considered deterministic parametrization allows to reproduce the eastwards jet for the coarse-resolution model, while the additional stochastic terms enhance the gyres circulation and improves the flow variability. Similarly, the random-forcing model proposed by [3] consists in a space-time correlated stochastic process to enhance the jet extension. The slow modes of the sub-grid scales can be provided by adequate high-pass filtering of high-resolution data on the coarse grid. We aim in this work at investigating the incorporation of such slow components within the LU framework. However, the derivation of LU models [10, 12, 1] relies on the martingale properties of the centered noise and we need hence to properly handle non centred Brownian terms. The Girsanov transformation [4] provides a theoretical tool that fully warrants such a superposition: by a change of the probability measure, the composed noise can be centered with respect to a new probability measure while the additional drift term appears, which pulls back time-correlated sub-grid-scale components into the dynamical system. The associated mathematical description is given as follows. Let $\boldsymbol{\Gamma}_t$ be an H-valued $\mathcal{F}_t$-predictable process satisfying the Novikov condition, $\mathbb{E}_{\mathbb{P}}\big[\exp(\frac{1}{2}\int_0^T \|\boldsymbol{\Gamma}_t\|_H^2\, dt)\big] < +\infty$, then the process $\{\widetilde{\boldsymbol{B}}_t := \boldsymbol{B}_t + \int_0^t \boldsymbol{\Gamma}_s\, ds\}_{0 \leq t \leq T}$ is an H-valued cylindrical Wiener process on $(\Omega, \mathcal{F}, \{\mathcal{F}_t\}_{0 \leq t \leq T}, \widetilde{\mathbb{P}})$ with Radon-Nikodym derivative

$$\frac{d\widetilde{\mathbb{P}}}{d\mathbb{P}} = \exp\left(-\int_0^T \langle \boldsymbol{\Gamma}_t, d\boldsymbol{B}_t \rangle_H - \frac{1}{2}\int_0^T \|\boldsymbol{\Gamma}_t\|_H^2\, dt\right). \tag{5a}$$

In this case, the SDE (1) under the probability measure $\widetilde{\mathbb{P}}$ reads:

$$d\boldsymbol{X}_t = \big(\boldsymbol{v}(\boldsymbol{X}_t, t) - \boldsymbol{\sigma}(\boldsymbol{X}_t, t)\boldsymbol{\Gamma}_t\big)\, dt + \boldsymbol{\sigma}(\boldsymbol{X}_t, t)\, d\widetilde{\boldsymbol{B}}_t. \tag{5b}$$

In the present work, we shall consider rather this modified stochastic flow defined on $(\Omega, \mathcal{F}, \{\mathcal{F}_t\}_{0 \leq t \leq T}, \widetilde{\mathbb{P}})$ with $\mathbb{E}_{\widetilde{\mathbb{P}}}[\boldsymbol{\sigma}\, d\widetilde{\boldsymbol{B}}_t] = \boldsymbol{0}$ as the physical solution. Hereafter, $\boldsymbol{\sigma}\boldsymbol{\Gamma}_t$ is referred to as the *Girsanov drift*.

2.2 Stochastic QG Model

The evolution law of a random tracer (function) Θ transported along the stochastic flow, $\Theta(\boldsymbol{X}_{t+\delta t}, t + \delta t) = \Theta(\boldsymbol{X}_t, t)$, is derived by [10, 1]. Under the probability measure $\widetilde{\mathbb{P}}$, this can be described by the following stochastic partial differential equation (SPDE), namely

$$\mathbb{D}_t\Theta := d_t\Theta + (\widetilde{\boldsymbol{v}}^*\, dt + \boldsymbol{\sigma}\, d\widetilde{\boldsymbol{B}}_t) \cdot \nabla\Theta - \frac{1}{2}\nabla \cdot (\boldsymbol{a}\nabla\Theta)\, dt = 0 \tag{6a}$$

$$\widetilde{\boldsymbol{v}}^* := \boldsymbol{v} - \frac{1}{2}\nabla \cdot \boldsymbol{a} + \boldsymbol{\sigma}^*(\nabla \cdot \boldsymbol{\sigma}) - \boldsymbol{\sigma}\boldsymbol{\Gamma}, \tag{6b}$$

In this SPDE, the first term $d_t\Theta(\boldsymbol{x}) := \Theta(\boldsymbol{x}, t + \delta t) - \Theta(\boldsymbol{x}, t)$ stands for the (forward) increment of Θ at a fixed point $\boldsymbol{x} \in \mathcal{D}$; the second term describes the tracer's advection by an *effective drift* $\widetilde{\boldsymbol{v}}^*$ and the noise $\boldsymbol{\sigma}\, d\widetilde{\boldsymbol{B}}_t$; the last term depicts the tracer's diffusion through the noise quadratic variation $\boldsymbol{a}$. The effective drift (6b) ensues from (*i*) the noise inhomogeneity, (*ii*) the possible unresolved flow divergence and (*iii*) the statistical correction due to the change of probability measures, respectively.

The derivation of the stochastic geophysical models under the LU framework follows exactly the same path as the deterministic derivation, together with a proper scaling of the noise and its amplitude. In particular, a continuously stratified QG model under LU has been derived by [12, 9] using an asymptotic approach. With horizontally moderate and vertically weak noises (see definitions in [12, 9]), the governing equations under the probability measure $\widetilde{\mathbb{P}}$ read:

Evolution of potential vorticity (PV):

$$\mathbb{D}_t q = \sum_{i=1,2} \mathrm{J}\Big((\widetilde{u}^\star)^i \, dt + (\sigma d\widetilde{B}_t)^i, u^i\Big) - \Big(\frac{1}{2}\nabla \cdot \big(\partial_{x_i}^\perp a \nabla u^i\big) + \beta \partial_{x_i} a_{i2}\Big) dt, \quad (7a)$$

From PV to streamfunction:

$$\nabla^2 \psi + \partial_z \Big(\frac{f_0^2}{N^2}\partial_z \psi\Big) = q - \beta y, \quad (7b)$$

Incompressible constraints:

$$u = \nabla^\perp \psi, \quad \nabla \cdot \sigma d\widetilde{B}_t = \nabla \cdot (\widetilde{u}^\star - u) = 0. \quad (7c)$$

Here, $\nabla = [\partial_x, \partial_y]^T$, $\nabla^\perp = [-\partial_y, \partial_x]^T$, $\nabla^2 = \partial_{xx}^2 + \partial_{yy}^2$ denote two-dimensional operators and $\mathrm{J}(f, g) = \partial_x f \partial_y g - \partial_x g \partial_y f$ stands for the Jacobian operator. The vector fields $u, \sigma d\widetilde{B}_t$ and the tensor field a are two-dimensional (2D) horizontal quantities. The horizontal effective drift is defined as $\widetilde{u}^\star := u - \nabla \cdot (a/2) - \sigma\Gamma$. The scalar fields q and ψ represent the PV and the streamfunction. In Eq. (7b), $N^2 = -(g/\rho_0)\partial_z \rho$ is the Brunt-Väisälä (or buoyancy) frequency with g the gravity value, ρ_0 the background density, ρ the density anomaly, and $f_0 + \beta y$ is the Coriolis parameter under a beta-plane approximation. As shown in [1], one important characteristic of the random model (7) is that it conserves the total energy of the resolved flow (under natural boundary condition) for any realization (i.e. pathwise). This property highlights a strong relation between the classical deterministic model and the stochastic formulation.

3 Numerical Parameterization of Unresolved Flow

Data-driven approaches are presented in this section to estimate the spatial correlation functions of the unresolved flow component based on the spectral decomposition (3). In practice, we work with a finite set of functions to represent the small-scale Eulerian velocity fluctuations rather than with the Lagrangian particles trajectory. We first review the empirical orthogonal functions (EOF) method for which the noise covariance is assumed quasi-stationary. We then propose an approach relying on the dynamic mode decomposition (DMD) to account for the temporal behavior of the spatial correlations.

3.1 EOF-Based Method

In the following, let $\{u_{\mathrm{HR}}(x, t_i)\}_{i=1,\ldots,N}$ be the set of velocity snapshots provided by a high-resolution (HR) simulation. We first build the spatial local fluctuations $u_f(x, t_i)$ of each snapshot on the coarse-grid points. In particular, for the QG system (7), one can first perform a high-pass filtering with a 2D Gaussian convolution kernel G on each HR streamfunction ψ_{HR}, to obtain the streamfunction fluctuations, $\psi_f(x, t_i) = ((I - G) \star \psi_{\mathrm{HR}})(x, t_i)$ (only for the coarse-grid points x). Then, the geostrophic velocity fluctuations can be derived by $u_f = \nabla^{\perp}_{\mathrm{LR}} \psi_f$. We next centre the data set by $u'_f = u_f - \overline{u_f}^t$ (with $\overline{\bullet}^t$ the temporal mean) and perform the EOF procedure [9] to get a set of orthogonal temporal modes $\{\alpha_m\}_{m=1,\ldots,N}$ and orthonormal spatial modes $\{\phi_m\}_{m=1,\ldots,N}$ satisfying

$$u'_f(x, t_i) = \sum_{m=1}^{N} \alpha_m(t_i)\phi_m(x), \quad \overline{\alpha_m \alpha_n}^t = \lambda_m \delta_{m,n}. \tag{8}$$

Truncating the modes (with $M \ll N$) and rescaling by a small-scale decorrelation time τ, the stationary noise and its quadratic variation can be build by

$$\sigma(x)\mathrm{d}\tilde{B}_t = \sqrt{\tau} \sum_{m=1}^{M} \sqrt{\lambda_m}\phi_m(x)\,\mathrm{d}\beta_m(t), \quad a(x) = \tau \sum_{m=1}^{M} \lambda_m \phi_m(x)\phi_m^T(x). \tag{9}$$

Note that this time scale τ is used to match the fact that the noise in (5b) has the physical dimension of a length. In practice, we often consider the coarse-grid simulation timestep Δt_{LR}. In addition, the Girsanov drift is set to be $\sigma(x)\Gamma_t = \overline{u_f}^t(x)$. It means that the Girsanov drift here is the projection of the temporal mean of the sub-grid scales onto the EOFs, i.e. $\sigma(x)\Gamma_t = \sum_{m=1}^{N} \gamma_m \phi_m(x)$ with $\gamma_m = \langle \overline{u_f}^t, \phi_m \rangle_H$ satisfying $\sum_{m=1}^{N} \gamma_m^2 < +\infty$.

3.2 DMD-Based Method

The DMD algorithm [13] seeks a spectral decomposition of the best-fit linear operator A that relates the two snapshots:

$$u'_f(x, t_{i+1}) \approx Au'_f(x, t_i). \tag{10a}$$

Applying the exact DMD procedure proposed by [14], the corresponding spectral expansion in continuous time reads

$$u'_f(x, t) = \sum_{m=1}^{N} b_m \exp\left((\sigma_m + \mathrm{i}\,\omega_m)t\right)\varphi_m(x), \tag{10b}$$

where $\varphi_m(x) \in \mathbb{C}^d$ are the DMD modes (eigenvectors of A) associated to the DMD eigenvalues $\mu_m \in \mathbb{C}$, $\sigma_m = \log(|\mu_m|)/\Delta t_d \in \mathbb{R}$ are the modes growth rate (with $\Delta t_s = t_{i+1} - t_i$ the sampling step of data), $\omega_m = \arg(\mu_m)/\Delta t_s \in \mathbb{R}$ are the modes frequencies (with i the imaginary unit) and $b_m \in \mathbb{C}$ are the modes amplitudes. In practice, our data set of velocity fluctuations is real valued, hence the DMD modes (also eigenvalues and amplitudes) are two-by-two complex conjugates, i.e. $\varphi_{2p} = \overline{\varphi}_{2p-1}$ ($p = 1, \ldots, N/2$).

We next propose to split the total set of DMD modes into two subsets, $\mathcal{M}^c$ and $\mathcal{M}^r$, to select separately adequate fast and slow modes for the noise (from $\mathcal{M}^r$) and the Girsanov drift (from $\mathcal{M}^c$), respectively, according to the following analysis of frequencies and amplitudes:

$$\mathcal{M}^c = \left\{ m \in [1, N] \,\middle|\, |\mu_m| \approx 1,\ |\omega_m| \le \frac{\pi}{\tau_c},\ |b_m| \ge C \right\}, \tag{11a}$$

$$\mathcal{M}^r = \left\{ m \in [1, N] \,\middle|\, |\mu_m| \approx 1,\ |\omega_m| > \frac{\pi}{\tau_c},\ |b_m| \ge C \right\}, \tag{11b}$$

where τ_c is a temporal-separation-scale that can be estimated by the spatial mean of the autocorrelation functions of data and C denotes an empirical cutoff of amplitudes. The DMD modes that are neither included in $\mathcal{M}^c$ nor in $\mathcal{M}^r$ are discarded. An example of spectrum and amplitudes of the selected DMD modes is shown in Fig. 1. In order to avoid spurious effects associated with the non-orthogonality of DMD modes, their amplitudes are rescaled such that the reconstructed data corresponds to

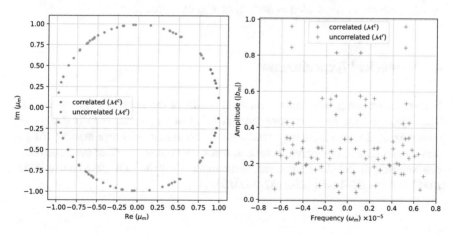

Fig. 1 Illustration of the selections of DMD modes used for the noise (orange) and the Girsanov drift (blue)

an orthogonal projection onto the subspace spanned by the modes in $\mathcal{M}^c$ or $\mathcal{M}^r$. In particular, we propose to rescale those truncated DMD modes as follows:

(i) Construct the Gramian $\boldsymbol{G} = (g_{m,n})_{m,n \in \mathcal{M}^c}$ with $g_{m,n} = \langle \boldsymbol{\varphi}_m, \boldsymbol{\varphi}_n \rangle_H$;

(ii) Inverse the Gramian $\boldsymbol{G}^{-1} := (g_{m,n}^{-1})_{m,n \in \mathcal{M}^c}$ and derive the dual set of the truncated DMD modes by $\boldsymbol{\varphi}_m^* = \sum_{n \in \mathcal{M}^c} g_{m,n}^{-1} \boldsymbol{\varphi}_n$;

(iii) Project the initial state of data on the dual set of modes to update the amplitudes: $\boldsymbol{\phi}_m := \langle \boldsymbol{u}_f'(\cdot, t_1), \boldsymbol{\varphi}_m^* \rangle_H \, \boldsymbol{\varphi}_m$.

Such procedure holds separately for the DMD modes of $\mathcal{M}^c$ and $\mathcal{M}^r$. Finally, the noise and the correction drift can be defined as

$$\boldsymbol{\sigma}(\boldsymbol{x}, t)\mathrm{d}\widetilde{\boldsymbol{B}}_t = \sqrt{\tau} \sum_{m \in \mathcal{M}^r} \exp(\mathrm{i}\,\omega_m t)\boldsymbol{\phi}_m(\boldsymbol{x})\,\mathrm{d}\beta_m(t), \tag{12a}$$

$$\boldsymbol{\sigma}(\boldsymbol{x}, t)\boldsymbol{\Gamma}_t = \overline{\boldsymbol{u}_f}^t(\boldsymbol{x}) + \sum_{m \in \mathcal{M}^c} \exp(\mathrm{i}\,\omega_m t)\boldsymbol{\phi}_m(\boldsymbol{x}), \tag{12b}$$

In particular, we assume that each pair of the complex Brownian motions are conjugates ($\beta_{2p} = \overline{\beta}_{2p-1}$) and their real and imaginary parts are independent. As such, both noise $\boldsymbol{\sigma}\mathrm{d}\widetilde{\boldsymbol{B}}_t$ and correction drift $\boldsymbol{\sigma}\boldsymbol{\Gamma}_t$ are real-valued fields. In addition, the joint quadratic variation of such noise remains stationary:

$$\boldsymbol{a}(\boldsymbol{x}) = \tau \sum_{m \in \mathcal{M}^r} \boldsymbol{\phi}_m(\boldsymbol{x})\boldsymbol{\phi}_m^\dagger(\boldsymbol{x}). \tag{12c}$$

In a similar way as in the EOF-based method, we could also construct the Girsanov drift by the projection of the RHS of (12b) onto the DMD modes. As we have dropped the unstable DMD modes, one can show that the predictability and the Novikov condition (presented in Sect. 2) of $\boldsymbol{\Gamma}$ hold in this case.

4 Numerical Experiments

In this section, we present some numerical results of the stochastic QG system (7). The objective consists to improve the variability of large-scale models defined on coarse grids. To that end, a high-resolution deterministic reference model (REF) is first simulated and compared to several coarse-resolution models: the benchmark deterministic model (DET), two stochastic models with an EOF-based noise (STO-EOF) and a DMD-based noise (STO-DMD).

4.1   Configurations

In this study, we consider a vertically discretized QG dynamical core proposed in [8] and extended in the stochastic setting [9]. This model consists in n isopycnal layers with constant thickness H_k and density ρ_k in each layer k. In this case, the prognostic variables such as ψ in (7) are assumed to be layer-averaged quantities. Homogeneous Dirichlet boundary conditions have been imposed for the term $f_0 \partial_z \psi / N^2$ in (7b) at the ocean surface and bottom. Moreover, external forcing and numerical dissipation are included in the evolution of PV (7a): the Ekman pumping $\nabla^\perp \cdot \tau$ due to the wind stress τ over ocean surface boundary, a linear drag $-(f_0 \eta_{ek}/2) \nabla^2 \psi_n$ at ocean bottom with a very thin thickness η_{ek}, and a biharmonic dissipation $-A_4 \nabla^4 (\nabla^2 \psi_k)$ in each layer with uniform coefficient A_4. In particular, we consider here a finite box ocean driven by an idealized (stationary and symmetric) wind stress $\tau = [-\tau_0 \cos(2\pi y)/L_y, 0]^T$. A mixed horizontal boundary condition is used for the k-th layer streamfunction: $\psi_k|_{\partial \mathcal{A}} = f_k(t)$ and $\partial_n^2 \psi_k|_{\partial \mathcal{A}} = -(\alpha_{bc}/\Delta x) \partial_n \psi_k|_{\partial \mathcal{A}}$ (same for the 4-th order derivative). Here, $\mathcal{A}$ denotes the 2D area, f_k is a time-dependent function constrained by mass conservation [7], Δx stands for the horizontal resolution and α_{bc} is a nondimensional coefficient associated to the slip conditions [7]. A quiescent initial condition is used for the REF, whereas a spin-up condition downsampled from REF (after 90-years integration) is adopted for all the coarse-resolution models. The common parameters for all the simulations are listed in Table 1, whereas resolution dependant parameters are presented separately in Table 2. Both EOF and DMD modes are calibrated from the REF data during 40 years (after the spin-up) with a 5-days sampling step. As for the numerical discretization, a conservative flux form [9] together with a stochastic Leapfrog scheme [5] is adopted for the evolution of PV (7a). The inversion of the modified Helmholtz equation (7b) is carried out with a discrete sine transform method [7].

Table 1 Common parameters for all the models. The buoyancy frequency N^2 in (7b) is approximated by $g'_{k+0.5}/(H_k + H_{k+1})/2$ on the interface between layers k and $k + 1$

Parameters	Value	Description
$X \times Y$	(3840×4800) km	Domain size
H_k	$(350, 750, 2900)$ m	Mean layer thickness
$g'_{k+0.5}$	$(0.025, 0.0125)$ m s^{-2}	Reduced gravity
η_{ek}	2 m	Bottom Ekman layer thickness
τ_0	2×10^{-5} m^2 s^{-2}	Wind stress magnitude
α_{bc}	0.2	Mixed boundary condition coefficient
f_0	9.375×10^{-5} s^{-1}	Mean Coriolis parameter
β	1.754×10^{-11} (m s)$^{-1}$	Coriolis parameter gradient
r_m	$(39, 22)$ km	Baroclinic Rossby radii

Table 2 Values of grid varying parameters. The energy proportion captured by the truncated EOF modes are given in the bracket. For DMD method, the first number stands for the size of $\mathcal{M}^c$ (11a) whereas the latter is the one of $\mathcal{M}^r$ (11b)

Resolution (km)	Timestep (s)	Viscosity ($m^4\,s^{-1}$)	EOF modes	DMD modes
5	600	2×10^9	–	–
40	1200	5×10^{11}	300 (83%)	14 + 46
80	1440	5×10^{12}	300 (92%)	16 + 74
120	1800	1×10^{13}	300 (97%)	16 + 110

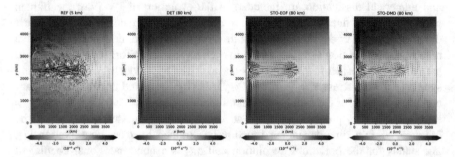

Fig. 2 Snapshots of surface PV provided by different simulations after 60-years integration. The black arrows are the interpolated geostrophic velocities

Snapshots of the surface PV provided by the different simulations are shown in Fig. 2. The dynamics of *REF* (5 km) model is mainly characterized by a meandering eastward jet with adjacent recirculations, which results from the most active mesoscale eddies effect through baroclinic instability. However, this effect cannot be properly resolved once the horizontal resolution exceeds the baroclinic deformation radius maximum (39 km here). For instance, the *DET* (80 km) simulation generates only a smooth symmetric field. On the other hand, both *STO-EOF* and *STO-DMD* models are able to reproduce the eastward jet on the coarse mesh (80 km) by including the non-linear effects carried both by the unresolved noise and the correction drift. In particular, the *STO-DMD* model produces a stronger meridional perturbation along the jet and is able to capture some of the large-wave structures predicted by the *REF* model. The improvements brought by these random models will be diagnosed and analyzed more precisely in the following.

4.2 Diagnostics

We first compare the long-term mean (over a 100-years interval) of the kinetic energy (KE) spectrum for both coarse models at different resolutions (40, 80,

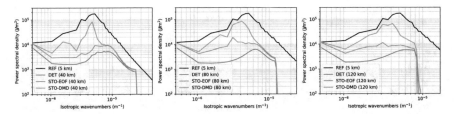

Fig. 3 Temporal mean of vertically integrated KE spectra for the different models

120 km). As shown in Fig. 3, introducing only a dissipation mechanism like the biharmonic viscosity in the *DET* coarse models leads to an excessive decrease of the resolved KE compared to the *REF* model. Both *STO-EOF* and *STO-DMD* models at different resolutions, recover a given amount of lost energy over all wavenumbers. In particular, the *STO-DMD* models provide higher KE backscattering at large scales and better spectrum slope in the inertial-range than the stationary unresolved models. This seems to highlight the importance of the non-stationary characteristic of the noise and Girsanov drift.

We then quantify the temporal variability (over the same 100-years interval) predicted by the different coarse models. In this work, we adopt the following three global metrics. The first one is the root-mean-square error (RMSE) between the standard deviation of the streamfunction of a coarse model (denoted by $\sigma[\psi^M]$) and the subsampled high-resolution one (denoted by $\sigma[\psi^R]$), $\|\sigma[\psi^M] - \sigma[\psi^R]\|_{L^2(\mathcal{D})}$, where $\mathcal{D} = \mathcal{A} \times [-H, 0]$ and H stands for the total depth of the ocean basin. The second criterion is the Gaussian relative entropy (GRE) [6] which assesses in a single measure the mean and variance reconstruction:

$$\text{GRE} = \frac{1}{|\mathcal{D}|} \int_{\mathcal{D}} \frac{1}{2} \left(\frac{\left(\overline{\psi^M}^t - \overline{\psi^R}^t \right)^2}{\sigma^2[\psi^M]} + \frac{\sigma^2[\psi^R]}{\sigma^2[\psi^M]} - 1 - \log\left(\frac{\sigma^2[\psi^R]}{\sigma^2[\psi^M]} \right) \right) \, d\mathbf{x}. \quad (13)$$

It is clear that a coarse model of high variability will have low RMSE and GRE, whereas a poor variability will lead to a large RMSE and GRE. The last metric measures the eddy kinetic energy (EKE), $(\rho_0/2)\|\mathbf{u}'\|^2_{(L^2(\mathcal{D}))^2}$, where $\mathbf{u}' := (I - \mathcal{F}_t)[\mathbf{u}]$ is the eddy velocity filtered out through a 2-years low-pass filter $\mathcal{F}_t$ at every point in space. For comparison reason, we show here only the time average of this metric $(\overline{\text{EKE}})$ for the different models.

These three criteria are shown in Fig. 4 as bar plots. The *DET* models show very high RMSE and GRE with a very low order of EKE, meaning that they produce poor variability along time and failed to represent the eddies effect. Compared to the *STO-EOF*, the *STO-DMD* models enable to increase significantly the internal variability and the eddy energy. Moreover, these improvements are resolution-aware. As shown

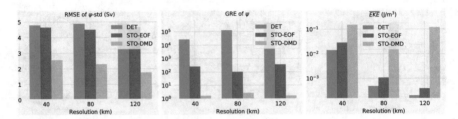

Fig. 4 Comparison of variability measures for different coarse models. The *y*-axis of the last two figures are in log-scales

in Table 2, under a similar level of captured energy, the *STO-DMD* models require much less modes than the *STO-EOF*, which reduces first the memory cost. Then, in terms of computational cost at each step, the former consists in generating less Gaussian variables than the latter, and reduces hence as well the dimension of the matrix-vector multiplication for the spectral decomposition (3).

4.3 Discussion

In order to distinguish the contribution of the correlated Girsanov drift and the uncorrelated noise, three additional benchmark runs (at resolution 80 km) have been further performed and compared to the proposed *STO-DMD* model, they are (i) *STO-DMD* without any correlation drift (i.e. $\sigma\Gamma_t = 0$); (ii) *STO-DMD* only with $\sigma\Gamma_t = \overline{u_f}^t$; (iii) a simplified deterministic version of the proposed *STO-DMD* model, denoted as *DET-DMD*, which only encodes the (full) correlated drift $\sigma\Gamma_t$ into the *DET* model. We remark that for the two first runs the DMD modes used for the correlated drift in the previous stochastic model are now included into the noise component. As shown in Fig. 5, run (i) fails to reproduce the eastwards jet on the coarse mesh, whereas the other runs succeed. However, run (ii) produces similar results as the *STO-EOF* model (see Fig. 2) with a lower improvement of variability, and run (iii) captures more waves than the others, yet leads to a reduction of the jet magnitude compared to the proposed *STO-DMD* model. In particular, by comparing the KE spectra of the different runs, Fig. 6 illustrates that the simplified *DET-DMD* model allows to produce backscattering of KE from small to large scales, and the proposed *STO-DMD* enhances this result with significantly higher KE at large-scales. We observe a consistent conclusion for the EKE budget (see Fig. 6). These comparisons demonstrate that the both correlated drift ($\sigma\Gamma_t$) and the uncorrelated noise ($\sigma d\widetilde{B}_t$) contribute on the prediction of large-scale patterns and on the improvement of the variability of the large-scale models.

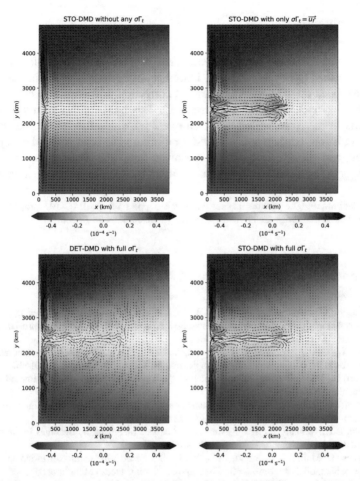

Fig. 5 Snapshots of surface PV provided by different simulations after 60-years integration. These four figures (from left to right) correspond to the benchmark runs (i), (ii), (iii) and the proposed *STO-DMD* model

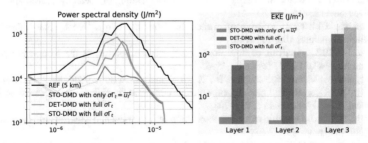

Fig. 6 Comparison of KE spectra and layered EKE (only horizontally integrated) for different coarse models

5 Conclusions

The proposed stochastic parameterization has been successfully implemented in a well established QG dynamical core. Different noises defined from high-resolution data have been considered. An additional correction drift ensuing from a change of probability measure has been introduced. This non-intuitive term seems quite important in the reproduction of the eastward jet within the wind-driven double-gyre circulation. Furthermore, the DMD procedure has been adopted to represent the quasi-periodic dynamic of the unresolved flow. The resulting random model enables us to improve the intrinsic variability of the large-scale resolved flow.

Acknowledgments The authors acknowledge the support of the ERC EU project 856408-STUOD. The source codes can be found in https://github.com/matlong/qgcm_lu.

References

1. Bauer, W., Chandramouli, P., Chapron, B., Li, L., Mémin, E., 2020a. Deciphering the role of small-scale inhomogeneity on geophysical flow structuration: a stochastic approach. Journal of Physical Oceanography 50, 983–1003.
2. Bauer, W., Chandramouli, P., Li, L., Mémin, E., 2020b. Stochastic representation of mesoscale eddy effects in coarse-resolution barotropic models. Ocean Modelling 151, 101646.
3. Berloff, P.,, 2005. Random-forcing model of the mesoscale oceanic eddies. Journal of Fluid Mechanics 529, 71–95.
4. Da Prato, G., Zabczyk, J., 2014. Stochastic equations in infinite dimensions. Encyclopedia of Mathematics and its Applications. 2 ed., Cambridge University Press.
5. Ewald, B.D., Témam, R., 2005. Numerical Analysis of Stochastic Schemes in Geophysics. SIAM Journal on Numerical Analysis 42, 2257–2276.
6. Grooms, I., Majda, A.J., Smith, K.S., 2015. Stochastic superparameterization in a quasi-geostrophic model of the Antarctic Circumpolar Current. Ocean Modelling 85, 1–15.
7. Hogg, A.M., Dewar, W.K., Killworth, P.D., Blundell, J.R., 2003. A quasi-geostrophic coupled model (Q-GCM). Monthly Weather Review 131, 2261–2278.
8. Hogg, A.M., Killworth, P.D., Blundell, J.R., 2004. Mechanisms of decadal variability of the wind-driven ocean circulation. Journal of Physical Oceanography 35.
9. Li, L., 2021. Stochastic modelling and numerical simulation of ocean dynamics. https://hal.archives-ouvertes.fr/tel-03207741.
10. Mémin, E., 2014. Fluid flow dynamics under location uncertainty. Geophysical & Astrophysical Fluid Dynamics 108, 119–146.
11. Mezić, I., 2013. Analysis of fluid flows via spectral properties of the Koopman operator. Annual Review of Fluid Mechanics 45, 357–378.
12. Resseguier, V., Mémin, E., Chapron, B., 2017b. Geophysical flows under location uncertainty, part II: Quasi-geostrophic models and efficient ensemble spreading. Geophysical & Astrophysical Fluid Dynamics 111, 177–208.
13. Schmid, P., 2010. Dynamic mode decomposition of numerical and experimental data. Journal of Fluid Mechanics 656, 5–28.
14. Tu, J.H., Rowley, C.W., Luchtenburg, D.M., 2014. On dynamic mode decomposition: Theory and applications. Journal of Computational Dynamics 1, 391–421.
15. Zanna, L., Porta Mana, P., Anstey, J., David, T., Bolton, T., 2017. Scale-aware deterministic and stochastic parametrizations of eddy-mean flow interaction. Ocean Modelling 111, 66–80.

Deep Learning for the Benes Filter

Alexander Lobbe

Abstract The filtering problem is concerned with the optimal estimation of a hidden state given partial and noisy observations. Filtering is extensively studied in the theoretical and applied mathematical literature. One of the central challenges in filtering today is the numerical approximation of the optimal filter. Here, accurate and fast methods are actively sought after, especially for such high-dimensional settings as numerical weather prediction, for example. In this paper we present a brief study of a new numerical method based on the mesh-free neural network representation of the density of the solution of the filtering problem achieved by deep learning. Based on the classical SPDE splitting method, our algorithm includes a recursive normalisation procedure to recover the normalised conditional distribution of the signal process. The present work uses the Benes model as a benchmark. The Benes filter is a well-known continuous-time stochastic filtering model in one dimension that has the advantage of being explicitly solvable. Within the analytically tractable setting of the Benes filter, we discuss the role of nonlinearity in the filtering model equations for the choice of the domain of the neural network. Further, we present the first study of the neural network method with an adaptive domain for the Benes model.

Keywords Nonlinear filtering · Deep learning · Stochastic PDE approximation

1 Introduction

Stochastic Filtering, i.e. the estimation of a signal process given only partial and noisy observations, is a well-studied problem, both in the theoretical and applied literature. It is relevant in many practical domains, for example in numerical weather prediction. Therefore, there is a high demand for efficient numerical methods to

A. Lobbe (✉)
Department of Mathematics, Imperial College London, London, UK
e-mail: alex.lobbe@imperial.ac.uk

© The Author(s) 2023
B. Chapron et al. (eds.), *Stochastic Transport in Upper Ocean Dynamics*,
Mathematics of Planet Earth 10, https://doi.org/10.1007/978-3-031-18988-3_12

approximate the optimal filter. Many such methods are known in the literature, among them the SPDE splitting method can be used to solve the filtering problem in low dimensions. The reason for the inefficiency of the splitting method in higher dimensions stems from the fact that the underlying state space must be explicitly discretised. This is problematic as the required number of discretisation points, known as the *mesh*, grows exponentially with the dimension of the state space. For this reason, the authors of [4] present a modified splitting method for the filtering problem which does not rely on the explicit space discretisation. The method developed in [4] is therefore called *mesh-free* and relies on a neural network representation of the solution. This means that, instead of approximating the values of the solution on a discrete mesh, we can optimize the parameters of a neural network defined on the state-space itself.

In this paper we present a further study of the deep learning method developed in [4] on the example of the Benes filter. The algorithm is derived from the classical splitting method for SPDEs which consists of a deterministic PDE approximation step and a normalisation step to incorporate the randomness of the SPDE. Our algorithm replaces the PDE approximation step of the splitting method by a neural network representation and learning algorithm. Combined with the Monte-Carlo method for the normalisation step, this method becomes completely mesh-free. Furthermore, an important property of the methodology in the filtering context is the ability to iterate it over several time steps. This allows the algorithm to be run *online* and to successively process observations arriving sequentially. In order to be computationally feasible, the domain of the neural network needs to be restricted. This restricted domain needs to cover the support of the density as well as possible in order to yield a sensible solution. In [4] the neural network domain is fixed a priori and does not move with the solution. This presents two problems. First, it is unnecessarily large to cover the support over all timesteps. Second, the solution may eventually move outside the computational domain, rendering the approximation inadequate. It was therefore noted in [4] that a possible extension of the approximation method would be given by an adaptive domain as the support of the neural network. We present in this work the first results obtained using an adaptive domain in the nonlinear and analytically tractable case of the Benes filter.

The paper is structured as follows. In Sect. 1.1 we briefly introduce the nonlinear, continuous-time stochastic filtering framework. The setting is identical to the one assumed in [4] and the reader may consult [1] for an in-depth treatment of stochastic filtering. Thereafter, in Sect. 2.2, we formulate the Benes filtering model used as a benchmark. Then, in Sect. 1.2 we introduce the filtering equation and the classical SPDE splitting method. This is the method upon which the new algorithm in [4] was built.

Next, in Sect. 2 we present an outline of the derivation of the new methodology. For details, the reader is referred to the original article [4]. The first idea of the algorithm, presented in Sect. 2.1 is to reformulate the solution of the PDE for the density of the unnormalised filter as an expected value. This is done using the Feynman–Kac formula, based on an auxiliary diffusion process derived from the model equations. Moreover, in Sect. 2.3 we briefly specify the neural network parameters used in the method, as well as the employed loss-function. The theoretical part of the paper is concluded with Sect. 2.4 where we show how to normalise the obtained neural network from the prediction step using Monte-Carlo approximation for linear sensor functions.

Section 3 contains the detailed parameter values and results of the numerical studies that we performed. Specifically, we perform two experiments, the first one, Sect. 3.1, is carried without any domain adaptation and highlights the limitations of ad-hoc parameterization of the domain. It is a simulation of the Benes filter using the deep learning method over a larger domain, as well as longer time interval than in the paper [4]. In particular, the size of the domain was estimated using the exact solution of the Benes model. This is necessary, as the nonlinearity of the Benes model makes it difficult to know the evolution of the posterior a priori. Thus we would be requiring a much larger domain, if chosen in an ad-hoc way. The second experiment, in Sect. 3.2, reports the performance of the proposed framework with domain adaptation. The adaptation was performed using precomputed estimates of the support of the filter by employing the solution formula for the Benes filter.

Finally, we formulate the conclusions from our experiments in Sect. 4. In short, the domain adapted method was more effective in resolving the bimodality in our study than the non-domain adapted one. However, this came at the cost of a linear trend in the error.

1.1 Nonlinear Stochastic Filtering Problem

The stochastic filtering framework consists of a pair of stochastic processes (X, Y) on a probability space $(\Omega, \mathcal{F}, P)$ with a normal filtration $(\mathcal{F}_t)_{t \geq 0}$ modelled, P-a.s., as

$$X_t = X_0 + \int_0^t f(X_s) \, ds + \int_0^t \sigma(X_s) \, dV_s \, , \tag{1}$$

and

$$Y_t = \int_0^t h(X_s) \, ds + W_t \, . \tag{2}$$

Here, the time parameter is $t \in [0, \infty)$, $d, p \in \mathbb{N}$ and $f : \mathbb{R}^d \to \mathbb{R}^d$ and $\sigma : \mathbb{R}^d \to \mathbb{R}^{d \times p}$ are the drift and diffusion coefficient functions of the signal. The processes V and W are $p-$ and m-dimensional independent, $(\mathcal{F}_t)_{t \geq 0}$-adapted Brownian motions. We call X the *signal process* and Y the *observation process*. The function $h : \mathbb{R}^d \to \mathbb{R}^m$ is often called the *sensor function*, or *link function*, because it models the possibly nonlinear connection of the signal and observation processes.

Further, consider the *observation filtration* $(\mathcal{Y}_t)_{t\geq 0}$ given as

$$\mathcal{Y}_t = \sigma(Y_s, s \in [0,t]) \vee \mathcal{N} \quad \text{and} \quad \mathcal{Y} = \sigma\left(\bigcup_{t\in[0,\infty)} \mathcal{Y}_t\right),$$

where $\mathcal{N}$ are the P-nullsets of $\mathcal{F}$. The aim of nonlinear filtering is to compute the probability measure valued $(\mathcal{Y}_t)_{t\geq 0}$-adapted stochastic process π that is defined by the requirement that for all bounded measurable test functions $\varphi : \mathbb{R}^d \to \mathbb{R}$ and $t \in [0,\infty)$ we have P-a.s. that

$$\pi_t\varphi = \mathbb{E}\left[\varphi(X_t)\,|\,\mathcal{Y}_t\right].$$

We call π the *filter*.

Furthermore, let the process Z be defined such that for all $t \in [0,\infty)$,

$$Z_t = \exp\{-\int_0^t h(X_s)\,dW_s - \frac{1}{2}\int_0^t h(X_s)^2\,ds\}.$$

Then, assumimg that

$$\mathbb{E}\left[\int_0^t h(X_s)^2\,ds\right] < \infty \quad \text{and} \quad \mathbb{E}\left[\int_0^t Z_s h(X_s)^2\,ds\right] < \infty,$$

we have that Z is an $(\mathcal{F}_t)_{t\geq 0}$-martingale and by the change of measure (for details, see [1]) given by $\frac{d\tilde{P}^t}{dP}\big|_{\mathcal{F}_t} = Z_t$, $t \geq 0$, the processes X and Y are independent under $\tilde{P}$ and Y is a $\tilde{P}$-Brownian motion. Here, $\tilde{P}$ is the consistent measure defined on $\bigcup_{t\in[0,\infty)} \mathcal{F}_t$. Finally, under $\tilde{P}$, we can define the measure valued stochastic process ρ by the requirement that for all bounded measurable functions $\varphi : \mathbb{R}^d \to \mathbb{R}$ and $t \in [0,\infty)$ we have P-a.s. that

$$\rho_t\varphi = \mathbb{E}\left[\varphi(X_t)\exp\{\int_0^t h(X_s)\,dY_s - \frac{1}{2}\int_0^t h(X_s)^2\,ds\}\,\Big|\,\mathcal{Y}_t\right]. \tag{3}$$

The Kallianpur–Striebel formula (see [1]) justifies the terminology to call ρ the *unnormalised filter*.

1.2 Filtering Equation and General Splitting Method

Note that under the conditions given in [4], X admits the infinitesimal generator $A : \mathcal{D}(A) \to B(\mathbb{R}^d)$ given, for all $\varphi \in \mathcal{D}(A)$, by

$$A\varphi = \langle f, \nabla\varphi \rangle + \mathrm{Tr}(a\,\mathrm{Hess}\,\varphi), \qquad (4)$$

where $\mathcal{D}(A)$ denotes the domain of the differential operator A and $a = \frac{1}{2}\sigma\sigma'$. The symbol $B(\mathbb{R}^d)$ denotes the set of real-valued, bounded, Borel-measurable functions defined on $\mathbb{R}^d$.

It is well-known (see, e.g., [1]), that the unnormalised filter ρ satisfies the *filtering equation*, i.e. for all $t \geq 0$, we have $\tilde{\mathbb{P}}$-a.s. that

$$\rho_t(\varphi) = \pi_0(\varphi) + \int_0^t \rho_s(A\varphi)\,\mathrm{d}s + \int_0^t \rho_s(\varphi h')\,\mathrm{d}Y_s. \qquad (5)$$

The classical splitting method for the filtering equation is given in [3] and seeks to approximate the following SPDE for the density p_t of the unnormalised filter given, for all $t \geq 0$, $x \in \mathbb{R}^d$, and P-a.s. as

$$p_t(x) = p_0(x) + \int_0^t A^* p_s(x)\,\mathrm{d}s + \int_0^t h'(x) p_s(x)\,\mathrm{d}Y_s$$

and relies on the splitting-up algorithm described in [9] and [10]. Here A^* is the formal adjoint of the infinitesimal generator A of the signal process X.

We summarise the splitting-up method below in Note 1.

Note 1 The splitting method for the filtering problem is defined by iterating the steps below with initial density $p^0(\cdot) = p_0(\cdot)$:

1. *(Prediction)* Compute an approximation $\tilde{p}^n$ of the solution to

$$\begin{aligned}
\frac{\partial q^n}{\partial t}(t, z) &= A^* q^n(t, z), \quad (t, z) \in (t_{n-1}, t_n] \times \mathbb{R}^d, \\
q^n(0, z) &= p^{n-1}(z), \qquad z \in \mathbb{R}^d,
\end{aligned} \qquad (6)$$

at time t_n and

2. *(Normalisation)* Compute the normalisation constant with $z_n = (Y_{t_n} - Y_{t_{n-1}})/(t_n - t_{n-1})$ and the function

$$\mathbb{R}^d \ni z \mapsto \xi_n(z) = \exp\left(-\frac{t_n - t_{n-1}}{2}\|z_n - h(z)\|^2\right),$$

so that we can set,

$$p^n(z) = \frac{1}{C_n}\xi_n(z)\tilde{p}^n(z); \; z \in \mathbb{R}^d,$$

where $C_n = \int_{\mathbb{R}^d} \xi_n(z)\tilde{p}^n(z)\,\mathrm{d}z$.

The deep learning method studied below replaces the predictor step of the splitting method above by a deep neural network approximation algorithm to avoid an explicit space discretisation. This is achieved by representing each $\tilde{p}^n(z)$ by a feed-forward neural network and approximating the initial value problem (6) based on its stochastic representation using a sampling procedure. The normalisation step may then be computed either using quadrature, or, to preserve the mesh-free characteristic, by Monte-Carlo approximation.

2 Derivation and Outline of the Deep Learning Algorithm

Here, we present a concise version of the derivation laid out in detail in [4].

2.1 Feynman–Kac Representation

Assuming sufficient differentiability of the coefficient functions, the operator A^* may be expanded such that for all compactly supported smooth test functions $\varphi \in C_c^\infty(\mathbb{R}^d, \mathbb{R})$ we have

$$A^*\varphi = \text{Tr}(a \, \text{Hess} \, \varphi) + \langle 2\overrightarrow{\text{div}}(a) - f, \text{grad} \, \varphi \rangle + \text{div}(\overrightarrow{\text{div}}(a) - f)\varphi. \tag{7}$$

Subtracting the zero-order term from (7), we obtain an operator that generates the auxiliary diffusion process, denoted $\hat{X}$, which is instrumental in the deep learning method.

Definition 1 Define the partial differential operator $\hat{A} : C_c^\infty(\mathbb{R}^d, \mathbb{R}) \to C_b(\mathbb{R}^d, \mathbb{R})$, with image in the set of bounded continuous function on $\mathbb{R}^d$, such that for all $\varphi \in C_c^\infty(\mathbb{R}^d, \mathbb{R})$,

$$\hat{A}\varphi = \text{Tr}(a \, \text{Hess} \, \varphi) + \langle 2\overrightarrow{\text{div}}(a) - f, \text{grad} \, \varphi \rangle$$

and the function $r : \mathbb{R}^d \to \mathbb{R}$ such that for all $x \in \mathbb{R}^d$,

$$r(x) = \text{div}(\overrightarrow{\text{div}}(a) - f)(x).$$

Lemma 1 For all $x \in \mathbb{R}^d$ the operator $\hat{A}$ defined in Definition 1 is the infinitesimal generator of the Itô diffusion $\hat{X} : [0, \infty) \times \Omega \to \mathbb{R}^d$ given, for all $t \geq 0$ and P-a.s. by

$$\hat{X}_t = x + \int_0^t b(\hat{X}_s)ds + \int_0^t \sigma(\hat{X}_s)d\hat{W}_s,$$

The deep learning method studied below replaces the predictor step of the splitting method above by a deep neural network approximation algorithm to avoid an explicit space discretisation. This is achieved by representing each $\tilde{p}^n(z)$ by a feed-forward neural network and approximating the initial value problem (6) based on its stochastic representation using a sampling procedure. The normalisation step may then be computed either using quadrature, or, to preserve the mesh-free characteristic, by Monte-Carlo approximation.

2　Derivation and Outline of the Deep Learning Algorithm

Here, we present a concise version of the derivation laid out in detail in [4].

2.1　Feynman–Kac Representation

Assuming sufficient differentiability of the coefficient functions, the operator A^* may be expanded such that for all compactly supported smooth test functions $\varphi \in C_c^\infty(\mathbb{R}^d, \mathbb{R})$ we have

$$A^*\varphi = \mathrm{Tr}(a\,\mathrm{Hess}\,\varphi) + \langle 2\overrightarrow{\mathrm{div}}(a) - f, \mathrm{grad}\,\varphi\rangle + \mathrm{div}(\overrightarrow{\mathrm{div}}(a) - f)\varphi. \qquad (7)$$

Subtracting the zero-order term from (7), we obtain an operator that generates the auxiliary diffusion process, denoted $\hat{X}$, which is instrumental in the deep learning method.

Definition 1 Define the partial differential operator $\hat{A} : C_c^\infty(\mathbb{R}^d, \mathbb{R}) \to C_b(\mathbb{R}^d, \mathbb{R})$, with image in the set of bounded continuous function on $\mathbb{R}^d$, such that for all $\varphi \in C_c^\infty(\mathbb{R}^d, \mathbb{R})$,

$$\hat{A}\varphi = \mathrm{Tr}(a\,\mathrm{Hess}\,\varphi) + \langle 2\overrightarrow{\mathrm{div}}(a) - f, \mathrm{grad}\,\varphi\rangle$$

and the function $r : \mathbb{R}^d \to \mathbb{R}$ such that for all $x \in \mathbb{R}^d$,

$$r(x) = \mathrm{div}(\overrightarrow{\mathrm{div}}(a) - f)(x).$$

Lemma 1 For all $x \in \mathbb{R}^d$ the operator $\hat{A}$ defined in Definition 1 is the infinitesimal generator of the Itô diffusion $\hat{X} : [0, \infty) \times \Omega \to \mathbb{R}^d$ given, for all $t \geq 0$ and P-a.s. by

$$\hat{X}_t = x + \int_0^t b(\hat{X}_s)\mathrm{d}s + \int_0^t \sigma(\hat{X}_s)\mathrm{d}\hat{W}_s,$$

$$A\varphi = \langle f, \nabla\varphi \rangle + \mathrm{Tr}(a\,\mathrm{Hess}\,\varphi), \tag{4}$$

where $\mathcal{D}(A)$ denotes the domain of the differential operator A and $a = \frac{1}{2}\sigma\sigma'$. The symbol $B(\mathbb{R}^d)$ denotes the set of real-valued, bounded, Borel-measurable functions defined on $\mathbb{R}^d$.

It is well-known (see, e.g., [1]), that the unnormalised filter ρ satisfies the *filtering equation*, i.e. for all $t \geq 0$, we have $\tilde{\mathrm{P}}$-a.s. that

$$\rho_t(\varphi) = \pi_0(\varphi) + \int_0^t \rho_s(A\varphi)\,\mathrm{d}s + \int_0^t \rho_s(\varphi h')\,\mathrm{d}Y_s. \tag{5}$$

The classical splitting method for the filtering equation is given in [3] and seeks to approximate the following SPDE for the density p_t of the unnormalised filter given, for all $t \geq 0$, $x \in \mathbb{R}^d$, and P-a.s. as

$$p_t(x) = p_0(x) + \int_0^t A^* p_s(x)\,\mathrm{d}s + \int_0^t h'(x)p_s(x)\,\mathrm{d}Y_s$$

and relies on the splitting-up algorithm described in [9] and [10]. Here A^* is the formal adjoint of the infinitesimal generator A of the signal process X.

We summarise the splitting-up method below in Note 1.

Note 1 The splitting method for the filtering problem is defined by iterating the steps below with initial density $p^0(\cdot) = p_0(\cdot)$:

1. *(Prediction)* Compute an approximation $\tilde{p}^n$ of the solution to

$$\frac{\partial q^n}{\partial t}(t, z) = A^* q^n(t, z), \quad (t, z) \in (t_{n-1}, t_n] \times \mathbb{R}^d,$$
$$q^n(0, z) = p^{n-1}(z), \quad z \in \mathbb{R}^d, \tag{6}$$

at time t_n and

2. *(Normalisation)* Compute the normalisation constant with $z_n = (Y_{t_n} - Y_{t_{n-1}})/(t_n - t_{n-1})$ and the function

$$\mathbb{R}^d \ni z \mapsto \xi_n(z) = \exp\left(-\frac{t_n - t_{n-1}}{2}||z_n - h(z)||^2\right),$$

so that we can set,

$$p^n(z) = \frac{1}{C_n}\xi_n(z)\tilde{p}^n(z); \ z \in \mathbb{R}^d,$$

where $C_n = \int_{\mathbb{R}^d} \xi_n(z)\tilde{p}^n(z)\,\mathrm{d}z$.

where $\hat{W} : [0, \infty) \times \Omega \to \mathbb{R}^d$ is a d-dimensional Brownian motion and $b : \mathbb{R}^d \to \mathbb{R}^d$ is the function

$$b = 2\overrightarrow{\mathrm{div}}(a) - f.$$

From the well-known Feynman–Kac formula (see Karatzas and Shreve [6, Chapter 5, Theorem 7.6]) we can deduce the Corollary 1 below for the initial value problem.

Corollary 1 Let $d \in \mathbb{N}$, $T > 0$, let $k : \mathbb{R}^d \to [0, \infty)$ be a continuous function, let $\hat{A}$ be the operator defined in Definition 1, and let $\psi : \mathbb{R}^d \to \mathbb{R}$ be an at most polynomially growing function. Suppose that $u \in C_b^{1,2}((0, T] \times \mathbb{R}^d, \mathbb{R})$ is continuously differentiable with bounded derivative in time and twice continuously differentiable with bounded derivatives in space, and satisfies the Cauchy problem

$$\frac{\partial u}{\partial t}(t, x) + k(x)u(t, x) = \hat{A}u(t, x), \quad (t, x) \in (0, T] \times \mathbb{R}^d,$$

$$u(0, x) = \psi(x), \quad x \in \mathbb{R}^d. \tag{8}$$

Then, for all $(t, x) \in (0, T] \times \mathbb{R}^d$, we have that

$$u(t, x) = \mathbb{E}\left[\psi(\hat{X}_t) \exp\left(-\int_0^t k(\hat{X}_\tau)\, d\tau \right) \Big| \hat{X}_0 = x \right],$$

where $\hat{X}$ is the diffusion generated by $\hat{A}$.

Recall that our aim is to approximate the Fokker–Planck equation (6). Assume from now on the discrete times $\{t_0 = 0, t_1, t_2 \ldots\}$, indexed by n. Written in the form as in Corollary 1, for any timestep $n = 1, 2, \ldots$, (6) reads as

$$\frac{\partial q^n}{\partial t}(t, z) = \hat{A}q^n(t, z) + r(z)q^n(t, z), \quad (t, z) \in (t_{n-1}, t_n] \times \mathbb{R}^d,$$

$$q^n(0, z) = p^{n-1}(z), \quad z \in \mathbb{R}^d.$$

Thus, with $k = -r$, and assuming that $-r$ is non-negative in (8), we obtain by Corollary 1 the representation, for all $n \in \{1, \ldots, N\}$, $t \in (t_{n-1}, t_n]$, $z \in \mathbb{R}^d$,

$$q^n(t, z) = \mathbb{E}\left[p^{n-1}(\hat{X}_t) \exp\left(\int_{t_{n-1}}^t r(\hat{X}_\tau)\, d\tau \right) \Big| \hat{X}_{t_{n-1}} = z \right]. \tag{9}$$

Note that [4, Proposition 2.4] shows that we have a feasible minimisation problem to approximate by the learning algorithm (see also [2, Proposition 2.7]).

2.2 The Benes Filtering Model

The Benes filter is a one-dimensional nonlinear model and is used as a benchmark in the numerical studies below. As we show below, it is one of the rare cases of explicitly solvable continuous-time stochastic filtering models. Here, we are considering a special case of the more general class of Benes filters, presented, for example, in [1, Chapter 6.1].

The signal is given by the coefficient functions

$$f(x) = \alpha\sigma \tanh(\beta + \alpha x/\sigma) \text{ and } \sigma(x) \equiv \sigma \in \mathbb{R},$$

where $\alpha, \beta \in \mathbb{R}$ and the observation is given by the affine-linear sensor function

$$h(x) = h_1 x + h_2,$$

with $h_1, h_2 \in \mathbb{R}$. The density p_B of the filter solving the Benes model is then given by two weighted Gaussians (see [1, Chapter 6.1]) as

$$p_B(z) = w^+ \Phi(\mu_t^+, v_t)(z) + w^- \Phi(\mu_t^-, v_t)(z), \qquad (10)$$

where $\mu_t^\pm = M_t^\pm/(2v_t)$, $v_t = 1/(2v_t)$, and

$$w^\pm = \frac{\exp((M_t^\pm)^2/(4v_t))}{\exp((M_t^+)^2/(4v_t))\exp((M_t^-)^2/(4v_t))}$$

with

$$M_t^\pm = \pm\frac{\alpha}{\sigma} + h_1 \int_0^t \frac{\sinh(s\zeta\sigma)}{\sinh(t\zeta\sigma)}\, dY_s + \frac{h_2 + h_1 x_0}{\sigma \sinh(t\zeta\sigma)} - \frac{h_2}{\sigma}\coth(t\zeta\sigma),$$

$v_t = h_1 \coth(t\zeta\sigma)/2\sigma$, and $\zeta = \sqrt{\alpha^2/\sigma^2 + h_1^2}$.

Further, for the Benes model, the auxiliary diffusion is given as

$$\hat{X}_t = \hat{X}_0 - \int_0^t \alpha\sigma \tanh(\beta + \alpha x/\sigma)\, ds + \int_0^t \sigma\, d\hat{W}_s,$$

and the coefficient

$$r(x) = -\operatorname{div} f(x) = -\alpha^2 \operatorname{sech}^2(\beta + \alpha x/\sigma).$$

Therefore the representation of the solution to the Fokker–Planck equation (6) in the Benes case reads

$$q^n(t, z) = \mathbb{E}\left[p^{n-1}(\hat{X}_t) \exp\left(-\int_{t_{n-1}}^t \alpha^2 \operatorname{sech}^2(\beta + \alpha\hat{X}_\tau/\sigma)\, d\tau \right) \middle| \hat{X}_{t_{n-1}} = z \right].$$

2.3 Neural Network Model for the Prediction Step

To solve the Fokker–Planck equation over a rectangular domain $\Omega_d = [\alpha_1, \beta_1] \times \cdots \times [\alpha_d, \beta_d]$, we employ the sampling based deep learning method from [2]. Using the representation (9), the solution of the Fokker–Planck equation is reformulated into an optimisation problem over function space given in [4, Proposition 2.4]. This in turn yields the loss functions for the learning algorithm. Writing $\hat{X}^\xi$ for the auxiliary diffusion with $\mathrm{Unif}(\Omega_d)$-random initial value ξ, the optimisation problem is approximated by the optimisation

$$\inf_{\theta \in \mathbb{R}^{\sum_{i=2}^{L} l_{i-1} l_i + l_i}} \mathbb{E}\left[\left| \psi(\hat{X}_T^\xi) \exp\left(-\int_0^T k(\hat{X}_\tau^\xi)\, d\tau \right) - \mathcal{N}\mathcal{N}_\theta(\xi) \right|^2 \right]$$

where the solution of the PDE is represented by a neural network $\mathcal{N}\mathcal{N}_\theta$ and the infinite-dimensional function space has been parametrised by θ. Here, L denotes the depth of the neural net, and the parameters l_i are the respective layer widths. Further details can be found in [4]. A comprehensive textbook on deep learning is [5]. We apply a modified gradient descent method, called ADAM [7], to determine the parameters in the model by minimising the *loss function*

$$\mathcal{L}(\theta; \{\xi^i, \{\hat{X}_{\tau_j}^{\xi,i}\}_{j=0}^J\}_{i=1}^{N_b}) =$$

$$\frac{1}{N_b} \sum_{i=1}^{N_b} \left| \psi(\hat{X}_T^{\xi,i}) \exp(-\sum_{j=0}^{J-1} k(\hat{X}_{\tau_j}^{\xi,i})(\tau_{j+1} - \tau_j)) - \mathcal{N}\mathcal{N}_\theta(\xi^i) \right|^2,$$

where N_b is the batch size and $\{\xi^i, \{\hat{X}_{\tau_j}^{\xi,i}\}_{j=0}^J\}_{i=1}^{N_b}$ is a training batch of independent identically distributed realisations ξ^i of $\xi \sim \mathcal{U}(\Omega_d)$ and $\{\hat{X}_{\tau_j}^{\xi,i}\}_{j=0}^J$ the approximate i.i.d. realisations of sample paths of the auxiliary diffusion started at ξ^i over the time-grid $\tau_0 = 0 < \tau_1 < \cdots < \tau_{J-1} < \tau_J = T$. For the approximation of the sample paths of the diffusion we use the Euler–Maruyama method [8]. Additionally, we augment the loss $\mathcal{L}$ by an additional term to encourage the positivity of the neural network. Thus, in practice, we use the loss

$$\tilde{\mathcal{L}}(\theta; \{\xi^i, \{\hat{X}_{\tau_j}^i\}_{j=0}^J\}_{i=1}^{N_b}) = \mathcal{L}(\theta; \{\xi^i, \{\hat{X}_{\tau_j}^i\}_{j=0}^J\}_{i=1}^{N_b}) + \lambda \sum_{i=1}^{N_b} \max\{0, -\mathcal{N}\mathcal{N}_\theta(\xi^i)\}$$

with the hyperparameter λ to be chosen.

Thus, in the notation of Sect. 1.2 we replace the Fokker–Planck solution by a neural network model, i.e. we *postulate* a neural network model

$$\tilde{p}_n(z) = \mathcal{N}\mathcal{N}(z),$$

with support on Ω_d. Therefore we require the a priori chosen domain to capture most of the mass of the probability distribution it is approximating.

2.4 Monte-Carlo Normalisation Step

We then realise the normalisation step via Monte-Carlo sampling over the bounded rectangular domain Ω_d to approximate the integral

$$\int_{\mathbb{R}^d} \xi_n(z) \mathcal{N} \mathcal{N}(z) \, dz = \int_{\Omega_d} \exp\left(-\frac{t_n - t_{n-1}}{2} ||z_n - h(z)||^2\right) \mathcal{N} \mathcal{N}(z) \, dz, \qquad (11)$$

where, as defined earlier, $z_n = \frac{1}{t_n - t_{n-1}}(Y_{t_n} - Y_{t_{n-1}})$. Note that, since Ω_d is the support of the neural network $\mathcal{N}\mathcal{N}$, the right-hand side above is indeed identical to the integral over the whole space.

The sensor function in the Benes model is given by $h(x) = h_1 x + h_2$. Then, the likelihood function becomes

$$\xi_n(z) = \frac{\sqrt{2\pi}}{\sqrt{(t_n - t_{n-1})h_1^2}} \mathcal{N}_{\text{pdf}}\left(\frac{z_n - h_2}{h_1}, \frac{1}{(t_n - t_{n-1})h_1^2}\right)(z),$$

where $\mathcal{N}_{\text{pdf}}(\mu, \sigma^2)$ denotes the probability density function of a normal distribution with mean μ and variance σ^2. Therefore, we can write the integral (11) as

$$\frac{\sqrt{2\pi}}{\sqrt{(t_n - t_{n-1})h_1^2}} \mathbb{E}_Z[\mathcal{N}\mathcal{N}(Z)]; \qquad Z \sim \mathcal{N}\left(\frac{z_n - h_2}{h_1}, \frac{1}{(t_n - t_{n-1})h_1^2}\right).$$

This is an implementable method to compute the normalisation constant C_n. Thus, we can express the approximate posterior density as

$$p^n(z) = \frac{1}{C_n} \xi_n(z) \tilde{p}^n(z).$$

Therefore, the methodology is fully recursive and can be applied sequentially.

Remark 1 In low-dimensions, the usage of the Monte-Carlo method to perform the normalisation is optional, since efficient quadrature methods are an alternative. We chose the sampling based method to preserve the grid-free nature of the algorithm.

3 Numerical Results for the Benes Filter

The neural network architecture for all our experiments below is a feed-forward fully connected neural network with a one-dimensional input layer, two hidden layers with a layer width of 51 neurons each and batch-normalisation, and an output layer of dimension one (a detailed illustration can be found in [4]). For the optimisation algorithm we chose the ADAM optimiser and performed the training over 6002 epochs with a batch size of 600 samples. The initial signal and observation values are $x_0 = y_0 = 0$ and the coefficients of the Benes model were chosen as $\alpha = 3$, $\beta = 0$, $\sigma = 0.5$, $h_1 = 3$, $h_2 = 0$, and timestep $\Delta t = 0.1$ over $N = 40$ steps. The initial condition is a Gaussian density with mean 0 and standard deviation 0.001. The posterior was calculated over the domain $[-9, 2.5]$. The domain boundaries were pre-estimated using a simulation of the exact Benes filter with fixed random seed. In the case of the domain adaptation we used the precomputed evolutions from the true solution to estimate the support of the posterior and set a fixed domain adaptation schedule. The spatial resolution is 1000 uniformly spaced values in the domain of definition of the neural network. At each time step, the training of the network consumes $6002 \cdot 600 = 3{,}601{,}200$ Monte-Carlo samples. Additionally we employ a piecewise constant learning rate schedule $lr(epoch) = 10^{-(2+epoch \bmod 2001)}$ and the normalisation constant is computed using 10^7 samples each timestep. The regularising parameter $\lambda = 1$.

3.1 No Domain Adaptation

Figure 1 shows the plots for the Benes filter without domain adaptation. In Fig. 1a we observe the drift of the posterior toward the left edge of the domain. The initial bimodality, reflecting the uncertainty due to few observed values, quickly resolves and the approximate posterior tracks the signal within the domain. In Fig. 1b the bimodality is mostly visible in the Monte-Carlo prior and smoothed out by the neural network. Figure 1c and d show snapshots of the progression of the filter. The absolute error in means with respect to the Benes reference solution is plotted in Fig. 2a and shows that as the posterior reaches the left domain boundary, the error increases. This is reflected as well in the drop of probability mass, Fig. 2c, and Monte-Carlo acceptance rate, Fig. 2d at later times. It is not clear from Fig. 2a if there is a trend in the error. Further experiments need to be performed to check this hypothesis. Figure 2b shows that the neural net training consistently succeeds as measured by the L_2 distance between the Monte-Carlo reference prior and the neural net prior.

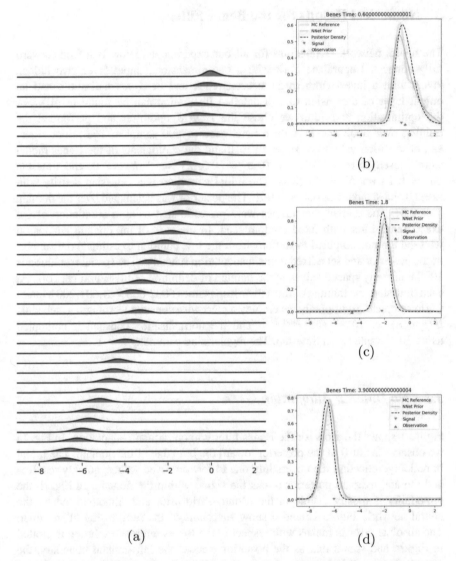

Fig. 1 Results of the combined splitting-up/machine-learning approximation applied iteratively to the Benes filtering problem (no domain adaptation). (**a**) The full evolution of the estimated posterior distribution produced by our method, plotted at all intermediate timesteps. (**b–d**) Snapshots of the approximation at times, $t = 0.6$, $t = 1.8$, and $t = 3.9$. The black dotted line in each graph shows the estimated posterior, the yellow line the prior estimate represented by the neural network, and the light-blue shaded line shows the Monte-Carlo reference solution for the prior

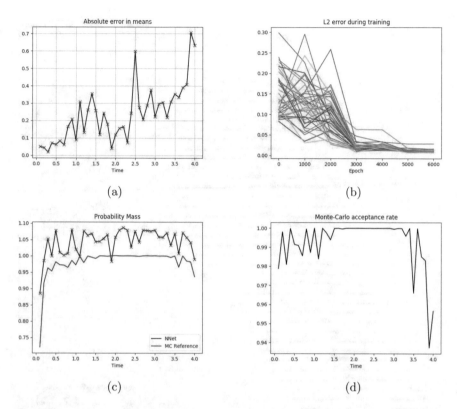

Fig. 2 Error and diagnostics for the Benes filter (no domain adaptation). (**a**) Absolute error in means between the approximated distribution and the exact solution. (**b**) L_2 error of the neural network during training with respect to the Monte-Carlo reference solution. (**c**) Probability mass of the neural network prior. (**d**) Monte-Carlo acceptance rate

3.2 With Domain Adaptation

Figure 3 shows the plots for the Benes filter with domain adaptation. In Fig. 3a we observe again the drift of the posterior toward the left edge of the domain. and the initial bimodality resolves. The approximate posterior tracks the signal within the domain. In Fig. 3b the bimodality is visible both in the prior an the posterior network. This shows that the domain adaptation helps resolve the bimodality in the nonlinear case by increasing the spatial resolution while keeping the computational cost equal. Figure 3c and d again show snapshots of the progression of the filter. The absolute error in means with respect to the Benes reference solution is plotted in Fig. 4a and shows a clear linear trend. This is an interesting phenomenon, likely due to the reduced domain size and subsequent error accumulation. The probability mass, Fig. 4c, and Monte-Carlo acceptance rate, Fig. 4d are stably fluctuating. Figure 4b shows here again that the neural net training consistently succeeds.

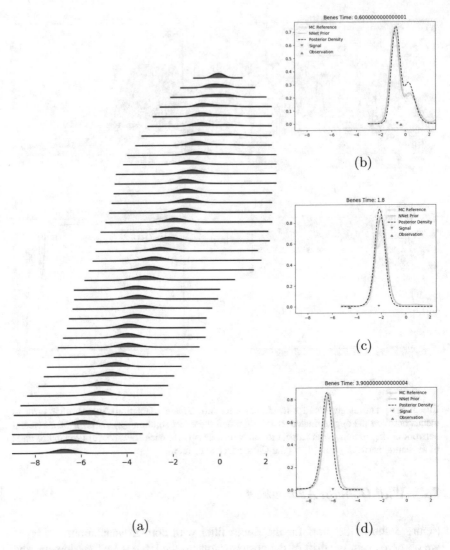

Fig. 3 Results of the combined splitting-up/machine-learning approximation applied iteratively to the Benes filtering problem (with domain adaptation). (**a**) The full evolution of the estimated posterior distribution produced by our method, plotted at all intermediate timesteps. (**b–d**) Snapshots of the approximation at times, $t = 0.6$, $t = 1.8$, and $t = 3.9$. The black dotted line in each graph shows the estimated posterior, the yellow line the prior estimate represented by the neural network, and the light-blue shaded line shows the Monte-Carlo reference solution for the prior

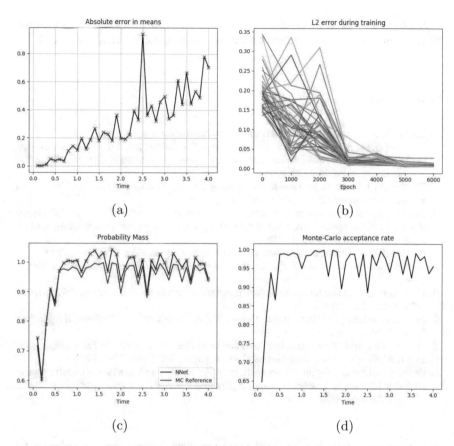

Fig. 4 Error and diagnostics for the Benes filter (with domain adaptation). (**a**) Absolute error in means between the approximated distribution and the exact solution. (**b**) L_2 error of the neural network during training with respect to the Monte-Carlo reference solution. (**c**) Probability mass of the neural network prior. (**d**) Monte-Carlo acceptance rate

4 Conclusion and Outlook

We have studied the domain adaptation in our method from [4] on the example of the Benes filter. We observed that the domain adapted method was more effective in resolving the bimodality than the non-domain adapted one. However, this came at the cost of a linear trend in the error. A possible direction for future work would thus be to investigate the optimal domain size more closely, in order to mitigate the error trend, and make full use of the increased resolution from the domain adaptation. This is subject of future research in connection with more general domain adaptation methods than the one employed here, which is specific to the Benes filter.

As already noted in the previous work [4], the possibility for *transfer learning* in our method should be explored.

A long-term goal in the development of neural network based numerical methods must of course be the rigorous error analysis, which remains a challenging task.

References

1. Alan Bain and Dan Crisan. *Fundamentals of Stochastic Filtering*. Springer, 2008.
2. Christian Beck, Sebastian Becker, Philipp Grohs, Nor Jaafari, and Arnulf Jentzen. Solving stochastic differential equations and Kolmogorov equations by means of deep learning. *arXiv preprint arXiv:1806.00421*, 2018.
3. Zhiqiang Cai, Francois Le Gland, and Huilong Zhang. *An adaptive local grid refinement method for nonlinear filtering*. PhD thesis, INRIA, 1995.
4. Dan Crisan, Alexander Lobbe, and Salvador Ortiz-Latorre. An application of the splitting-up method for the computation of a neural network representation for the solution for the filtering equations. *Preprint arXiv 2201.03283*, 2022.
5. Ian Goodfellow, Yoshua Bengio, and Aaron Courville. *Deep learning*. MIT press, 2016.
6. Ioannis Karatzas and Steven E Shreve. *Brownian Motion and Stochastic Calculus*. Springer, 1998.
7. Diederik P Kingma and Jimmy Ba. Adam: A method for stochastic optimization. *arXiv preprint arXiv:1412.6980*, 2014.
8. Peter E. Kloeden and Eckhard Platen. *Numerical Solution of Stochastic Differential Equations*. Springer, 1992.
9. François Le Gland. Time discretization of nonlinear filtering equations. In *Proceedings of the 28th IEEE Conference on Decision and Control,*, pages 2601–2606. IEEE, 1989.
10. François LeGland. Splitting-up approximation for SPDE's and SDE's with application to nonlinear filtering. In *Stochastic partial differential equations and their applications*, pages 177–187. Springer, 1992.

End-to-End Kalman Filter in a High Dimensional Linear Embedding of the Observations

Said Ouala, Pierre Tandeo, Bertrand Chapron, Fabrice Collard, and Ronan Fablet

Abstract Data assimilation techniques are the state-of-the-art approaches in the reconstruction of a spatio-temporal geophysical state such as the atmosphere or the ocean. These methods rely on a numerical model that fills the spatial and temporal gaps in the observational network. Unfortunately, limitations regarding the uncertainty of the state estimate may arise when considering the restriction of the data assimilation problems to a small subset of observations, as encountered for instance in ocean surface reconstruction. These limitations motivated the exploration of reconstruction techniques that do not rely on numerical models. In this context, the increasing availability of geophysical observations and model simulations motivates the exploitation of machine learning tools to tackle the reconstruction of ocean surface variables. In this work, we formulate sea surface spatio-temporal reconstruction problems as state space Bayesian smoothing problems with unknown augmented linear dynamics. The solution of the smoothing problem, given by the Kalman smoother, is written in a differentiable framework which allows, given some training data, to optimize the parameters of the state space model.

Keywords Kalman filter · Machine learning · Spatio-temporal interpolation

1 Introduction

Data assimilation in a broad sense can be considered as the inference of a hidden state, based on several sources of information. When considering data assimilation

S. Ouala (✉) · P. Tandeo · R. Fablet
IMT Atlantique, Lab-STICC, Brest, France
e-mail: said.ouala@imt-atlantique.fr

B. Chapron
Ifremer, LOPS, Plouzané, France

F. Collard
ODL, Locmaria-Plouzané, France

© The Author(s) 2023
B. Chapron et al. (eds.), *Stochastic Transport in Upper Ocean Dynamics*,
Mathematics of Planet Earth 10, https://doi.org/10.1007/978-3-031-18988-3_13

in the context of oceanography, these schemes exploit, in addition to some given observations, a dynamical model to perform simulations from given ocean states [1]. Unfortunately, realistic analytic parameterizations of the dynamical model, in the context of sea surface variables reconstruction, lead to computationally demanding representations [2]. Furthermore, when associated to a small subset of observations (as encountered for instance when assimilating sea surface variables with a global ocean model), these realistic models may result in modeling and inversion uncertainties. On the other hand, the analytic derivation of computationally-efficient, low-order models involves theoretical assumptions, which may not be fulfilled by real observations. These limitations motivated the exploration of interpolation techniques that do not require an explicit dynamical representation. Among other methods, Optimal Interpolation (OI) became the state-of-the-art framework [3, 4]. This technique does not need an explicit formulation of the dynamical model and rather relies on the modelization of the covariance of the spatio-temporal fields. Despite the success of OI, this technique tends to smooth the fine scale structures which motivates the development of new spatio-temporal interpolation schemes, mainly based on machine learning representations [5–10].

From the perspective of the machine learning community, state-of-the-art reconstruction techniques are usually formulated as inverse problems, where one searches to maximize the reconstruction performance of an inversion model, given the observed field as an input. Several methods were developed for this purpose in the fields of signal denoising [11, 12] and image inpainting [13] where the inversion model typically relies on a deep learning architecture. This *end-to-end* learning strategy, differs from classical inversion techniques used in geosciences, where the state-space representations (specifically the dynamical models) and the inversion schemes are a priori unrelated. The recent exploration of machine learning representations in the context of sea surface fields reconstruction was inspired by the latter methodological viewpoint, where a data-driven dynamical model is optimized based on the minimization of a forecasting cost. This data-driven prior is then plugged into a data assimilation framework to perform reconstruction based on classical (Kalman based, variational formulations) inversion schemes [7, 14, 8].

Recently, several works investigated *end-to-end* deep learning architectures in the resolution of reconstruction issues in geosciences [15–17, 10]. However, this tools, although relevant, were naturally explored in the context of image denoising and inpainting applications due to the lack of methodological formulation. When considering geosciences applications, a huge effort was carried within the geosciences community to derive reconstruction algorithms that, beyond being efficient with respect to a given metric, are robust and rely on a solid methodological formulation. From this point of view, we believe that *end-to-end* deep learning techniques should build on such methodological knowledge to propose new reconstruction solutions that can achieve both a decent performance score, and remain theoretically relevant which helps the understanding and generalization of these algorithms. From this point of view, we exploit ideas from machine learning and Bayesian filtering to propose a framework that is able to provide a relevant reconstruction of a spatio-temporal state. Specifically, we formulate a new state space model for ocean surface observations based on an augmented linear dynamical system. Assuming that the model and observation errors are Gaussian, the solution of the filtering/smoothing problem on this new state space model is given by the Kalman filter/smoother. Inspired by deep learning architectures, the Kalman recursion is written in a differentiable framework, which allows for the derivation of the parameters of the new state-space model based on a reconstruction cost of the observations.

2 Method

Motivation Let us assume the following state-space model

$$\dot{\mathbf{x}}_t = f(\mathbf{x}_t) + \boldsymbol{\eta}_t \tag{1}$$

$$\mathbf{y}_t = \mathcal{H}_t(\mathbf{x}_t) + \boldsymbol{\epsilon}_t \tag{2}$$

where $t \in [0, +\infty]$ is time. The variables $\mathbf{x}_t \in \mathbb{R}^s$ and $\mathbf{y}_t \in \mathbb{R}^n$ represent the state variables and the observations respectively. f and $\mathcal{H}_t$ are the dynamical and observation operators. $\boldsymbol{\eta}_t$ and $\boldsymbol{\epsilon}_t$ are random processes accounting for the uncertainties. They are defined as centered Gaussian processes with covariances $\mathbf{Q}_t$ and $\mathbf{R}_t$ respectively.

In the context of geosciences, and when considering the resolution of filtering and smoothing problems using data assimilation, the dynamical and observation models f and $\mathcal{H}$, the model and observation error covariances $\mathbf{Q}_t$ and $\mathbf{R}_t$ as well as the true state $\mathbf{x}_t$ of Eqs. (1) and (2) are either unavailable or too complicated to handle. In this context, we show in this work how to exploit observations $\mathbf{y}_t$ sampled from time t_1 to time t_f to learn a Bayesian scheme that allows for reconstruction applications given new observations (i.e., at time $t > t_f$).

Definition of a New State Space Model In this work, we consider an embedding of the observations as proposed in [18]. Specifically, we project our observations (or a reduced order version of our observations) into a higher dimensional space where the dynamics of the observations are assumed to be linear. Formally, in order to derive our new state-space model, we first start by writing an augmented state $\mathbf{u}_t$ such as $\mathbf{u}_t^T = [(\mathbf{M}\mathbf{y}_t)^T, \mathbf{z}_t^T]$ with $\mathbf{z}_t \in \mathbb{R}^l$ is the unobserved component of the augmented state $\mathbf{u}_t$ and $\mathbf{M} \in \mathbb{R}^{r \times n}$ with $r \leq n$ a linear projection operator (that can be used for instance in the context of reduced order modeling). The matrix $\mathbf{M}$ is assumed to have r orthogonal lines so that the matrix $\mathbf{M}^{-1} = \mathbf{M}^T$ verifies $\mathbf{M}\mathbf{M}^{-1} = \mathbf{I}$. We used in this work an Empirical Orthogonal Functions (EOF) projection. This constraints $\mathbf{M}$ to be a matrix of orthogonal eigenvectors of the covariance matrix of the centered data. The augmented state $\mathbf{u}_t \in \mathbb{R}^{d_E}$, with $d_E = l + r$, evolves in time according to the following state-space model:

$$\dot{\mathbf{u}}_t = \mathbf{A}_\sigma \mathbf{u}_t + \boldsymbol{\eta}_t \tag{3}$$

$$\mathbf{y}_t = \mathbf{M}^{-1}\mathbf{G}\mathbf{u}_t + \boldsymbol{\epsilon}_t \tag{4}$$

where the dynamical operator $\mathbf{A}_\sigma$ is a $d_E \times d_E$ matrix with coefficients σ. $\mathbf{G}$ is a projection matrix that satisfies $\mathbf{M}\mathbf{y}_t = \mathbf{G}\mathbf{u}_t$. The eigenvalues of the matrix $\mathbf{A}_\sigma$ encode the decaying and oscillating modes of the dynamics that are learned from data. Furthermore, the matrix $\mathbf{A}_\sigma$ can be constrained to be skew-symmetric (simply by imposing $\mathbf{A}_\sigma = 0.5(\mathbf{B}_\sigma - \mathbf{B}_\sigma^T)$ with $\mathbf{B}_\sigma$ a trainable matrix) so the solution of (3) will be written as a weighted sum of $d_E/2$ trainable oscillations, where the corresponding frequencies been encoded in the imaginary parts of the eigenvalues of $\mathbf{A}_\sigma$. This formulation is highly suitable for Hamiltonian (conservative) dynamical systems since the energy of the system is conserved and allows guaranteeing long term boundedness of the model. Furthermore, this formulation differs fundamentally from classical Auto Regressive (AR) models written in the space of the observations. Indeed, simple AR models only have a number of $r < d_E$ eigenvalues, which limits their expressivity.

It is worth noting that this formulation closely relates to the Koopman operator [19] where the augmented state $\mathbf{u}_t$ can be seen as a finite dimensional approximation of the infinite dimensional Hilbert space of measurements of the hidden state $\mathbf{x}_t$. This model takes advantage of a linear formulation of the dynamics in a space of observables, where the resulting model is perfectly linear for a category of dynamical regimes (typically periodic and quasi-periodic ones), and can provide a decent short-term approximation of chaotic regimes. It can also be seen as a generalization of the Dynamic Mode Decomposition (DMD) method, in which $\mathbf{u}_t = \mathbf{M}\mathbf{y}_t$.

Model and Observations Error Covariances The model and observation errors $\boldsymbol{\eta}_t$ and $\boldsymbol{\epsilon}_t$ are assumed to follow Gaussian distributions with zero mean and covariance matrices $\mathbf{Q}_{\lambda,t}$ and $\mathbf{R}_{\phi,t}$, respectively. These covariance models can be parameterized as neural networks with parameter vectors λ and ϕ.

Smoothing Scheme A Kalman smoother, based on the above state-space model, is written in a differentiable framework. The idea is to derive an analytical solution of the posterior distribution $p(\mathbf{u}_t|\mathbf{y}_{t_1:t_f})$, based on the Kalman recursion. Formally, given a regular time discretization $t \in [t_1, \ldots, t_N]$ where N is a positive integer and given the initial moments $\mathbf{u}_{t_1}^a$ and $\mathbf{P}_{t_1}^a$, the mean $\mathbf{u}^s$ and covariance $\mathbf{P}^s$ of the posterior distribution $p(\mathbf{u}_t|\mathbf{y}_{t_1:t_f})$ can be computed as follows:

$$\mathbf{u}_{t+1}^f = \mathbf{F}\mathbf{u}_t^a \tag{5}$$

$$\mathbf{P}_{t+1}^f = \mathbf{F}\mathbf{P}_t^a\mathbf{F}^T + \mathbf{Q}_{\lambda,t} \tag{6}$$

$$\mathbf{K}_{t+1} = \mathbf{P}_{t+1}^f\mathbf{H}^T[\mathbf{H}\mathbf{P}_{t+1}^f(\mathbf{H})^T + \mathbf{R}_{\phi,t}]^{-1} \tag{7}$$

$$\mathbf{u}_{t+1}^a = \mathbf{u}_{t+1}^f + \mathbf{K}_{t+1}[\mathbf{y}_{t+1} - \mathbf{H}\mathbf{u}_{t+1}^f] \tag{8}$$

$$\mathbf{P}_{t+1}^a = \mathbf{P}_{t+1}^f - \mathbf{K}_{t+1}\mathbf{H}\mathbf{P}_{t+1}^f \tag{9}$$

$$\mathbf{K}_{t+1}^s = \mathbf{P}_{t+1}^a \mathbf{F}^T (\mathbf{P}_{t+2}^f)^{-1} \tag{10}$$

$$\mathbf{u}_{t+1}^s = \mathbf{u}_{t+1}^a + \mathbf{K}_{t+1}^s [\mathbf{u}_{t+1}^s - \mathbf{u}_{t+1}^f] \tag{11}$$

$$\mathbf{P}_{t+1}^s = \mathbf{P}_{t+1}^a - \mathbf{K}_{t+1}^s (\mathbf{P}_{t+1}^f - \mathbf{P}_{t+2}^s)(\mathbf{K}_{t+1}^s)^T \tag{12}$$

where $\mathbf{F} = e^{dt\mathbf{A}_\sigma}$ with dt the prediction time step and $\mathbf{H} = \mathbf{M}^{-1}\mathbf{G}$. The smoothing (Eqs. (10), (11) and (12)) is carried backward in time with $\mathbf{P}_{t_f}^s = \mathbf{P}_{t_f}^a$ and $\mathbf{u}_{t_f}^s = \mathbf{u}_{t_f}^a$.

Learning Scheme The tuning of the trainable parameters vector $\boldsymbol{\theta} = [\sigma, \lambda, \phi]^T$ is carried using the following loss function: $\hat{\theta} = \arg\min_\theta \{\gamma_1 \mathcal{L}_1 + \gamma_2 \mathcal{L}_2\}$ where $\mathcal{L}_1 = \sum_{t=t_0}^{t_N} \|\mathbf{y}_t - \mathbf{H}\mathbf{u}_t^s\|^2$ and $\mathcal{L}_2 = \frac{1}{2}\log(|\mathbf{H}\mathbf{P}_{t+1}^f \mathbf{H}^T + \mathbf{R}_{\phi,t}|)$ $+ \frac{1}{2}\sum_{t=1}^{t=t_N} \|\mathbf{y}_t - \mathbf{H}\mathbf{u}_t^f\|_{\mathbf{H}\mathbf{P}_{t+1}^f \mathbf{H}^T + \mathbf{R}_{\phi,t}}^2$ and γ_1 and γ_2 are weighting parameters. The first term $\mathcal{L}_1$ is simply the quadratic reconstruction error of the observation. The minimization of this error helps to recover an initial guess of the trainable parameters. The second term, $\mathcal{L}_2$ is the negative log likelihood of the observations. This likelihood is derived from the likelihood of the innovation, i.e. $p(\mathbf{y}_{1:T}) = \prod_{t=1}^{t=T} p(\mathbf{y}_t|\mathbf{y}_{t-1})$ [20].

3 Numerical Experiments

3.1 Preliminary Analysis on SST Anomaly Data

As an illustration of the proposed framework, we consider scalar measurements of the anomaly of the Sea Surface Temperature (SST) in the Mediterranean Sea (8.6°N and 43.8°E). The data are computed based on of the annual 99th percentile of Sea Surface Temperature (SST) from model data [21]. The time series consists of daily measurements of the SST anomaly from 1987 to 2019. The training data is composed of a sparse sampling of the original time series, as highlighted in Fig. 1a. The proposed framework is tested with the following configuration: The augmented state space model is built with $\mathbf{M} = I_1$, and $\mathbf{z} \in \mathbb{R}^5$. The model error covariance is a constant matrix of size, $d_E \times d_E$ and the observation error covariance is a scalar parameter that corresponds to the variance of the SST anomaly measurement error. Finally, the training is carried with $\gamma_1 = 0$ and $\gamma_2 = 1$.

Figure 1b highlights the reconstruction performance of the smoothing Probability Density Function (PDF) with respect to the true (unobserved) state. Interestingly, and despite the fact that the observations used to train the parameters of the Kalman filtering scheme were extremely sparse, the proposed framework is able to catch the correct underlying frequencies. Furthermore, the coverage probability of the PDF highlights the effectiveness of the estimated model and observations error covariances.

Fig. 1 Performance of the proposed framework in the reconstruction of the smoothing PDF of SST anomaly data. (**a**) Sparse training data (**b**) Reconstructed smoothing PDF on the test set (We only visualize the standard deviation of the SST anomaly measurements)

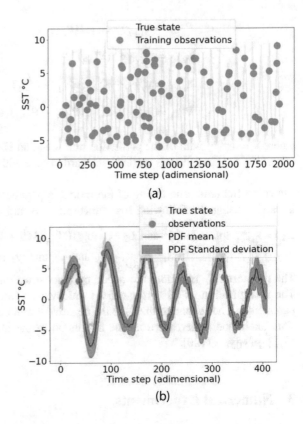

(a)

(b)

3.2   Shallow Water Equation (SWE) Case-Study

Dataset Description We consider the SWE without wind stress and bottom friction. The momentum equations are taken to be linear, and the continuity equation is solved in its nonlinear form. The direct numerical simulation is carried using a finite difference method. The size of the domain is set to $1000\,\text{km} \times 1000\,\text{km}$ with a corresponding regular discretization of 80×80. The temporal step size was set to satisfy the Courant–Friedrichs–Lewy condition ($h = 40.41\,\text{s}$). The data were subsampled to $h = 40.41 \times 10$ and 500 time-steps were used as training data. The models were validated on a series of length 100. As observations, we randomly sample 1% of the pixels with a temporal coverage given in Fig. 2.

Parametrization of the Data-Driven Models The application of the above framework in the spatio-temporal reconstruction of sea surface fields should be considered with care to account for the underlying dimensionality. In this context, and following several related works [14, 9], a patch based representations is considered in order to reduce the computational complexity of the model. Specifically, this patch based representations allows a block diagonal modelization of the covariance

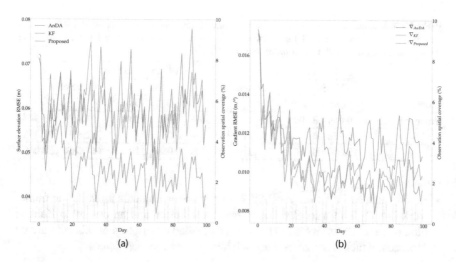

Fig. 2 Daily performance time series: we report the reconstruction performance of the sea surface elevation and its gradient in (**a**) and (**b**) respectively

matrices, which significantly reduces the computational and memory complexity of the model. This patch-based representation is fully embedded in the considered architecture to make explicit both the extraction of the patches from a 2D field and the reconstruction of a 2D field from the collection of patches. The latter involves a reconstruction operator $\mathcal{F}_r$ which is learned from data.

This patch-level representation is carried with a fixed shape of 35×35 pixels and a 10 pixels overlap between neighboring patches, resulting in a total of 16 overlapping patches. For each patch $\mathcal{P}_i$, $i = 1, \ldots, 16$ we learn an EOF basis $\mathbf{M}_{\mathcal{P}_i}$ from the training data. We keep the first 20 EOF components, which amount on average to 95% of the total variance. This patch-based decomposition is shared among all the tested models. The end-to-end Kalman filter architecture (E2EKF) is applied on a patch level with an augmented linear model operating on an embedding of dimension $d_E = 60$. The reconstructed patches are combined through the reconstruction model $\mathcal{F}_r$. This model is implemented as a residual, two blocks, convolutional neural network. The first block of the network contains four layers with 6 filters of size $k \times k$ (with k ranging from 3 to 17). The second block involves 5 layers, the first four containing 24 filters and a similar kernel size distribution as the ones in the first block, the last layer is a linear convolution with a single filter.

The proposed technique is compared in this work to the following schemes:

– Data-driven plug-and-play Kalman filter (KF): In order to show the relevance of the proposed end-to-end architecture, its plug-and-play counterpart is also tested. This model exploits the same patch based augmented linear formulation as the end-to-end one, however, the parameters of the dynamical model are trained based on a forecasting criterion and plugged into a Kalman filtering scheme.

Table 1 Surface elevation (η) interpolation experiment: reconstruction correlation coefficient and root mean squared error (RMSE) over the elevation time series and their gradient. Bold values denote smallest RMSE and highest percentage correlation

	Entire map				Missing data areas			
	RMSE		Correlation		RMSE		Correlation	
Model	$\eta(m)$	$\nabla\eta(m/°)$	η	$\nabla\eta$	$\eta(m)$	$\nabla\eta(m/°)$	η	$\nabla\eta$
Proposed, E2EKF	**0.046**	**0.009**	**73.10%**	**41.89%**	**0.047**	**0.010**	**73.80%**	**41.90%**
AnDA	0.058	0.011	52.74%	35.91%	0.060	0.011	52.82%	21.25%
KF	0.060	0.010	64.57%	21.21%	0.059	0.010	64.68%	36.06%

– Analog data assimilation (AnDA): We apply the analog data assimilation framework [14, 7] with a locally linear dynamical kernel and an ensemble Kalman filter scheme. Please refer to [14, 7] for a detailed description of this data-driven approach, which relies on nearest-neighbor regression techniques.

Following [14], an EOF based post-processing step is applied to all the reconstructions. Furthermore, in this experiment, we only report the reconstruction performance of the mean component as a relevant benchmark of the uncertainty of the above data-driven models would be out of the scope of this paper. Thus, the model and observation error covariances are assumed to be known matrices with appropriate dimensions, and the training of the proposed model is carried with $\gamma_1 = 1$ and $\gamma_2 = 0$.

Reconstructing Performance of the Proposed Data-Driven Models A quantitative analysis of the benchmark is given in Table 1 based on (i) a mean RMSE criterion and (ii) a mean correlation coefficient criterion of the interpolated fields as well as their gradients. The RMSE and correlation coefficient time series, as well as the spatial coverage of the observations are also reported in Fig. 2. Overall, the proposed end-to-end architecture leads to very significant improvements with respect to the state-of-the-art AnDA technique, as well as to its plug-and-play counterpart both in terms of RMSE and correlation coefficients. These results emphasize the importance of the end-to-end methodology with respect to classical plug-and-play techniques since, when considering data-assimilation applications, and as shown by [16, 10], the reconstruction performance depends, in addition to the quality of the dynamical prior, on the provided measurements and their sampling. Classical plug-and-play techniques, in the opposite to end-to-end strategies, ignore the latter source of information which explains the performance of our framework.

Qualitative Analysis of the Proposed Schemes the conclusions of the quantitative analysis are also illustrated through the visual analysis of the reconstructed surface elevation and its gradient in Fig. 3. Interestingly, this visual analysis reveals that the AnDA technique tend to smooth out fine-scale patterns. By contrast, the Kalman filter based schemes (in both its end-to-end and plug and play versions) achieve a better reproduction of fine scale structures, illustrated for instance by the gradients

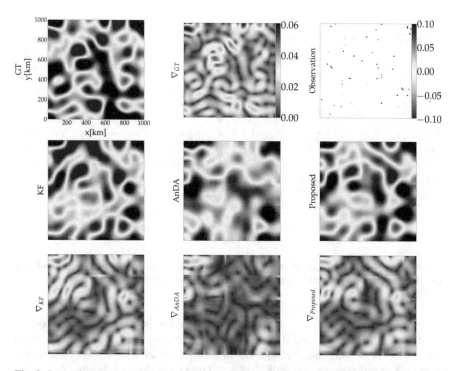

Fig. 3 Interpolation example of the surface elevation field: first row, the reference surface elevation, its gradient and the observation with missing data; second row, interpolation results using respectively the plug-and-play Augmented Koopman Kalman filter, AnDA, and the proposed E2EKF; third row, gradient of the reconstructed fields

of the field. The analysis of the spectral signatures in Fig. 4 leads to similar conclusions since, when compared to the state-of-the-art AnDA technique, as well as to its plug and play counterpart, the proposed end-to-end architecture leads to significant improvements especially regarding the reproduction of the gradient energy-level.

4 Conclusion

Spatio-temporal interpolation applications are important in the context of ocean surface modeling. For this reason, deriving new data assimilation architectures that can perfectly exploit the observations and the current advances in signal processing, modeling and artificial intelligence is crucial. In this context, this work investigated the ability of augmented linear state space models in solving smoothing issues of ocean surface observations using the Kalman filter.

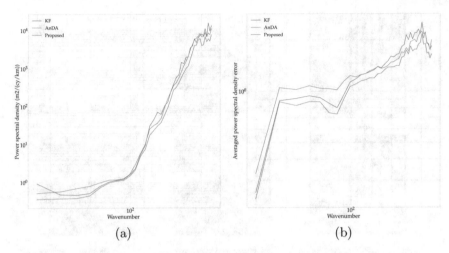

Fig. 4 Spectral comparison of the tested models: the averaged power spectral densities and their error with respect to the ground truth are given in (**a**) and (**b**) respectively

Beyond filtering and smoothing applications, we believe that the proposed framework provides an initial playground for learning approximate linear state space models of real observations. Given a sequence of sparse observations, the proposed framework may be able to unfold large scale frequencies that are useful for prediction. Interesting case studies include sea level rise and the increase of the anomaly of the sea surface temperature.

References

1. C. Gordon, C. Cooper, C. A. Senior, H. Banks, J. M. Gregory, T. C. Johns, J. F. B. Mitchell, and R. A. Wood. The simulation of SST, sea ice extents and ocean heat transports in a version of the Hadley Centre coupled model without flux adjustments. *Climate Dynamics*, 16(2–3):147–168, feb 2000.
2. van Leeuwen P. J. Nonlinear data assimilation in geosciences: an extremely efficient particle filter. *Quarterly Journal of the Royal Meteorological Society*, 136(653):1991–1999, dec 2010.
3. L. Gandin. Objective analysis of meteorological fields. 1963.
4. F Bouttier and P Courtier. Data assimilation concepts and methods march 1999. *Meteorological training course lecture series. ECMWF*, 718:59, 2002.
5. Aïda ALVERA-AZCÁRATE, Alexander Barth, Damien Sirjacobs, Fabian Lenartz, and Jean-Marie Beckers. Data interpolating empirical orthogonal functions (dineof): a tool for geophysical data analyses. *Mediterranean Marine Science*, 12(3):5–11, 2011.
6. Bo Ping, Fenzhen Su, and Yunshan Meng. An Improved DINEOF Algorithm for Filling Missing Values in Spatio-Temporal Sea Surface Temperature Data. *PLOS ONE*, 11(5):e0155928, may 2016.
7. Redouane Lguensat, Pierre Tandeo, Pierre Ailliot, Manuel Pulido, and Ronan Fablet. The Analog Data Assimilation. *Monthly Weather Review*, aug 2017.

8. Said OUALA, Cédric Herzet, and Ronan Fablet. Sea surface temperature prediction and reconstruction using patch-level neural network representations. In *IGARSS*, Italy, 2018. IEEE.
9. Said Ouala, Ronan Fablet, Cédric Herzet, Bertrand Chapron, Ananda Pascual, Fabrice Collard, and Lucile Gaultier. Neural network based kalman filters for the spatio-temporal interpolation of satellite-derived sea surface temperature. *Remote Sens.*, 10(12):1864, Nov 2018.
10. Ronan Fablet, Lucas Drumetz, and Francois Rousseau. Joint learning of variational representations and solvers for inverse problems with partially-observed data, 2020.
11. Kai Zhang, Wangmeng Zuo, Yunjin Chen, Deyu Meng, and Lei Zhang. Beyond a gaussian denoiser: Residual learning of deep cnn for image denoising. *IEEE transactions on image processing*, 26(7):3142–3155, 2017.
12. Chunwei Tian, Yong Xu, and Wangmeng Zuo. Image denoising using deep cnn with batch renormalization. *Neural Networks*, 121:461–473, 2020.
13. Zhen Qin, Qingliang Zeng, Yixin Zong, and Fan Xu. Image inpainting based on deep learning: A review. *Displays*, page 102028, 2021.
14. R. Fablet, P. H. Viet, and R. Lguensat. Data-Driven Models for the Spatio-Temporal Interpolation of Satellite-Derived SST Fields. *IEEE Transactions on Computational Imaging*, 3(4):647–657, dec 2017.
15. Qiang Zhang, Qiangqiang Yuan, Chao Zeng, Xinghua Li, and Yancong Wei. Missing data reconstruction in remote sensing image with a unified spatial–temporal–spectral deep convolutional neural network. *IEEE Transactions on Geoscience and Remote Sensing*, 56(8):4274–4288, 2018.
16. Ronan Fablet, Lucas Drumetz, and François Rousseau. End-to-end learning of energy-based representations for irregularly-sampled signals and images, 2019.
17. Bo Ping, Fenzhen Su, Xingxing Han, and Yunshan Meng. Applications of deep learning-based super-resolution for sea surface temperature reconstruction. *IEEE Journal of Selected Topics in Applied Earth Observations and Remote Sensing*, 14:887–896, 2020.
18. S Ouala, D Nguyen, L Drumetz, B Chapron, A Pascual, F Collard, L Gaultier, and R Fablet. Learning latent dynamics for partially observed chaotic systems. *Chaos: An Interdisciplinary Journal of Nonlinear Science*, 30(10):103121, 2020.
19. Bernard O Koopman. Hamiltonian systems and transformation in hilbert space. *Proceedings of the national academy of sciences of the united states of america*, 17(5):315, 1931.
20. Alberto Carrassi, Marc Bocquet, Alexis Hannart, and Michael Ghil. Estimating model evidence using data assimilation. *Quarterly Journal of the Royal Meteorological Society*, 143(703):866–880, 2017.
21. Karina von Schuckmann et al. Copernicus marine service ocean state report, issue 3. *Journal of Operational Oceanography*, 12(sup1):S1–S123, 2019.

Dynamical Properties of Weather Regime Transitions

Paul Platzer, Bertrand Chapron, and Pierre Tandeo

Abstract Large-scale weather can often be successfully described using a small amount of patterns. A statistical description of reanalysed pressure fields identifies these recurring patterns with clusters in state-space, also called "regimes". Recently, these weather regimes have been described through instantaneous, local indicators of dimension and persistence, borrowed from dynamical systems theory and extreme value theory. Using similar indicators and going further, we focus here on weather regime transitions. We use 60 years of winter-time sea-level pressure reanalysis data centered on the North-Atlantic ocean and western Europe. These experiments reveal regime-dependent behaviours of dimension and persistence near transitions, although in average one observes an increase of dimension and a decrease of persistence near transitions. The effect of transition on persistence is stronger and lasts longer than on dimension. These findings confirm the relevance of such dynamical indicators for the study of large-scale weather regimes, and reveal their potential to be used for both the understanding and detection of weather regime transitions.

Keywords Weather · Regime · Transition · Shift · Dynamical systems · Dimension · Persistence

1 Introduction

The concept of weather regime was introduced in 1949 by [1]. Broadly speaking, weather regimes are recurring, quasi-stationary states of the atmosphere, which allow to describe most of the subseasonal variability of atmospheric states, the

P. Platzer (✉) · B. Chapron
Laboratoire d'Océanographie Physique et Spatiale (LOPS), Ifremer, Plouzané, France
e-mail: paul.platzer@ifremer.fr

P. Tandeo
Lab-STICC, UMR CNRS 6285, IMT Atlantique, Plouzané, France

© The Author(s) 2023   223
B. Chapron et al. (eds.), *Stochastic Transport in Upper Ocean Dynamics*,
Mathematics of Planet Earth 10, https://doi.org/10.1007/978-3-031-18988-3_14

latter being defined through large-scale maps of either mean sea-level pressure or geopotential height. The study of weather regimes has numerous potential applications as a tool to understand subseasonal atmospheric dynamics [2]. The understanding and correct representation of weather regimes is also paramount for adequate climate projections [3].

Vautard [4] defines weather regimes through stationarity and searches for geopotential fields with a quasi-vanishing time-derivative. Others (see e.g. [5]) use cluster analysis (i.e. k-means or Gaussian Mixture Models) to find recurring patterns. To perform such analyses, one usually uses a low-order description of the atmospheric state, through empirical orthogonal functions (EOFs). Some authors simply rely on projection on a low number of EOFs (two in the case of [6]), and on forecaster's empirical knowledge of the recurrence of regimes defined through positive and negative phases of dominant EOFs.

A natural concern is not only the definition of weather regime, but also the study of their transition [5]. Statistical tools such as random forest can be used to perform such a task [7]. The performance of physics-based weather forecasts can also be assessed through their ability to predict weather regime transitions [6]. Our study of weather regime transition is noticeably motivated by the relevance and difficulty of their forecast.

We aim to focus on the time-evolution of two dynamical indicators (local dimension and persistence) around transitions between winter-time, North-Atlantic weather regimes. These indicators are relevant to the study of Atlantic-European weather regimes, as each weather regime can be associated with specific values of these indicators [8]. From this static study of weather regimes, we carry on with a dynamic study of transitions.

Note, [9] already investigated the temporal behaviour of local dimension and persistence at the mature stage of seven regimes, used to define round-year sub-seasonal variability of weather over the North-Atlantic and western Europe. These mature stages were identified as local minima of the weather regime index defined by [10] as the projection of the instantaneous atmospheric state on the atmospheric state associated with each regime. Hochman et al. [9] showed that the so-defined mature stages of weather regimes coincided with locally low values of the dimension and inverse persistence, and that these mature stages were both preceded and followed by higher relative values of these indicators. The present paper is concerned with weather regime transitions, which are located between weather regime mature stages. We therefore expect to confirm the relatively higher values of dimension and persistence observed by [9] before and after regime mature stages. However, our study could reveal varying behaviours as we focus on transitions from one specific regime to another, while the study of [9] does not specify which regime precedes or follows a given mature stage.

Our analysis also bears similarity with the one of [11], in which the temporal behaviour of local dimension and persistence during Eastern Mediterranean cold spells was examined. The main difference with the present study is the nature of the event of interest: we are interested in transitions between weather regimes, while cold spells could be viewed as a special type of weather regime (a particular case of Cyprus Lows which is the dominant regime responsible for precipitation in the Eastern Mediterranean region).

The next section is the core of our paper and reviews the results of our study, describing salient features of the time-evolution of dimension and persistence around transitions between four winter-time North-Atlantic weather regimes. The following section draws perspectives and proposes potential applications to real-world meteorological issues. Appendix sections provide details to the tools and data used in the present study.

2 European-Atlantic Weather Regime Transitions

An EOF-decomposition is performed (see section "Empirical Orthogonal Functions") of winter-time, reanalysed sea-level pressure fields described in Appendix 1. A weather-regime analysis follows using a Gaussian Mixture Model with four modes, corresponding to four weather regimes, in a reduced-space spanned by the three first EOFs (see section "Gaussian Mixture Model" for a discussion). The resulting regimes are shown in Fig. 1 in EOF space and there centroids are shown in Fig. 2 as SLP-anomaly maps.

Figure 1 illustrates that the four regimes are mostly defined through EOF1 and EOF2, as the centroids' EOF3-coordinates are close to zero. Two regimes are associated with positive-negative phases of the first EOF, corresponding to a strong north-south pressure gradient (see Fig. 2), and we label these regimes NAO+ and NAO− to match previous works in the litterature. The two other regimes are

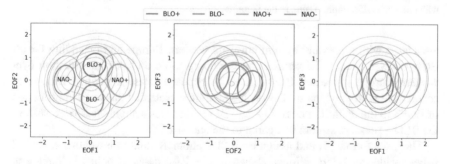

Fig. 1 Weather regimes as cluster distributions from the fit of a Gaussian Mixture Model to winter-time sea-level-pressure anomaly (SLPa) from reanalysis data. The fit is performed in reduced space through projection of SLPa maps on three leading empirical orthogonal functions (EOF). Colored contours show the 0.75σ (thick lines) and 1.25σ (thin lines) ellipses of each distribution around their centroids, with σ denoting standard deviation. Grey contours show the whole GMM distribution through marginal distributions in two-dimenisonal EOF-subspaces. Regime names are assigned from comparison with other scientific studies found in the litterature (see Fig. 2)

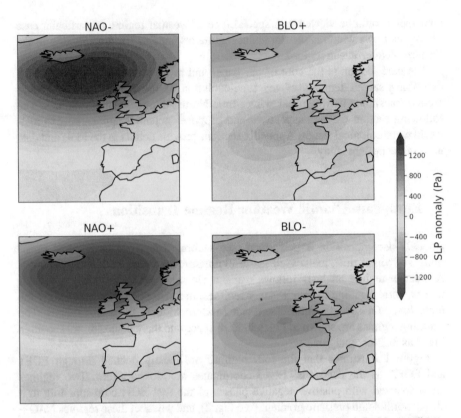

Fig. 2 Weather regimes as sea-level-pressure anomalies in (longitude, latitude) coordinates (coastlines are shown), defined by the distributions' centroids from a Gaussian Mixture Model (see Fig. 1 and section "Gaussian Mixture Model"). Regime names are assigned from comparison with other scientific studies found in the literature

associated with a pressure system covering western Europe and extending far-off Europe's west-coast. The regime corresponding to an anticyclonic situation over western Europe is termed BLO+, and its opposite phase is termed BLO−, in accordance with previous studies on such regimes. Note that the small contribution of EOF3 to the definition of BLO+ and BLO− induces a slight west-ward shift of the BLO− pressure system compared to the one of BLO+.

Then, we follow [5] and assign each SLP-anomaly field to a weather regime if it lies inside the 1.25σ ellipses, shown in Fig. 1 (in cases of points belonging to two regimes, we assign the regime with highest probability), otherwise no regime is assigned. Next, for any regimes "A" and "B", a transition from regime "A" to regime "B" is defined as either the consecutive passing from "A" to "B" or the consecutive passing from "A" to "no regime" and then to "B" (note that this allows transitions from a regime to itself). As we are interested in the behaviour of dynamical indicators around transitions, we discard transitions of the type "A"→"no regime"→"B" if the "no regime" phase exceeds 24 h.

3 Dimensionality Around Transitions

The local dimension of sea-level pressure fields is used as an indicator of the state of the atmosphere. Details on this indicator and how is it computed can be found in section "Local Dimensions".

In Fig. 3, one observes statistics of dimension-versus-time profiles centered on transitions. The number of transitions on which the statistics were computed is also mentioned, showing preferred transitions in agreement with [5]. Several behaviours can be observed.

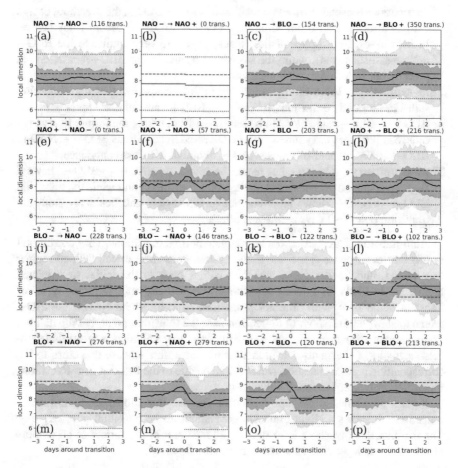

Fig. 3 Typical profiles of local dimension versus time, centered at transition point, for each possible transitions. Light (resp. dark) greys fill between the 0.05 and 0.95 (resp. 0.25 and 0.75) quantiles, while the dark lines show the average dimension profile around transition from regime "A" to regime "B". In red, statistics over each regime (with no restriction to transitions) are shown. Red dotted (resp. dashed) lines show the 0.05 and 0.95 (resp. 0.25 and 0.75) quantiles, while the full red lines show the average dimension of regime "A" and "B"

Smooth transition   The transition BLO+ → NAO− shows a smooth transition from the dimension statistics of regime BLO+ to the statistics of regime NAO− over a transition period of ∼1 day, starting after the transition, with no particular behaviour at the transition itself.

Dimension overshoot   Right before, after, or during transitions NAO− → BLO−, BLO− → BLO+, BLO+ → BLO−, and BLO+ →NAO+, the local dimension statistics exceed what is expected from statistics computed over each regime. Transitions BLO− → BLO+ and BLO+ → BLO− show the highest intensity of dimension overshoot (around +1 in dimension), with the average dimension near transition (black, full) reaching the 0.75 quantile of the regime distributions (dashed, red). For transition BLO− → BLO+, the overshoot occurs ∼1 day after the transition, while for BLO+ → BLO− it occurs 1 day before. In both cases, transition-statistics (black, grey) are very similar to the BLO− regime-statistics (red), while the overshoot occurs in the BLO+ phase, and is preceded or followed by an undershoot.

Time-symmetry   From the previous description, it appears that the dimension statistics around transition BLO− →BLO+ are almost symmetric to BLO+ →BLO−: the latter can be recovered from taking the former in reverse-time. Similar types of symmetry can be observed in transitions BLO+ ↔NAO+, BLO− ↔NAO+, and BLO+ ↔NAO−, although with less confidence.

Time-asymmetry   On the other hand, the transition NAO− →BLO− shows a slight overshoot of dimension statistics at the transition point while the transition BLO− →NAO− shows an overshoot of dimension statistics away from the transition point (∼2 days before and after).

Auto-transitions are harder to interpret than normal transitions. They correspond to trajectories in phase-space where the system goes from a well-defined regime to a mixed, undefined regime, and then comes back to the initial well-defined regime. It is likely that these auto-transitions actually mix different types of transient behaviours, with different properties. Auto-transition NAO+ →NAO+ seems to show an overshoot of dimension near the transition point, but the number of transitions (57) is small and therefore only low confidence is attributed to these statistics. Other auto-transition statistics are rather smooth and close to the corresponding regime-statistics, which might be due to the fact that auto-transitions mix different types of transient behaviours.

Figure 5b shows dimension statistics for all transitions, excluding auto-transitions. It shows a slight dimension overshoot at the transition point ±1 day. The fact that this overshoot is so small is an indicator of the variety of behaviours near transition, depending on which regimes are involved.

4 Persistence Around Transitions

We now use the inverse persistence θ (also called extremal index) of sea-level pressure fields as an indicator of the state of the atmosphere. Details on this indicator and how is it computed can be found in section "Inverse Persistence θ".

In Fig. 4, we show the result of the same procedure followed in the previous section, but replacing the local dimension by the inverse persistence. As these two variables are correlated, the behaviour of inverse persistence resembles the one of dimension around much of the observed transitions. However, the difference between transition-statistics and regime-statistics appear to be more significant for θ than for the dimension, with some special behaviours described below.

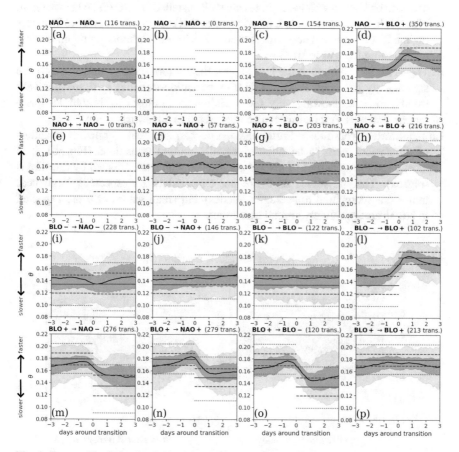

Fig. 4 Same as Fig. 3, but for the inverse persistence θ (also called extremal index). High values indicate a rapidly changing dynamical system

Transitions to/from BLO+ The BLO+ regime-statistics of θ are much higher than the ones of other regimes, with most values concentrated between 0.17 and 0.19, and almost all values above 0.16. We therefore see high variations of θ around transitions from or to BLO+. However, when one is in the regime BLO+, either after or before a transition, we do not observe an overshoot as with the dimension. Rather, we see that the transition-statistics match the BLO+ statistics very near the transition point, while they are much lower 2–3 days away from the transition. This means that, in the regime BLO+, the inverse persistence is much lower either 2–3 days before or 2–3 days after any transition. Also, the values of θ in regimes NAO± and BLO−, up to at least three days around a transition from or to BLO+, are much higher than expected from intra-regime statistics.

We can interpret these fact using the results of [9] who observed a strong decrease of θ when weather regimes are well-installed. Therefore, what we see in Figs. 3d, h, l, m–p and 4d, h, l, m–p indicates that the systems rapidly exits/enters regime BLO+, while it needs more time to exit/enter neighbouring regimes when transitioning from or to BLO+.

BLO−↔NAO+ Although the NAO+ and BLO− intra-regime statistics of θ are significantly different, BLO− ↔NAO+ transition-statistics of θ are relatively smooth in time, showing very few variations, and closer to the NAO+ intra-regime statistics. Again, this can be interpreted as a slow transition.

Low-quantiles overshoot From Fig. 5, one can see that, while all quantiles of dimension seem to be affected equally around transitions (Fig. 5b), it is mostly the low quantiles of inverse persistence which are affected by transitions (Fig. 5a). That is, values of θ are not expected to be especially large near transitions (compared to average statistics), but small values of θ are expected to be extremely unlikely around transitions.

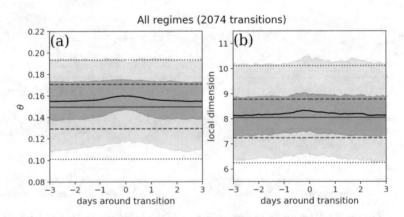

Fig. 5 In grey: statistics (0.05, 0.25, 0.75 and 0.95 quantiles, as well as mean) of inverse persistence (**a**) and local dimension (**b**) over all transitions, discarding auto-transitions (from regime "A" to "A"). In red: statistics (0.05, 0.25, 0.75 and 0.95 quantiles, as well as mean) over all values from the dataset (winter-time from 1956 to 2015), without restriction to transitions

Already mentioned earlier, we discard transitions "A→B" if the "no regime" phase between regimes "A" and "B" exceeds 24 h. Raising the maximum length of this "no regime" phase allows to find more transitions, and results in a slight smoothing of the profiles of Figs. 3 and 5, but the observed tendencies remain. Reducing the maximum length of the "no regime" phase between regimes "A" and "B" results in slightly sharper, yet noisier profiles (not shown).

5 Conclusion and Perspectives

The analysis of reanalysed sea-level pressure maps covering a large part of the North-Atlantic ocean and western Europe, demonstrates that local dynamical indicators of dimension and persistence display great sensitivity to transitions between weather regimes. In particular, we observe higher values of dimension and lower values of persistence near transitions, which is in agreement both with the early definition of weather regimes (as quasi-stationary, low-order recurring states) and with recent studies of weather regimes through these same two dynamical indicators. The study reveals non-homogeneous behaviour of these indicators near transitions, meaning that different transition show different signatures in terms of time-variation of dimension and persistence. Furthermore, we observe that the fingerprint of transitions is more pronounced for persistence than for dimension, and that it spreads over a larger duration (more than ± 3 days for persistence but around ± 1.5 day for dimension).

This study, combined with recent studies on weather regimes and dynamical indicators, confirm the relevance of these indicators for the understanding of weather regimes, and even reveal the potential for these indicators to be used in the definition of weather regimes. Present findings also indicate that each transition could be identified through the time-behaviour of dimension and persistence. This has great implications and shall motivate further investigations on how to use these indicators for the purpose of detecting regime transitions. However, for each transition we still observe a great variability of time-profiles of dimension and persistence. This suggests to use a variety of related indicators, and not only these two. Recent studies have used these indicators on separated scales, allowing to explore variations in dimensionality and persistence of small-scale variables [23]. Our current analyses also reveal a signature of large-scale weather regime transitions in the time-variation of small-scale dimension and persistence, however with less intensity than for large-scale dynamical indicators (not shown). We interpret this as a hint that small-scale organization may be necessary to large-scale transitions. Other local indicators also based on analogues such as the ones used by [24] and [25] shall also be considered in an attempt to predict transitions.

Acknowledgments We thank Pierre Ailliot for fruitful discussions on Gaussian Mixture Models. This work was financially supported by the ERC project 856408-STUOD. Support for the Twentieth Century Reanalysis Project version 3 dataset is provided by the U.S. Department of Energy, Office of Science Biological and Environmental Research (BER), by the National Oceanic and Atmospheric Administration Climate Program Office, and by the NOAA Physical Sciences Laboratory. We thank the anonymous reviewer for helpful comments and suggestions.

Appendix 1: Data Description: Twentieth Century Reanalysis

We use data from the 3rd version of the twentieth Century Reanalysis, which combines surface observations of synoptic pressure and NOAA's Global Forecast System, and prescribes sea surface temperature and sea ice distribution [12].

From this reanalysis we extract the ensemble-mean, sea-Level pressure maps from year 1956 to 2015, at 3h-intervals. We do not use preceding years in order to avoid inconsistency between past, observation-scarce data, and more recent data, better constrained by observations. We could also have selected only data from the satellite era starting in 1979, but this would have diminished the statistical significance of our work.

We focus on a 41×41 grid at $1°$-resolution covering longitudes $30W \leq LON \leq 10E$ and latitudes $30N \leq LAT \leq 70N$, including western Europe and the eastern part of the North-Atlantic Ocean (see Fig. 2). We use only extended-winter data, from October to March, as is typical in North-Atlantic weather-regime studies (see e.g., [9, 6, 8]).

Appendix 2: Statistical Descriptors

Empirical Orthogonal Functions

To study winter-time SLP fields, we use the empirical orthogonal function decomposition, also called principal component analysis [13]. It allows to decompose any spatial field (snapshot) of SLP-anomaly (SLPa) onto orthogonal maps (EOFs), ordered by their respective contribution to the total variability in time of SLPa fields. To compute SLPa, we remove a moving seasonal-average using data from ± 10 years and ± 5 calendar-days, with a Gaussian kernel to give more weight to neighbouring years and calendar days.

In our case, EOFs n°1–7 contribute respectively to 41%, 24%, 14%, 5.5%, 4.8%, 2.2% and 1.5% of the total signal variance. No that, for our analyses of weather regimes, we use only EOFs n°1–3, which contribute collectively to 79% of the total variance.

Gaussian Mixture Model

A Gaussian Mixture Model (GMM) assumes that the random variable it describes is the result of pooling from a finite number of sub-populations (in our case, regimes) whose distributions are Gaussian [14]. Expectation-maximization (EM) allows to find optimal parameters (averages and covariances) of the Gaussian distributions, once the number of regimes has been fixed.

We follow [5], and make a GMM EM-fit using a finite number of EOFs. As we allow the covariances to have any possible shape, the number of parameters to be optimized depends exponentially on the number of EOFs kept, we therefore have not tried using more than 5 EOFs. Then, once the number of EOFs is fixed, a trade-off between the number of parameters (dictated by the number of regimes) and the model adequacy to the data can be found by computing either the Bayesian Information Criterion or the average log-likelihood over an independent set [16]. However, as in the study by [5], we find a very low sensitivity of these indicators to the number of regimes chosen (not shown). We also compute the Silhouette score proposed by [15] to estimate the degree of overlapping between regimes, and find that using more EOFs always leads to more overlapping, and so does using more regimes but to a lesser extent (not shown).

In the end, we make the choice of keeping 3 EOFs and 4 regimes. The choice of 3 EOFs is motivated by the fact that each of the three first EOFs account for more than 10% of the total variance, while EOFs n°4 and further only represent up to ~5%. This has the consequence that, even when we retain more than 3 EOFs, the regime centroids found through GMM EM-fits are mostly defined by their projection on the 3 first EOFs, as projections on EOFs 4 and 5 are always closer to 0 then one of the other projections (not shown). The choice of 4 regimes is motivated by the adequacy with other studies [6] and operational weather-forecasting services such as ECMWF who divide into 4 quadrants the reduced-space formed by the projection of geopotential height fields onto their corresponding first-2 EOFs.

Appendix 3: Dynamical Indicators

Local Dimensions

We use the same estimator of local dimension as [8], borrowing the python code from the Chaotic Dynamical Systems Kit (https://github.com/yrobink/CDSK). This estimator is based on a definition of local dimension at any point z in state-space through the extreme-value distribution of the observable $g_z : x \rightarrow g_z(x) = -\log \text{dist}(z, x)$ for any other state-space vector x (where "dist" is any distance in the mathematical sense). Large values of this observable are found for points x close to z: these points are called "analogues" of z in the atmospheric- and ocean-sciences community. Then, the probability that $g(x)$ exceeds a given threshold ρ is exponential (see, for instance, [17]):

$$P\left(g_z(x) > \rho\right) \propto \exp(-\rho\, d(z))\,, \tag{A.1}$$

where $d(z)$ is the local-dimension that we estimate here. The geometric interpretation of this dimension is that in a space of dimension d, the typical number of points inside a sphere of radius r scales as r^d. Although such an interpretation of dimension has been connected to the distances to analogues for a long time (see for instance [18] and the famous Grassberger-Proccacia algorithm [19]), only recent works have used extreme-value theory to provide instantaneous, local estimators of dimension [20]. These recent tools are particularly suited for the study of local behaviours, while previous works focused on average, global indicators.

Recently, distances between analogues x and their target z have been shown to follow distributions whose parameters are given by the length of the available dataset, the analogue rank, and the local dimension as estimated in this paper [21]. This indicator is thus both relevant from a dynamical systems point of view and for practical use of data-based methods.

Inverse Persistence θ

However, Eq. A.1 is not valid when the system passes close to a fixed point, as this causes trajectories to slow down. In this case, another parameter called the extremal index, or inverse persistence, comes into play:

$$P\left(g_z(x) > \rho\right) \propto \exp(-\rho\, \theta(z) d(z))\,, \tag{A.2}$$

with $0 < \theta(z) \le 1$. Low values of θ correspond to highly persistent areas of state-space. It can be interpreted as the inverse mean residence time within a sphere centered on z (if divided by the time-increment between two consecutive points in the dataset, which is 3 h in our case). We estimate this parameter with the Süveges likelihood estimator [22]. It is based on counting consecutive points inside a ball centered on z (i.e., analogues of the same point z that are also consecutive points in the time-ordered dataset).

References

1. Levick, R. B. M. (1949). Fifty years of English weather. Weather, 4, 206–211.
2. White, C. J., Carlsen, H., Robertson, A. W., Klein, R. J., Lazo, J. K., Kumar, A., et al. (2017). Potential applications of sub seasonal-to-seasonal (S2S) predictions. Meteorological Applications, 24, 315–325.
3. Dorrington, J., Strommen, K., & Fabiano, F. (2021). How well does CMIP6 capture the dynamics of Euro-Atlantic weather regimes, and why?. Weather and Climate Dynamics Discussions, 1–41.

4. Vautard, R. (1990). Multiple weather regimes over the North Atlantic: Analysis of precursors and successors. Monthly weather review, 118(10), 2056–2081.
5. Kondrashov, D., Ide, K., & Ghil, M. (2004). Weather regimes and preferred transition paths in a three-level quasigeostrophic model. Journal of the atmospheric sciences, 61(5), 568–587.
6. Ferranti, L., Magnusson, L., Vitart, F., & Richardson, D. S. (2018). How far in advance can we predict changes in large-scale flow leading to severe cold conditions over Europe? Quarterly Journal of the Royal Meteorological Society, 144(715), 1788–1802.
7. Kondrashov, D., Shen, J., Berk, R., D'Andrea, F., & Ghil, M. (2007). Predicting weather regime transitions in Northern Hemisphere datasets. Climate Dynamics, 29(5), 535–551.
8. Faranda, D., Messori, G., & Yiou, P. (2017). Dynamical proxies of North Atlantic predictability and extremes. Scientific reports, 7(1), 1–10.
9. Hochman, A., Messori, G., Quinting, J. F., Pinto, J. G., & Grams, C. M. (2021). Do Atlantic-European Weather Regimes Physically Exist?. Geophysical Research Letters, 48(20), e2021GL095574.
10. Michel, C., & Rivière, G. (2011). The link between Rossby wave breakings and weather regime transitions. Journal of the Atmospheric Sciences, 68(8), 1730–1748.
11. Hochman, A., Scher, S., Quinting, J., Pinto, J. G., & Messori, G. (2020). Dynamics and predictability of cold spells over the Eastern Mediterranean. Climate Dynamics, 1–18.
12. Slivinski, L. C., Compo, G. P., Whitaker, J. S., Sardeshmukh, P. D., Giese, B. S., McColl, C., ...& Wyszyński, P. (2019). Towards a more reliable historical reanalysis: Improvements for version 3 of the Twentieth Century Reanalysis system. Quarterly Journal of the Royal Meteorological Society, 145(724), 2876–2908.
13. Abdi, H., & Williams, L. J. (2010). Principal component analysis. Wiley interdisciplinary reviews: computational statistics, 2(4), 433–459.
14. Reynolds, D. A. (2009). Gaussian mixture models. Encyclopedia of biometrics, 741(659–663).
15. Kaufman, L., & Rousseeuw, P. J. (2009). Finding groups in data: an introduction to cluster analysis. John Wiley & Sons.
16. McLachlan, G. J., & Rathnayake, S. (2014). On the number of components in a Gaussian mixture model. Wiley Interdisciplinary Reviews: Data Mining and Knowledge Discovery, 4(5), 341–355.
17. Caby, T., Faranda, D., Mantica, G., Vaienti, S., & Yiou, P. (2019). Generalized dimensions, large deviations and the distribution of rare events. Physica D: Nonlinear Phenomena, 400, 132143.
18. Farmer, J. D., & Sidorowich, J. J. (1987). Predicting chaotic time series. Physical review letters, 59(8), 845.
19. Grassberger, P., & Procaccia, I. (1983). Characterization of strange attractors. Physical review letters, 50(5), 346.
20. Lucarini, V., Faranda, D., de Freitas, J. M. M., Holland, M., Kuna, T., Nicol, M., ...& Vaienti, S. (2016). Extremes and recurrence in dynamical systems. John Wiley & Sons.
21. Platzer, P., Yiou, P., Naveau, P., Filipot, J. F., Thiébaut, M., & Tandeo, P. (2021). Probability distributions for analog-to-target distances. Journal of the Atmospheric Sciences, 78(10), 3317–3335.
22. Süveges, M. (2007). Likelihood estimation of the extremal index. Extremes, 10(1), 41–55.
23. Alberti, T., Daviaud, F., Donner, R. V., Dubrulle, B., Faranda, D., & Lucarini, V. (2021). Chameleon attractors in a turbulent flow. arXiv preprint arXiv:2112.10488.
24. Blanc, A., Blanchet, J., & Creutin, J. D. (2021). Past Evolution and Recent Changes in Western Europe Large-scale Circulation. Weather and Climate Dynamics Discussions, 1–27.
25. Yiou, P., Cattiaux, J., Ribes, A., Vautard, R., & Vrac, M. (2018). Recent Trends in the Recurrence of North Atlantic Atmospheric Circulation Patterns. Complexity, 2018.

Frequentist Perspective on Robust Parameter Estimation Using the Ensemble Kalman Filter

Sebastian Reich

Abstract Standard maximum likelihood or Bayesian approaches to parameter estimation for stochastic differential equations are not robust to perturbations in the continuous-in-time data. In this paper, we give a rather elementary explanation of this observation in the context of continuous-time parameter estimation using an ensemble Kalman filter. We employ the frequentist perspective to shed new light on two robust estimation techniques; namely subsampling the data and rough path corrections. We illustrate our findings through a simple numerical experiment.

Keywords Parameter estimation · Stochastic differential equations · Ensemble Kalman filter · Frequentist approach · Rough path theory

1 Introduction

In this note, we consider the well-studied problem of parameter estimation for stochastic differential equations (SDEs) from continuous-time observations $X_t^\dagger$, $t \in [0, T]$ [25]. It is well-known that the corresponding maximum likelihood estimator does not depend continuously on the observations $X_t^\dagger$, $t \in [0, T]$, which can result in a systematic estimation bias [27, 14]. In other words, the maximum likelihood estimator is not robust with respect to perturbations in the observations. Here, we revisit this problem from the perspective of online (time-continuous) parameter estimation [6, 11] using the popular ensemble Kalman filter (EnKF) and its continuous-time ensemble Kalman-Bucy filter (EnKBF) formulations [15, 10, 26]. As for the corresponding maximum likelihood approaches, the EnKBF does not depend continuously on the incoming observations $X_t^\dagger$, $t \geq 0$, with respect to the uniform norm topology on the space of continuous functions. This fact has been first investigated in [9] using rough path theory [16]. In particular, as already

S. Reich (✉)
Institute of Mathematics, University of Potsdam, Potsdam, Germany
e-mail: sebastian.reich@uni-postdam.de

© The Author(s) 2023   237
B. Chapron et al. (eds.), *Stochastic Transport in Upper Ocean Dynamics*,
Mathematics of Planet Earth 10, https://doi.org/10.1007/978-3-031-18988-3_15

demonstrated for the related maximum likelihood estimator in [14], rough path theory allows one to specify an appropriately generalised topology which leads to a continuous dependence of the EnKBF estimators on the observations. Here we expand the analysis of [9] to a frequentist analysis of the EnKBF in the spirit of [29], where the primary focus is on the expected behaviour of the EnKBF estimators over all admissible observation paths. One recovers that the discontinuous dependence of the EnKBF estimators on the driving observations results in a systematic bias from a frequentist perspective. This is also a well known fact for SDEs driven by multiplicative noise [23].

The proposed frequentist perspective naturally enables the study of known bias correction methods, such as subsampling the data [27], as well as novel de-biasing approaches in the context of the EnKBF.

In order to facilitate a rather elementary mathematical analysis, we consider only the very much simplified problem of parameter estimation for linear SDEs. This restriction allows us to avoid certain technicalities from rough path theory and enables a rather straightforward application of the numerical rough path approach put forward in [13]. As a result we are able to demonstrate that the popular approach of subsampling the data [2, 27, 5] can be well justified from a frequentist perspective. The frequentist perspective also suggests a rather natural approach to the estimation of the required correction term in the case an EnKBF is implemented without subsampling.

We end this introductory paragraph with a reference to [1], which includes a broad survey on alternative estimation techniques. We also point to [9] for an in-depth discussion of rough path theory in connection to filtering and parameter estimation.

The remainder of this paper is structured as follows. The problem setting and the EnKBF are introduced in the subsequent Sect. 2. The frequentist perspective and its implications on the specific implementations of an EnKBF in the context of low and high frequency data assimilation are laid out in Sect. 3. The importance of these considerations becomes transparent when applying the EnKBF to perturbed data in Sect. 4. Here again, we restrict attention to a rather simple model setting taken from [17] and also used in [9]. As a result we build a clear connection between subsampling and the necessity for a correction term in the case high frequency data is assimilated directly. A brief numerical demonstration is provided in Sect. 5, which is followed by a concluding remark in Sect. 6.

2   Ensemble Kalman Parameter Estimation

We consider the SDE parameter estimation problem

$$dX_t = f(X_t, \theta)dt + \gamma^{1/2}dW_t \tag{1}$$

subject to observations $X_t^\dagger, t \in [0, T]$, which arise from the reference system

$$dX_t^\dagger = f^\dagger(X_t^\dagger)dt + \gamma^{1/2}dW_t^\dagger, \tag{2}$$

where the unknown drift function $f^\dagger(x)$ typically satisfies $f^\dagger(x) = f(x, \theta^\dagger)$ and $\theta^\dagger$ denotes the true parameter value. Here we assume for simplicity that the unknown parameter is scalar-valued and that the state variable is d-dimensional with $d \geq 1$. Furthermore, W_t and $W_t^\dagger$ denote independent standard d-dimensional Brownian motions and $\gamma > 0$ is the (known) diffusion constant.

Following the Bayesian paradigm, we treat the unknown parameter as a random variable Θ. Furthermore, we apply a sequential approach and update Θ with the incoming data $X_t^\dagger$ as a function of time. Hence we introduce the random variable Θ_t which obeys the Bayesian posterior distribution given all observations $X_\tau^\dagger$, $\tau \in [0, t]$, up to time $t > 0$. Furthermore, instead of exactly solving the time-continuous Bayesian inference problem as specified by the associated Kushner–Stratonovitch equation [6, 26], we define the time evolution of Θ_t by an application of the (deterministic) ensemble Kalman–Bucy filter (EnKBF) mean-field equations [10, 26], which take the form

$$d\Theta_t = \gamma^{-1}\pi_t\left[(\theta - \pi_t[\theta]) \otimes f(X_t^\dagger, \theta)\right]dI_t, \tag{3a}$$

$$dI_t = dX_t^\dagger - \frac{1}{2}\left(f(X_t^\dagger, \Theta_t) + \pi_t[f(X_t^\dagger, \theta)]\right)dt, \tag{3b}$$

where π_t denotes the probability density function (PDF) of Θ_t and $\pi_t[g]$ the associated expectation value of a function $g(\theta)$. The column vector I_t, defined by (3b), is called the innovation, while the row vector

$$K_t(\pi_t) = \gamma^{-1}\pi_t\left[(\theta - \pi_t[\theta]) \otimes f(X_t^\dagger, \theta)\right], \tag{4}$$

premultiplying the innovation in (3a) is called the gain. Here the notation $a \otimes b = ab^T$, where a, b can be any two column vectors, has been used. The initial condition $\Theta_0 \sim \pi_0$ is provided by the prior PDF of the unknown parameter.

A Monte-Carlo implementation of the mean-field equations (3) leads to the interacting particle system

$$d\Theta_t^{(i)} = \gamma^{-1}\pi_t^M\left[(\theta - \pi_t^M[\theta]) \otimes f(X_t^\dagger, \theta)\right]dI_t^{(i)}, \tag{5a}$$

$$dI_t^{(i)} = dX_t^\dagger - \frac{1}{2}\left(f(X_t^\dagger, \Theta_t^{(i)}) + \pi_t^M[f(X_t^\dagger, \theta)]\right)dt, \tag{5b}$$

$i = 1, \ldots, M$, where expectations are now taken with respect to the empirical measure. That is,

$$\pi_t^M[g] = \frac{1}{M}\sum_{i=1}^{M}g(\Theta_t^{(i)}) \tag{6}$$

for given function $g(\theta)$, and all Monte-Carlo samples are driven by the same (fixed) observations $X_t^\dagger$. The initial samples $\Theta_0^{(i)}$, $i = 1, \ldots, M$, are drawn identically and independently from the prior distribution π_0.

We note in passing that there is also a stochastic variant of the innovation process [26] defined by

$$dI_t = dX_t^\dagger - f(X_t^\dagger, \Theta_t)dt - \gamma^{1/2}dW_t, \tag{7}$$

which leads to the Monte-Carlo approximation

$$dI_t^{(i)} = dX_t^\dagger - f(X_t^\dagger, \Theta_t^{(i)})dt - \gamma^{1/2}dW_t^{(i)} \tag{8}$$

of the innovation in (5).

Remark 1 There is an intriguing connection to the stochastic gradient descent approach to the estimation of $\theta^\dagger$, as proposed in [30], which is written as

$$d\theta_t = \frac{\alpha_t}{\gamma}\nabla_\theta f(X_t^\dagger, \theta_t)d\tilde{I}_t, \tag{9a}$$

$$d\tilde{I}_t = dX_t^\dagger - f(X_t^\dagger, \theta_t)dt \tag{9b}$$

in our notation, where $\alpha_t > 0$ denotes the learning rate. We note that (9) shares with (3) the gain times innovation structure. However, while (3) approximates the Bayesian inference problem, formulation (9) treats the parameter estimation problem from an optimisation perspective. Both formulations share, however, the discontinuous dependence on the observation path $X_t^\dagger$, and the proposed frequentist analysis of the EnKBF (3) also applies in simplified form to (9). We also point out that (3) is affine invariant [18] and does not require the computation of partial derivatives.

We now state a numerical implementation with step-size $\Delta t > 0$ and denote the resulting numerical approximations at $t_n = n\Delta t$ by $\Theta_n \sim \pi_n$, $n \geq 1$. While a standard Euler–Maruyama approximation could be applied, the following stable discrete-time mean-field formulation of the EnKBF

$$\Theta_{n+1} = \Theta_n + K_n \left\{ (X_{t_{n+1}}^\dagger - X_{t_n}^\dagger) - \frac{1}{2}\left(f(X_{t_n}^\dagger, \Theta_n) + \pi_n[f(X_{t_n}^\dagger, \theta)]\right)\Delta t\right\} \tag{10}$$

is inspired by [3] with Kalman gain

$$K_n = \pi_n\left[(\theta - \pi_n[\theta]) \otimes f(X_{t_n}^\dagger, \theta)\right] \times \tag{11a}$$

$$\left(\gamma + \Delta t\pi_n\left[\left(f(X_{t_n}^\dagger, \theta) - \pi_n[f(X_{t_n}^\dagger, \theta)]\right) \otimes f(X_{t_n}^\dagger, \theta)\right]\right)^{-1}. \tag{11b}$$

It is straightforward to combine this time discretisation with the Monte-Carlo approximation (5) in order to obtain a complete numerical implementation of the EnKBF.

Remark 2 The rough path analysis of the EnKBF presented in [9] is based on a Stratonovich reformulation of (3) and its appropriate time discretisation. Here we follow the Itô/Euler–Maruyama formulation of the data-driven term in (3),

$$\int_0^T g(X_t^\dagger, t)\, dX_t^\dagger = \lim_{\Delta t \to 0} \sum_{i=1}^L g(X_{t_n}^\dagger, t_n)(X_{t_{n+1}}^\dagger - X_{t_n}^\dagger) \tag{12}$$

for any continuous function $g(x, t)$ and $\Delta t = T/L$, as it corresponds to standard implementation of the EnKBF and is easier to analyse in the context of this paper.

The EnKBF provides only an approximate solution to the Bayesian inference problem for general nonlinear $f(x, \theta)$. However, it becomes exact in the mean-field limit for affine drift functions $f(x, \theta) = \theta Ax + Bx + c$.

Example 1 Consider the stochastic partial differential equation

$$\partial_t u = -U \partial_y u + \rho \partial_y^2 u + \dot{\mathcal{W}} \tag{13}$$

over a periodic spatial domain $y \in [0, L)$, where $\mathcal{W}(t, y)$ denotes space-time white noise, $U \in \mathbb{R}$, and $\rho > 0$ are given parameters. A standard finite-difference discretisation in space with d grid points and mesh-size Δy leads to a linear system of SDEs of the form

$$d\mathbf{u}_t = -(UD + \rho DD^T)\mathbf{u}_t dt + \Delta y^{-1/2} dW_t, \tag{14}$$

where $\mathbf{u}_t \in \mathbb{R}^d$ denotes the vector of grid approximations at time t, $D \in \mathbb{R}^{d \times d}$ a finite difference approximation of the spatial derivative ∂_y, and W_t the standard d-dimensional Brownian motion. We can now set $X_t = \mathbf{u}_t$, $\gamma = \Delta y^{-1}$ and identify either $\theta = U$ or $\theta = \rho$ as the unknown parameter in order to obtain an SDE of the form (1).

In this note, we further simplify our given inference problem to the case

$$f(x, \theta) = \theta Ax, \tag{15}$$

where $A \in \mathbb{R}^{d \times d}$ is a normal matrix with eigenvalues in the left half plane. That is $\sigma(A) \subset \mathbb{C}_-$. The reference parameter value is set to $\theta^\dagger = 1$. Hence the SDE (2) possesses a Gaussian invariant measure with mean zero and covariance matrix

$$C = -\gamma(A + A^T)^{-1}. \tag{16}$$

We assume from now on that the observations $X_t^\dagger$ are realisations of (2) with initial condition $X_0^\dagger \sim N(0, C)$.

Under these assumptions, the EnKBF (3) simplifies drastically, and we obtain

$$d\Theta_t = \frac{\sigma_t}{\gamma}(AX_t^\dagger)^T dI_t, \tag{17a}$$

$$dI_t = dX_t^\dagger - \frac{1}{2}\left(\Theta_t + \pi_t[\theta]\right) AX_t^\dagger dt, \tag{17b}$$

with variance

$$\sigma_t = \pi_t\left[(\theta - \pi_t[\theta])^2\right]. \tag{18}$$

Remark 3 For completeness, we state the corresponding formulation for the stochastic gradient descent approach (9):

$$d\theta_t = \frac{\alpha_t}{\gamma}(AX_t^\dagger)^T d\tilde{I}_t, \tag{19a}$$

$$d\tilde{I}_t = dX_t^\dagger - \theta_t AX_t^\dagger dt. \tag{19b}$$

We find that the learning rate α_t takes the role of the variance σ_t in (17). However, we emphasise again that the same pathwise stochastic integrals arise from both formulations, and therefore, the same robustness issue of the resulting estimators $\theta_t, t > 0$, arises.

Similarly, the discrete-time mean-field EnKBF (10) reduces to

$$\Theta_{n+1} = \Theta_n + K_n\left\{(X_{t_{n+1}}^\dagger - X_{t_n}^\dagger) - \frac{1}{2}\left(\Theta_n + \pi_n[\theta]\right) AX_{t_n}^\dagger \Delta t\right\} \tag{20}$$

with Kalman gain

$$K_n = \sigma_n(AX_{t_n}^\dagger)^T\left(\gamma + \Delta t \sigma_n(AX_{t_n}^\dagger)^T AX_{t_n}^\dagger\right)^{-1}. \tag{21}$$

Furthermore, since $X_t^\dagger \sim N(0, C)$,

$$(AX_t^\dagger)^T AX_t^\dagger = (A^T A) : (X_t^\dagger \otimes X_t^\dagger) \approx (A^T A) : C \tag{22}$$

for $d \gg 1$, and we may simplify the Kalman gain to

$$K_n = \sigma_n(AX_{t_n}^\dagger)^T\left(\gamma + \Delta t \sigma_n(A^T A) : C\right)^{-1}. \tag{23}$$

Here we have used the notation $A : B = \mathrm{tr}(A^T B)$ to denote the Frobenius inner product of two matrices $A, B \in \mathbb{R}^{d \times d}$. The approximation (22) becomes exact in the limit $d \to \infty$, which we will frequently assume in the following section. Please note that

$$K_n = \frac{\sigma_n}{\gamma} (A X_{t_n}^\dagger)^T + \mathcal{O}(\Delta t) \tag{24}$$

under the stated assumptions.

Remark 4 The Stratonovitch reformulation of (17) replaces (17a) by

$$d\Theta_t = \frac{\sigma_t}{\gamma} \left\{ (A X_t^\dagger)^T \circ dI_t - \frac{\gamma}{2} \mathrm{tr}(A) \, dt \right\}. \tag{25}$$

The innovation I_t remains as before. See Appendix B of [9] for more details. An appropriate time discretisation of the innovation-driven term replaces the Kalman gain (21) by

$$K_{n+1/2} = \sigma_n (A X_{t_{n+1/2}}^\dagger)^T \left(\gamma + \Delta t \sigma_n (A X_{t_{n+1/2}}^\dagger)^T A X_{t_{n+1/2}}^\dagger \right)^{-1}, \tag{26}$$

where

$$X_{t_{n+1/2}}^\dagger = \frac{1}{2} (X_{t_n}^\dagger + X_{t_{n+1}}^\dagger). \tag{27}$$

Please note that a midpoint discretisation of the data-driven term in (25) results in

$$(A X_{t_{n+1/2}}^\dagger)^T (X_{t_{n+1}}^\dagger - X_{t_n}^\dagger) = (A X_{t_n}^\dagger)^T (X_{t_{n+1}}^\dagger - X_{t_n}^\dagger) + \tag{28a}$$

$$\frac{1}{2} A^T : (X_{t_{n+1}}^\dagger - X_{t_n}^\dagger) \otimes (X_{t_{n+1}}^\dagger - X_{t_n}^\dagger) \tag{28b}$$

and that

$$\frac{1}{2} A^T : (X_{t_{n+1}}^\dagger - X_{t_n}^\dagger) \otimes (X_{t_{n+1}}^\dagger - X_{t_n}^\dagger) \approx \frac{\Delta t \, \gamma}{2} \mathrm{tr}(A), \tag{29}$$

which justifies the additional drift term in (25). A precise meaning of the approximation in (29) will be given in Remark 5 below.

Alternatively, if one wishes to explicitly utilise the availability of continuous-time data $X_t^\dagger$, one could apply the following variant of (20):

$$\Theta_{n+1} = \Theta_n + \frac{\sigma_n}{\gamma} \int_{t_n}^{t_{n+1}} (A X_t^\dagger)^T dX_t^\dagger - \frac{1}{2} K_n A X_{t_n}^\dagger (\Theta_n + \pi_n[\theta]) \, \Delta t, \tag{30}$$

and following the Itô/Euler–Maruyama approximation (12), discretise the integral with a small inner step-size $\Delta\tau = \Delta t/L$, $L \gg 1$; that is,

$$\int_{t_n}^{t_{n+1}} (AX_t^\dagger)^T dX_t^\dagger \approx \sum_{l=0}^{L-1} (AX_{\tau_l}^\dagger)^T (X_{\tau_{l+1}}^\dagger - X_{\tau_l}^\dagger) \tag{31}$$

with $\tau_l = t_n + l\Delta\tau$. We note that

$$\sum_{l=0}^{L-1} (AX_{\tau_l}^\dagger)^T (X_{\tau_{l+1}}^\dagger - X_{\tau_l}^\dagger) = (AX_{t_n}^\dagger)^T (X_{t_{n+1}}^\dagger - X_{t_n}^\dagger) + \tag{32a}$$

$$A^T : \left(\sum_{l=0}^{L-1} (X_{\tau_l}^\dagger - X_{t_n}^\dagger) \otimes (X_{\tau_{l+1}}^\dagger - X_{\tau_l}^\dagger) \right), \tag{32b}$$

which is at the heart of rough path analysis [13] and which we utilise in the following section.

3 Frequentist Analysis

It is well-known that the second-order contribution in (32) leads to a discontinuous dependence of the integral on the observed $X_t^\dagger$ in the uniform norm topology on the space of continuous functions. Rough path theory fixes this problem by defining appropriately extended topologies and has been extended to the EnKBF in [9]. In this section, we complement the path-wise analysis from [9] by an analysis of the impact of second-order contribution on the EnKBF (17) from a frequentist perspective, which analyses the behaviour of EnKBF over all possible observations $X_t^\dagger$ subject to (2). In other words, one switches from a strong solution concept to a weak one. While we assume that the observations satisfy (2), throughout this section, we will analyse the impact of a perturbed observation process on the EnKBF in Sect. 4.

We first derive evolution equations for the conditional mean and variance under the assumption that Θ_0 is Gaussian distributed with given prior mean m_{prior} and variance σ_{prior}. It follows directly from (17) that the conditional mean $\mu_t = \pi_t[\theta]$, that is the mean of Θ_t, satisfies the SDE

$$d\mu_t = \frac{\sigma_t}{\gamma} \left((AX_t^\dagger)^T dX_t^\dagger - \mu_t (A^T A) : (X_t^\dagger \otimes X_t^\dagger) dt \right), \tag{33}$$

which simplifies to

$$d\mu_t = \frac{\sigma_t}{\gamma} \left((AX_t^\dagger)^\mathrm{T} dX_t^\dagger - \mu_t (A^\mathrm{T}A) : C\, dt \right), \tag{34}$$

under the approximation (22). The initial condition is $\mu_0 = m_{\mathrm{prior}}$. The evolution equation for the conditional variance, that is the variance of Θ_t, is given by

$$\frac{d}{dt}\sigma_t = -\frac{\sigma_t^2}{\gamma} (A^\mathrm{T}A) : (X_t^\dagger \otimes X_t^\dagger) \tag{35}$$

with initial condition $\sigma_0 = \sigma_{\mathrm{prior}}$ and which again reduces to

$$\frac{d}{dt}\sigma_t = -\frac{\sigma_t^2}{\gamma} (A^\mathrm{T}A) : C \tag{36}$$

under the approximation (22).

We now perform a frequentist analysis of the estimator μ_t defined by (34) and (36), that is, we perform a weak analysis of the SDE (34) in terms of the first two moments of μ_t [29]. In the first step, we take the expectation of (34) over all realisations $X_t^\dagger$ of the SDE (2), which we denote by

$$m_t := \mathbb{E}^\dagger[\mu_t]. \tag{37}$$

The associated evolution equation is given by

$$\frac{d}{dt}m_t = \frac{\sigma_t}{\gamma} (A^\mathrm{T}A) : \mathbb{E}^\dagger \left[X_t^\dagger \otimes X_t^\dagger \right] - \frac{\sigma_t}{\gamma} (A^\mathrm{T}A) : C\, m_t, \tag{38}$$

which reduces to

$$\frac{d}{dt}m_t = \frac{\sigma_t}{\gamma} (A^\mathrm{T}A) : C\,(1 - m_t) = \sigma_t (A^\mathrm{T}A) : (A + A^\mathrm{T})^{-1} (1 - m_t). \tag{39}$$

In the second step, we also look at the frequentist variance

$$p_t := \mathbb{E}^\dagger[(\mu_t - m_t)^2]. \tag{40}$$

Using

$$d(\mu_t - m_t) = \frac{\sigma_t}{\gamma} \left\{ (A^\mathrm{T}A) : \left(X_t^\dagger \otimes X_t^\dagger - C \right) dt + \gamma^{1/2}(AX_t^\dagger)^\mathrm{T} dW_t^\dagger \right\} - \tag{41a}$$

$$\frac{\sigma_t}{\gamma}(A^\mathrm{T}A) : C\,(\mu_t - m_t)dt, \tag{41b}$$

we obtain

$$\frac{d}{dt} p_t = -\frac{\sigma_t}{\gamma} (A^{\mathrm{T}} A) : C (2p_t - \sigma_t) + \tag{42a}$$

$$\frac{2\sigma_t}{\gamma} (A^{\mathrm{T}} A) : \mathbb{E}^{\dagger} \left[(X_t^{\dagger} \otimes X_t^{\dagger} - C) (\mu_t - m_t) \right], \tag{42b}$$

which we simplify to

$$\frac{d}{dt} p_t = \frac{\sigma_t}{\gamma} (A^{\mathrm{T}} A) : C (\sigma_t - 2p_t) = \sigma_t (A^{\mathrm{T}} A) : (A + A^{\mathrm{T}})^{-1} (\sigma_t - 2p_t) \tag{43}$$

under the approximation (22). The initial conditions are $m_0 = m_{\mathrm{prior}}$ and $p_0 = 0$, respectively. We note that the differential equations (36) and (43) are explicitly solvable. For example, it holds that

$$\sigma_t = \frac{\sigma_0}{1 + (A^{\mathrm{T}} A) : (A^{\mathrm{T}} + A)^{-1} \sigma_0 t} \tag{44}$$

and one finds that $\sigma_t \sim 1/((A^{\mathrm{T}} A) : (A^{\mathrm{T}} + A)^{-1} t)$ for $t \gg 1$. It can also be shown that $p_t \le \sigma_t$ for all $t \ge 0$. Furthermore, this analysis suggests that the learning rate in the stochastic gradient descent formulation (19) should be chosen as

$$\alpha_t = \min \left\{ \bar{\alpha}, \frac{1}{(A^{\mathrm{T}} A) : (A^{\mathrm{T}} + A)^{-1} t} \right\}, \tag{45}$$

where $\bar{\alpha} > 0$ denotes an initial learning rate; for example $\bar{\alpha} = \sigma_0$.

We finally conduct a formal analysis of the ensemble Kalman filter time-stepping (20) and demonstrate that the method is first-order accurate with regard to the implied frequentist mean m_t. We recall (24) and conclude from (20) that the implied update on the variance σ_n satisfies

$$\sigma_{n+1} = \sigma_n - \frac{\sigma_n^2}{\gamma} (A^{\mathrm{T}} A) : C \Delta t + \mathcal{O}(\Delta t^2), \tag{46}$$

which provides a first-order approximation to (36).

We next analyse the evolution equation (34) for the conditional mean μ_t and its numerical approximation

$$\mu_{n+1} = \mu_n + K_n \left\{ (X_{t_{n+1}}^{\dagger} - X_{t_n}^{\dagger}) - \mu_n A X_{t_n}^{\dagger} \Delta t \right\} \tag{47}$$

arising from (20). Here we follow [13] in order to analyse the impact of the data $X_t^{\dagger}$ on the estimator. An in-depth theoretical treatment can be found in [9].

Comparing (47) to (34) and utilising (24), we find that the key quantity of interest is

$$J^\dagger_{t_n,t_{n+1}} := \int_{t_n}^{t_{n+1}} (AX^\dagger_t)^T dX^\dagger_t, \tag{48}$$

which we can rewrite as

$$J^\dagger_{t_n,t_{n+1}} = A^T : (X^\dagger_{t_n} \otimes X^\dagger_{t_n,t_{n+1}}) + A^T : \mathbb{X}^\dagger_{t_n,t_{n+1}}. \tag{49}$$

Here, motivated by (32) and following standard rough path notation, we have used

$$X^\dagger_{t_n,t_{n+1}} := X^\dagger_{t_{n+1}} - X^\dagger_{t_n} \tag{50}$$

and the second-order iterated Itô integral

$$\mathbb{X}^\dagger_{t_n,t_{n+1}} := \int_{t_n}^{t_{n+1}} (X^\dagger_t - X^\dagger_{t_n}) \otimes dX^\dagger_t. \tag{51}$$

The difference between the integral (48) and its corresponding approximation in (47) is provided by $A^T : \mathbb{X}^\dagger_{t_n,t_{n+1}}$ plus higher-order terms arising from (24). The iterated integral $\mathbb{X}^\dagger_{t_n,t_{n+1}}$ becomes a random variable from the frequentist perspective. Taking note of (2), we find that the drift, $f(x) = Ax$, contributes with terms of order $\mathcal{O}(\Delta t^2)$ to $\mathbb{X}^\dagger_{t_n,t_{n+1}}$ and the expected value of $\mathbb{X}^\dagger_{t_n,t_{n+1}}$ therefore satisfies

$$\mathbb{E}^\dagger[\mathbb{X}^\dagger_{t_n,t_{n+1}}] = \mathcal{O}(\Delta t^2), \tag{52}$$

since $\mathbb{E}^\dagger[W^\dagger_{t_n,\tau}] = 0$ for $\tau > t_n$, and

$$\mathbb{E}^\dagger[\mathbb{W}^\dagger_{t_n,t_{n+1}}] = \frac{1}{2}\mathbb{E}^\dagger[W^\dagger_{t_n,t_{n+1}} \otimes W^\dagger_{t_n,t_{n+1}} - [W^\dagger_{t_n}, W^\dagger_{t_n,t_{n+1}}]] - \frac{\Delta t}{2}I = 0, \tag{53}$$

where we have introduced the commutator

$$[W^\dagger_{t_n}, W^\dagger_{t_n,t_{n+1}}] := W^\dagger_{t_n} \otimes W^\dagger_{t_n,t_{n+1}} - W^\dagger_{t_n,t_{n+1}} \otimes W^\dagger_{t_n}. \tag{54}$$

Hence we find that, while (47) is not a first-order (strong) approximation of the SDE (34), the approximation becomes first-order in m_t when averaged over realisations $X^\dagger_t$ of the SDE (2). More precisely, one obtains

$$\mathbb{E}^\dagger[J^\dagger_{t_n,t_{n+1}}] = (A^T A) : C\Delta t + \mathcal{O}(\Delta t^2). \tag{55}$$

We note that the modified scheme (30) leads to the same time evolution in the variance σ_n while the update in μ_n is changed to

$$\mu_{n+1} = \mu_n + \frac{\sigma_n}{\gamma} \int_{t_n}^{t_{n+1}} (AX_t^\dagger)^\mathrm{T} \mathrm{d}X_t^\dagger - K_n AX_{t_n}^\dagger \mu_n \Delta t. \tag{56}$$

This modification results in a more accurate evolution in the conditional mean μ_n, but because of (52) it does not impact to leading order the evolution of the underlying frequentist mean, $m_n = \mathbb{E}^\dagger[\mu_n]$. We summarise our findings in the following proposition.

Proposition 1 *The discrete-time EnKBF implementations (20) and (30) both provide first-order approximations to the time evolution of the frequentist mean, m_t, and the frequentist variance, p_t. In other words, both methods converge weakly with order one.*

We also note that the frequentist uncertainty is essentially data-independent and depends only on the time window $[0, T]$ over which the data gets observed. Hence, for fixed observation interval $[0, T]$, it makes sense to choose the step-size Δt such that the discretisation error (bias) remains on the same order of magnitude as $p_T^{1/2} \approx \sigma_T^{1/2}$. Selecting a much smaller step-size would not significantly reduce the frequentist estimation error in the conditional estimator μ_T.

Remark 5 We can now give a precise reformulation of the approximation (29):

$$\frac{1}{2}\mathbb{E}^\dagger \left[A^\mathrm{T} : (X_{t_n,t_{n+1}}^\dagger \otimes X_{t_n,t_{n+1}}^\dagger) \right] = \frac{\Delta t\, \gamma}{2} \mathrm{tr}\,(A) + \mathcal{O}(\Delta t^2), \tag{57}$$

which is at the heart of the Stratonovich formulation (25) of the EnKFB [9].

4 Multi-Scale Data

We now have all the material in place to study the dependency of the EnKBF estimator on a set of observations $X_t^{(\epsilon)}$, $\epsilon > 0$, which approach the theoretical $X_t^\dagger$ with respect to the uniform norm topology on the space of continuous functions as $\epsilon \to 0$. Since the second-order contribution in (32), that is (51), does not depend continuously on such perturbations, we demonstrate in this section that a systematic bias arises in the EnKBF. Furthermore, we show how the bias can be eliminated either via subsampling the data, which effectively amounts to ignoring these second-order contributions, or via an appropriate correction term, which ensures a continuous dependence on observations $X_t^{(\epsilon)}$ with respect to the uniform norm topology. More specifically, we investigate the impact of a possible discrepancy between the SDE model (1), for which we aim to estimate the parameter θ, and the data generating SDE (2). We therefore replace (2) by the following two-scale SDE [17]:

$$dX_t^{(\epsilon)} = AX_t^{(\epsilon)}\,dt + \frac{\gamma^{1/2}}{\epsilon}MP_t^{(\epsilon)}\,dt, \tag{58a}$$

$$dP_t^{(\epsilon)} = -\frac{1}{\epsilon}MP_t^{(\epsilon)}\,dt + dW_t^{\dagger}, \tag{58b}$$

where

$$M = \begin{pmatrix} 1 & \beta \\ -\beta & 1 \end{pmatrix}, \tag{59}$$

$\beta = 2$ and $\epsilon = 0.01$. The dimension of state space is $d = 2$ throughout this section. While we restrict here to the simple two-scale model (58), similar scenarios can arise from deterministic fast-slow systems [24, 7].

The associated EnKBF mean-field equations in the parameter Θ_t, which we now denote by $\Theta_t^{(\epsilon)}$ in order to explicitly record its dependence on the scale parameter $\epsilon \ll 1$, become

$$d\Theta_t^{(\epsilon)} = \frac{\sigma_t^{(\epsilon)}}{\gamma}(AX_t^{(\epsilon)})^{\mathrm{T}}dI_t^{(\epsilon)}, \tag{60a}$$

$$dI_t^{(\epsilon)} = dX_t^{(\epsilon)} - \frac{1}{2}\left(\Theta_t^{(\epsilon)} + \pi_t^{(\epsilon)}[\theta]\right)AX_t^{(\epsilon)}dt, \tag{60b}$$

with variance

$$\sigma_t^{(\epsilon)} = \pi_t^{(\epsilon)}\left[(\theta - \pi_t^{(\epsilon)}[\theta])^2\right] \tag{61}$$

and $\Theta_t^{\epsilon} \sim \pi_t^{(\epsilon)}$. The discrete-time mean-field EnKBF (20) turns into

$$\Theta_{n+1}^{(\epsilon)} = \Theta_n^{(\epsilon)} + K_n^{(\epsilon)}\left\{\left(X_{t_{n+1}}^{(\epsilon)} - X_{t_n}^{(\epsilon)}\right) - \frac{1}{2}\left(\Theta_n^{(\epsilon)} + \pi_n^{(\epsilon)}[\theta]\right)AX_{t_n}^{(\epsilon)}\Delta t\right\} \tag{62}$$

with Kalman gain

$$K_n^{(\epsilon)} = \sigma_n^{(\epsilon)}(AX_{t_n}^{(\epsilon)})^{\mathrm{T}}\left(\gamma + \Delta t\sigma_n^{(\epsilon)}(AX_{t_n}^{(\epsilon)})^{\mathrm{T}}AX_{t_n}^{(\epsilon)}\right)^{-1}. \tag{63}$$

We also consider the appropriately modified scheme (30):

$$\Theta_{n+1}^{(\epsilon)} = \Theta_n^{(\epsilon)} + \frac{\sigma_n^{(\epsilon)}}{\gamma}\int_{t_n}^{t_{n+1}}(AX_t^{(\epsilon)})^{\mathrm{T}}dX_t^{(\epsilon)} - \frac{1}{2}K_n^{(\epsilon)}AX_{t_n}^{(\epsilon)}\left(\Theta_n^{(\epsilon)} + \pi_n^{(\epsilon)}[\theta]\right)\Delta t. \tag{64}$$

In order to understand the impact of the modified data generating process on the two mean-field EnKBF formulations (62) and (64), respectively, we follow [17] and investigate the difference between $X_t^{(\epsilon)}$ and $X_t^{\dagger}$:

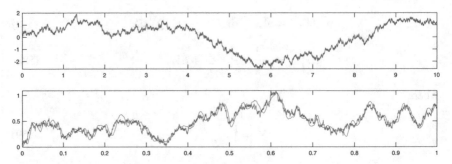

Fig. 1 SDE driven by mathematical vs. physical Brownian motion ($\epsilon = 0.01$). The top panel displays both $X_t^\dagger$ (blue) and $X_t^{(\epsilon)}$ (red) over the long time interval $t \in [0, 10]$, while the lower panel provides a zoomed in perspective over the interval $t \in [0, 1]$

$$d(X_t^{(\epsilon)} - X_t^\dagger) = A(X_t^{(\epsilon)} - X_t^\dagger)dt + \frac{\gamma^{1/2}}{\epsilon}MP_t^{(\epsilon)}dt - \gamma^{1/2}dW_t^\dagger \tag{65a}$$

$$= A(X_t^{(\epsilon)} - X_t^\dagger)dt - \gamma^{1/2}dP_t^{(\epsilon)}. \tag{65b}$$

When $P_t^{(\epsilon)}$ is stationary, it is Gaussian with mean zero and covariance

$$\mathbb{E}_{\text{stat}}\left[P_t^{(\epsilon)} \otimes P_t^{(\epsilon)}\right] = \epsilon\,(M + M^{\text{T}})^{-1} = \frac{\epsilon}{2}I. \tag{66}$$

Hence $P_t^{(\epsilon)} \to 0$ as $\epsilon \to 0$ and also

$$X_t^{(\epsilon)} \to X_t^\dagger \tag{67}$$

in L^2 uniformly in t, provided $\sigma(A) \subset \mathbb{C}_-$ and $X_0^{(\epsilon)} = X_0^\dagger$. This is illustrated in Fig. 1.

In order to investigate the problem further, we study the integral

$$J_{t_n,t_{n+1}}^{(\epsilon)} := \int_{t_n}^{t_{n+1}} (AX_t^{(\epsilon)})^{\text{T}}dX_t^{(\epsilon)} \tag{68}$$

and its relation to (48). As for (48), we can rewrite (68) as

$$J_{t_n,t_{n+1}}^{(\epsilon)} = A^{\text{T}} : (X_{t_n}^{(\epsilon)} \otimes X_{t_n,t_{n+1}}^{(\epsilon)}) + A^{\text{T}} : \mathbb{X}_{t_n,t_{n+1}}^{(\epsilon)}. \tag{69}$$

We now investigate the limit of the second-order iterated integral

$$\mathbb{X}_{t_n,t_{n+1}}^{(\epsilon)} = \int_{t_n}^{t_{n+1}} X_{t_n,t}^{(\epsilon)} \otimes dX_t^{(\epsilon)} \tag{70a}$$

$$= \frac{1}{2} X_{t_n,t_{n+1}}^{(\epsilon)} \otimes X_{t_n,t_{n+1}}^{(\epsilon)} - \frac{1}{2} \int_{t_n}^{t_{n+1}} [X_{t_n,t}^{(\epsilon)}, dX_t^{(\epsilon)}] \tag{70b}$$

as $\epsilon \to 0$ [17]. Here $[.,.]$ denotes the commutator defined by (54).

Proposition 2 *The second-order iterated integral* $\mathbb{X}_{t_n,t_{n+1}}^{(\epsilon)}$ *satisfies*

$$\lim_{\epsilon \to 0} \mathbb{X}_{t_n,t_{n+1}}^{(\epsilon)} = \mathbb{X}_{t_n,t_{n+1}}^{\dagger} + \frac{\Delta t \, \gamma}{2} M \tag{71}$$

Proof The proof follows [17] and can be summarised as follows:

$$\mathbb{X}_{t_n,t_{n+1}}^{(\epsilon)} = \int_{t_n}^{t_{n+1}} X_{t_n,t}^{(\epsilon)} \otimes dX_t^{(\epsilon)} \tag{72a}$$

$$\to \int_{t_n}^{t_{n+1}} X_{t_n,t}^{\dagger} \otimes dX_t^{\dagger} - \gamma^{1/2} \int_{t_n}^{t_{n+1}} X_{t_n,t}^{(\epsilon)} \otimes dP_t^{(\epsilon)} \tag{72b}$$

$$= \mathbb{X}_{t_n,t_{n+1}}^{\dagger} - \gamma^{1/2} X_{t_n,t_{n+1}}^{(\epsilon)} \otimes P_{t_{n+1}}^{(\epsilon)} + \gamma^{1/2} \int_{t_n}^{t_{n+1}} dX_t^{(\epsilon)} \otimes P_t^{(\epsilon)} \tag{72c}$$

$$\to \mathbb{X}_{t_n,t_{n+1}}^{\dagger} + \gamma^{1/2} \int_{t_n}^{t_{n+1}} \left\{ A X_t^{(\epsilon)} + \frac{\gamma^{1/2}}{\epsilon} M P_t^{(\epsilon)} \right\} \otimes P_t^{(\epsilon)} dt \tag{72d}$$

$$\to \mathbb{X}_{t_n,t_{n+1}}^{\dagger} + \frac{\Delta t \, \gamma}{\epsilon} M \, \mathbb{E}_{\text{stat}} \left[P_{t_n}^{(\epsilon)} \otimes P_{t_n}^{(\epsilon)} \right] \tag{72e}$$

$$= \mathbb{X}_{t_n,t_{n+1}}^{\dagger} + \frac{\Delta t \, \gamma}{2} M. \tag{72f}$$

As discussed in detail in [9] already, Proposition 2 implies that the scheme (64) does not, in general, converge to the scheme (64) as $\epsilon \to 0$ since

$$J_{t_n,t_{n+1}}^{\dagger} = \lim_{\epsilon \to 0} J_{t_n,t_{n+1}}^{(\epsilon)} - \frac{\Delta t \, \gamma}{2} A^{\mathrm{T}} : M. \tag{73}$$

This observation suggests the following modification

$$\Theta_{n+1}^{(\epsilon)} = \Theta_n^{(\epsilon)} + \frac{\sigma_n^{(\epsilon)}}{\gamma} \int_{t_n}^{t_{n+1}} (A X_t^{(\epsilon)})^{\mathrm{T}} dX_t^{(\epsilon)} - \frac{\Delta t}{2} \sigma_n^{(\epsilon)} A^{\mathrm{T}} : M - \tag{74a}$$

$$\frac{1}{2} K_n^{(\epsilon)} A X_{t_n}^{(\epsilon)} \left(\Theta_n^{(\epsilon)} + \pi_n^{(\epsilon)}[\theta] \right) \Delta t \tag{74b}$$

to (64). Please note that it follows from (70) that

$$\int_{t_n}^{t_{n+1}} (AX_t^{(\epsilon)})^{\mathrm{T}} dX_t^{(\epsilon)} = A^{\mathrm{T}} : \left(X_{t_{n+1/2}}^{(\epsilon)} \otimes X_{t_n,t_{n+1}}^{(\epsilon)} - \frac{1}{2} \int_{t_n}^{t_{n+1}} [X_{t_n,t}^{(\epsilon)}, dX_t^{(\epsilon)}] \right).$$

(75)

Proposition 3 *The discrete-time EnKBF (62) converges to (20) for fixed Δt as $\epsilon \to$ 0. Similarly, (74) converges to (30) under the same limit.*

Proof The first statement follows from $\sigma_n^{(\epsilon)} = \sigma_n$, the limiting behaviour (67), and

$$\lim_{\epsilon \to 0} K_n^{(\epsilon)} = K_n.$$

(76)

The second statement additionally requires (73) to be substituted into (74) when taking the limit $\epsilon \to 0$.

Remark 6 The analogous adaptation of (74) to the gradient descent formulation (19) with $X_t^{\dagger}$ replaced by $X_t^{(\epsilon)}$ becomes

$$\theta_{n+1}^{(\epsilon)} = \theta_n^{(\epsilon)} + \frac{\alpha_{t_n}}{\gamma} \left(\int_{t_n}^{t_{n+1}} (AX_t^{(\epsilon)})^{\mathrm{T}} dX_t^{(\epsilon)} - \frac{\gamma \Delta t}{2} A^{\mathrm{T}} : M - \right.$$

(77a)

$$\left. \theta_n^{(\epsilon)} (AX_{t_n}^{(\epsilon)})^{\mathrm{T}} A X_{t_n}^{(\epsilon)} \Delta t \right).$$

(77b)

Alternatively, subsampling the data can be applied which leads to the simpler formulation

$$\theta_{n+1}^{(\epsilon)} = \theta_n^{(\epsilon)} + \frac{\alpha_{t_n}}{\gamma} (AX_{t_n}^{(\epsilon)})^{\mathrm{T}} \left((X_{t_{n+1}}^{(\epsilon)} - X_{t_n}^{(\epsilon)}) - \theta_n^{(\epsilon)} A X_{t_n}^{(\epsilon)} \Delta t \right).$$

(78)

Remark 7 A two-scale SDE, closely related to (58), has been investigated in [8] in terms of the time integrated autocorrelation function of $P_t^{(\epsilon)}$ and modified stochastic integrals. In our case, the modified quadrature rule, here denoted by $\diamond$, has to satisfy

$$\int_{t_n}^{t_{n+1}} (AX_t^{\dagger})^{\mathrm{T}} \diamond dX_t^{\dagger} = \lim_{\epsilon \to 0} \int_{t_n}^{t_{n+1}} (AX_t^{(\epsilon)})^{\mathrm{T}} dX_t^{(\epsilon)},$$

(79)

and it is therefore related to the standard Itô integral via

$$\int_{t_n}^{t_{n+1}} (AX_t^{\dagger})^{\mathrm{T}} \diamond dX_t^{\dagger} = \int_{t_n}^{t_{n+1}} (AX_t^{\dagger})^{\mathrm{T}} dX_t^{\dagger} + \frac{\Delta t \gamma}{2} A^{\mathrm{T}} : M.$$

(80)

Hence M playes the role of the integrated autocorrelation function of $P_t^{(\epsilon)}$ in our approach. We note that the modified quadrature rule reduces to the standard Stratonovitch integral if either $\beta = 0$ in (59) or A is symmetric. While the results from [8] could, therefore, also be used as a starting point for discussing the induced estimation bias, practical implementations would still require knowledge of the integrated autocorrelation function of $P_t^{(\epsilon)}$ or, equivalently, the estimation of M in addition to observing $X_t^{(\epsilon)}$. We address this aspect next.

The numerical implementation of (74) requires an estimator for the generally unknown M in (73). This task is challenging as we only have access to $X_t^{(\epsilon)}$ without any explicit knowledge of the underlying generating process (58). While the estimator proposed in [9] is based on the idea of subsampling the data, the frequentist perspective taken in this note suggests the alternative estimator M_{est} defined by

$$\frac{\Delta t\, \gamma}{2} M_{est} = \mathbb{E}^\dagger[\mathbb{X}_{t_n,t_{n+1}}^{(\epsilon)}], \tag{81}$$

which follows from (72f) and (52). That is, $\mathbb{E}^\dagger[\mathbb{X}_{t_n,t_{n+1}}^\dagger] = \mathcal{O}(\Delta t^2)$ for Δt sufficiently small. Note that second-order iterated integral $X_{t_n,t_{n+1}}^{(\epsilon)}$ satisfies (70) and is therefore easy to compute. In practice, the frequentist expectation value can be replaced by an approximation along a given single observation path $X_t^{(\epsilon)}, t \in [0, T]$, under the assumption of ergodicity.

An appropriate choice of the outer or sub-sampling step-size Δt [27] constitutes an important aspect for the practical implementation of the EnKBF formulation (62) for finite values of $\epsilon > 0$ [26]. Consistency of the second-order iterated integrals [13] implies

$$\mathbb{X}_{t_n,t_{n+2}}^{(\epsilon)} = \mathbb{X}_{t_n,t_{n+1}}^{(\epsilon)} + \mathbb{X}_{t_{n+1},t_{n+2}}^{(\epsilon)} + X_{t_n,t_{n+1}}^{(\epsilon)} \otimes X_{t_{n+1},t_{n+2}}^{(\epsilon)}. \tag{82}$$

A sensible choice of Δt is dictated by

$$\mathbb{E}^\dagger\left[X_{t_n,t_{n+1}}^{(\epsilon)} \otimes X_{t_{n+1},t_{n+2}}^{(\epsilon)}\right] = \mathcal{O}(\Delta t^2), \tag{83}$$

that is, the sub-sampled data $X_{t_n}^{(\epsilon)}$ behaves to leading order like solution increments from the reference model (2) at scale Δt independent of the specific value of ϵ. Note that, on the other hand,

$$\mathbb{E}^\dagger\left[X_{\tau_l,\tau_{l+1}}^{(\epsilon)} \otimes X_{\tau_{l+1},\tau_{l+2}}^{(\epsilon)}\right] = \mathcal{O}(\epsilon^{-1}\Delta\tau^2) \tag{84}$$

for an inner step-size $\Delta\tau \sim \epsilon$. In other words, a suitable step-size $\Delta t > 0$ can be defined by making

$$h(\Delta t) := \Delta t^{-2} \left\| \mathbb{E}^\dagger\left[X_{t_n,t_{n+1}}^{(\epsilon)} \otimes X_{t_{n+1},t_{n+2}}^{(\epsilon)}\right]\right\| \tag{85}$$

as small as possible while still guaranteeing an accurate numerical approximation in (62).

Remark 8 The choice of the outer time step Δt is less critical for the EnKBF formulation (74) since it does not rely on sub-sampling the data and is robust with regard to perturbations in the data provided the appropriate M is explicitly available or has been estimated from the available data using (81). Furthermore, if A is symmetric, then it follows from (75) and the skew-symmetry of the commutator $[., .]$ that

$$\int_{t_n}^{t_{n+1}} (A X_t^{(\epsilon)})^{\mathsf{T}} \mathrm{d} X_t^{(\epsilon)} = A : \left(X_{t_{n+1/2}}^{(\epsilon)} \otimes X_{t_n, t_{n+1}}^{(\epsilon)} \right), \tag{86}$$

which can be used in (74). The same simplification arises when M is symmetric. This insight is at the heart of the geometric rough path approach followed in [9] and which starts from the Stratonovich formulation (25) of the EnKBF. See also [28] on the convergence of Wong–Zakai approximations for stochastic differential equations. In all other cases, a more refined numerical approximation of the data-driven integral in (74) is necessary; such as, for example, (31). For that reason, we rely on the Itô/Euler–Maruyama interpretation of (68) in this note instead, that is the approximation (12).

5   Numerical Example

We consider the linear SDE (2) with $\gamma = 1$ and

$$A = \frac{-1}{2} \begin{pmatrix} 1 & -1 \\ 1 & 1 \end{pmatrix}. \tag{87}$$

We find that $C = I$ and $A^{\mathsf{T}} A = 1/2 I$. Hence $(A^{\mathsf{T}} A) : C = 1$, and the posterior variance simply satisfies $\sigma_t = \sigma_0/(1 + \sigma_0 t)$ according to (44). We set $m_{\mathrm{prior}} = 0$ and $\sigma_{\mathrm{prior}} = 4$ for the Gaussian prior distribution of Θ_0, and the observation interval is $[0, T]$ with $T = 6$. We find that $\sigma_T = 0.16$. Solving (39) for given σ_t with initial condition $m_0 = 0$ yields

$$m_t = 1 - \frac{\sigma_t}{\sigma_0} \tag{88}$$

and $m_T = 0.96$. The corresponding curves are displayed in red in Fig. 2.

  We implement the EnKBF schemes (20) and (30) with $t_n = n \Delta t$. The inner time-step is $\Delta \tau = 10^{-4}$ while $\Delta t = 0.06$, that is, $L = 600$. We repeat the experiment $N = 10^4$ times and compare the outcome with the predicted mean value of $m_T = 0.96$ and the posterior variance of $\sigma_T = 0.16$ in Fig. 2. The differences in the computed time evolutions of m_t and p_t are rather minor and support the idea that it is not necessary to assimilate continuous-time data beyond Δt. We

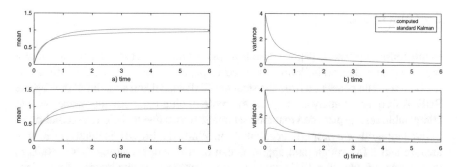

Fig. 2 (**a–b**) Frequentist mean, m_t and variance, p_t, from EnKBF implementation (20) with step-size $\Delta t = 0.06$; (**c–d**) Same results from EnKBF implementation (30) with inner time-step $\Delta \tau = \Delta t / 600$. We also display the curves arising for σ_t and m_t from the standard Kalman theory using the approximation (22). Note that the posterior variance, σ_t, should provide an upper bound on the frequentist uncertainty p_t

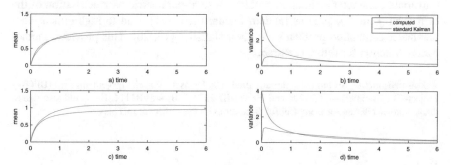

Fig. 3 Same experimental setting as in Fig. 2 but with the data now generated from the multi-scale SDE (58). Again, subsampling the data in intervals of $\Delta t = 0.06$ and high-frequency assimilation with step-size $\Delta \tau = 10^{-4}$ lead to very similar results in terms of their frequentist means and variances

also find that the simple prediction (88), based on standard Kalman filter theory, is not very accurate for this low-dimensional problem ($d = 2$). The corresponding approximation for σ_t provides, however, a good upper bound for p_t.

We now replace the data generating SDE model (2) by the multi-scale formulation (58) with $\epsilon = 0.01$ and $\beta = 2$. This parameter choice agrees with the one used in [9]. We again find that assimilating the data at the slow time-scale $\Delta t = 0.06$ leads to very similar results obtained from an assimilation at the fast time-scale $\Delta \tau = 10^{-4}$ with the EnKBF formulation (74), provided the correction term resulting from the second-order iterated integral (73) is included (See Fig. 3). We also verified numerically that $\Delta t = 0.06$ constitutes a nearly optimal step-size in the sense of making (85) sufficiently small while maintaining numerical accuracy. For example, reducing the outer step-size to $\Delta t = 0.02$ leads to $h(0.02) - h(0.06) \approx 10$ in (85).

6 Conclusions

In this follow-up note to [9], we have investigated the impact of subsampling and/or high-frequency data assimilation on the corresponding conditional mean estimators, μ_t, both for data generated from the standard SDE model and a modified multi-scale SDE. A frequentist analysis supports the basic finding that both approaches lead to comparable results provided that the systematic biases due to different second-order iterated integrals are properly accounted for. While the EnKBF is relatively easy to analyse and a full rough path approach can be avoided, extending these results to the nonlinear feedback particle filter [26, 9] will prove more challenging. Extensions to systems without a strong scale separation [4, 31] and applications to geophysical fluid dynamics [22, 12] are also of interest. In this context, the approximation quality of the proposed estimator (81) and the choice of the step-size Δt following (85) (and potentially $\Delta \tau$) will be of particular interest. Finally, while we have investigated the univariate parameter estimation problem, a semi-parametric parametrisation of the drift term f in (1), such as random feature maps [21], lead to high-dimensional parameter estimation problems and their statistics [19, 20]. This provides another fertile direction for future research.

Acknowledgments SR has been partially funded by Deutsche Forschungsgemeinschaft (DFG)— Project-ID 318763901—SFB1294 and Project-ID 235221301—SFB1114. He would also like to thank Nikolas Nüsken for many fruitful discussions on the subject of this paper.

References

1. A. Abdulle, G. Garegnani, G. A. Pavliotis, A. M. Stuart, and A. Zanoni. Drift estimation of multiscale diffusions based on filtered data. *Foundations of Computational Mathematics*, published online 2021/10/13: in press, 2021. https://doi.org/10.1007/s10208-021-09541-9.
2. Y. Ait-Sahalia, P. A. Mykland, and L. Zhang. How often to sample a continuous-time process in the presence of market microstructure noise. *The Review of Financial Studies*, 18: 351–416, 2005.
3. J. Amezcua, E. Kalnay, K. Ide, and S. Reich. Ensemble transform Kalman-Bucy filters. *Q.J.R. Meteor. Soc.*, 140: 995–1004, 2014.
4. L. Arnold. Hasselmann's program revisited: The analysis of stochasticity in deterministic climate models. In *Stochastic Climate Models*, pages 141–158. Birkhäuser Basel, 2001. https://doi.org/10.1007/978-3-0348-8287-3.
5. R. Azencott, A. Beri, A. Jain, and I. Timofeyev. Sub-sampling and parametric estimation for multiscale dynamics. *Communications in Mathematical Sciences*, 11: 939–970, 2013.
6. A. Bain and D. Crisan. *Fundamentals of Stochastic Filtering*, volume 60 of *Stoch. Model. Appl. Probab.* Springer, New York, 2009. https://doi.org/10.1007/978-0-387-76896-0.
7. P. Bálint and I. Melbourne. Statistical properties for flows with unbounded roof function, including the Lorenz attractor. *Journal of Statistical Physics*, 172: 1101–1126, 2018. https://doi.org/10.1007/s10955-018-2093-y.
8. S. Bo and A. Celani. White-noise limit of nonwhite nonequilibrium processes. *Physical Review E*, 88: 062150, 2013. https://doi.org/10.1103/PhysRevE.88.062150.

9. M. Coghi, T. Nilssen, N. Nüsken, and S. Reich. Rough McKean–Vlasov dynamics for robust ensemble Kalman filtering, 2021. arXiv:2107.06621.

10. C. Cotter and S. Reich. Ensemble filter techniques for intermittent data assimilation. *Radon Ser. Comput. Appl. Math.*, 13: 91–134, 2013. https://doi.org/10.1515/9783110282269.91.

11. D. Crisan, J. Diehl, P. K. Friz, H. Oberhauser, et al. Robust filtering: correlated noise and multidimensional observation. *The Annals of Applied Probability*, 23: 2139–2160, 2013.

12. J. Culina, S. Kravtsov, and A. H. Monahan. Stochastic parameterization schemes for use in realistic climate models. *Journal of the Atmospheric Sciences*, 68: 284 – 299, 2011. https://doi.org/10.1175/2010JAS3509.1.

13. A. M. Davie. Differential equations driven by rough paths: An approach via discrete approximation. *Applied Mathematics Research eXpress*, 2008, 2008. https://doi.org/10.1093/amrx/abm009. abm009.

14. J. Diehl, P. Friz, and H. Mai. Pathwise stability of likelihood estimators for diffusion via rough paths. *The Annals of Applied Probability*, 26: 2169–2192, 2016. https://doi.org/10.1214/15-AAP1143.

15. G. Evensen. *Data assimilation*. Springer-Verlag, Berlin, second edition, 2009. ISBN 978-3-642-03710-8. https://doi.org/10.1007/978-3-642-03711-5.

16. P. Friz and M. Hairer. *A course on rough paths*. Springer-Verlag, 2020.

17. P. Friz, P. Gassiat, and T. Lyons. Physical Brownian motion in a magnetic field as a rough path. *Transactions of the American Mathematical Society*, 367: 7939–7955, 2015.

18. A. Garbuno-Inigo, N. Nüsken, and S. Reich. Affine invariant interacting Langevin dynamics for Bayesian inference. *SIAM J. Appl. Dyn. Syst.*, 19: 1633–1658, 2020. https://doi.org/10.1137/19M1304891.

19. S. Ghosal and A. van der Vaart. *Fundamentals of Nonparametric Bayesian Inference*. Cambridge Series in Statistical and Probabilistic Mathematics. Cambridge University Press, 2017. https://doi.org/10.1017/9781139029834.

20. E. Giné and R. Nickl. *Mathematical Foundations of Infinite-Dimensional Statistical Models*. Cambridge University Press, Cambridge, 2016. https://doi.org/10.1017/CBO9781107337862.

21. G. A. Gottwald and S. Reich. Supervised learning from noisy observations: Combining machine-learning techniques with data assimilation. *Physica D: Nonlinear Phenomena*, 423: 132911, 2021. ISSN 0167-2789. https://doi.org/10.1016/j.physd.2021.132911.

22. K. Hasselmann. Stochastic climate models Part I. Theory. *Tellus*, 28: 473–485, 1976. https://doi.org/10.1111/j.2153-3490.1976.tb00696.x.

23. N. Ikeda and S. Watanabe. *Stochastic differential equations and diffusion processes*. North Holland Publishing Company, Amsterdam-New York, 2nd edition, 1989.

24. D. Kelly and I. Melbourne. Deterministic homogenization for fast-slow systems with chaotic noise. *Journal of Functional Analysis*, 272: 4063–4102, 2017. https://doi.org/10.1016/j.jfa.2017.01.015.

25. Y. A. Kutoyants. *Statistical inference for ergodic diffusion processes*. Springer Science & Business Media, 2013.

26. N. Nüsken, S. Reich, and P. J. Rozdeba. State and parameter estimation from observed signal increments. *Entropy*, 21 (5): 505, 2019. https://doi.org/10.3390/e21050505.

27. A. Papavasiliou, G. Pavliotis, and A. Stuart. Maximum likelihood estimation for multiscale diffusions. *Stochastic Processes and their Applications*, 19: 3173–3210, 2009.

28. S. Pathiraja. L^2 convergence of smooth approximations of stochastic differential equations with unbounded coefficients, 2020. arXiv:2011.13009.

29. S. Reich and P. Rozdeba. Posterior contraction rates for non-parametric state and drift estimation. *Foundation of Data Science*, 2: 333–349, 2020. https://doi.org/10.3934/fods.2020016.

30. J. Sirignano and K. Spiliopoulos. Stochastic gradient descent in continuous time. *SIAM J. Financial Math.*, 8: 933–961, 2017. https://doi.org/10.1137/17M1126825.

31. J. Wouters and G. A. Gottwald. Stochastic model reduction for slow-fast systems with moderate time scale separation. *Multiscale Modeling & Simulation*, 17: 1172–1188, 2019.

Random Ocean Swell-Rays: A Stochastic Framework

Valentin Resseguier, Erwan Hascoët, and Bertrand Chapron

1 Introduction

Originating from distant storms, swell systems radiate across all ocean basins (Snodgrass et al., 1966; Collard et al., 2009; Ardhuin et al., 2009). Far from their sources, emerging surface waves have low steepness characteristics, with very slow amplitude variations. Swell propagation then closely follows principles of geometrical optics, i.e. the eikonal approximation to the wave equation, with a constant wave period along geodesics, when following a wave packet at its group velocity. The phase averaged evolution of quasi-linear wave fields is then dominated by interactions with underlying current and/or topography changes (Phillips, 1977). Comparable to the propagation of light in a slowly varying medium, over many wavelengths, cumulative effects can lead to refraction, i.e. change of the direction of propagation of a given wave packet, so that it departs from its initial ray-propagation direction. This opens the possibility of using surface swell systems as probes to estimate turbulence along their propagating path.

For a single progressive swell wave train, a description of the form

$$h(\boldsymbol{x}, t) = a(\boldsymbol{x}, t)e^{i\phi(\boldsymbol{x},t)}, \tag{1}$$

V. Resseguier (✉)
Lab, SCALIAN DS, Rennes, France
e-mail: valentin.resseguier@scalian.com
https://sites.google.com/view/valentinresseguier

E. Hascoët
OceanDataLab, Locmaria-Plouzané, France

B. Chapron
Laboratoire d'Océanographie Physique et Spatiale (LOPS), Ifremer, Plouzané, France

© The Author(s) 2023
B. Chapron et al. (eds.), *Stochastic Transport in Upper Ocean Dynamics*,
Mathematics of Planet Earth 10, https://doi.org/10.1007/978-3-031-18988-3_16

is locally possible for most wave properties, i.e. the surface elevation, slope, orbital velocities. If the wave-ray propagation is to be followed, or predicted, the phase, $\phi(x, t)$, must vary smoothly along the wave's path. Mathematically, $\phi(x, t)$ is required to be differentiable, to define the relative frequency

$$\omega = -\partial_t \phi(x, t), \tag{2}$$

and the wave number vector

$$k = \nabla \phi(x, t). \tag{3}$$

These partial derivatives of $\phi(x, t)$ being independent of the differentiation order, the kinematical conservation equation for the density of waves writes

$$-\nabla \omega = \partial_t k, \tag{4}$$

with the irrotational condition

$$\nabla \times k = 0, \tag{5}$$

to serve as an initial condition for use with Kelvin's circulation theorem. The rate of change of the wave-number is balanced by the convergence of the frequency, the number of wave crests passing a fixed point.

Let us now consider an ocean moving with velocity v, slowly varying with respect to time and space. The frequency of wave crests passing a fixed point, i.e. the apparent frequency, becomes

$$\omega = \omega_0 + v \cdot k, \tag{6}$$

with $\omega_0 = f(k, H)$, H the depth, the intrinsic frequency, whose functional dependence on k is known. For gravity waves, this dispersion relationship is

$$\omega_0 = \sqrt{g \|k\| \tanh \|k\| H}, \tag{7}$$

and thus

$$\partial_t k + \partial_k \omega_0 \nabla k + \partial_H \omega_0 \nabla H + l \cdot v \nabla \|k\| + \|k\| \nabla (l \cdot v) = 0, \tag{8}$$

with l is a unit vector in the direction of k and $k = \|k\|$. Consequently, for a steady wave train, the variation of the wave-number magnitude along the propagation s is

$$\partial_s \|k\| = -(c_g + l \cdot v)^{-1} [\partial_H \omega_0 \partial_s H + \|k\| \partial_s (l \cdot v)], \tag{9}$$

with $c_g = \partial_k \omega_0$, the local group velocity. Using the irrotational condition, the evolution of the ray direction, $\theta(s)$, follows

$$\partial_s \theta = -(c_g + \boldsymbol{l} \cdot \boldsymbol{v})^{-1} [\frac{1}{\|\boldsymbol{k}\|} \partial_H \omega_0 \partial_v H + \partial_v (\boldsymbol{l} \cdot \boldsymbol{v})], \qquad (10)$$

where v is unit vector normal to the direction of the ray. Accordingly, wave trajectories will bend with depth variations. For deep water, the dispersion relationship reduces to $\omega_0 = \sqrt{g\|\boldsymbol{k}\|}$, and $\theta(s)$ solely depends upon the ratio between the cross-ray current gradient and the local group velocity. More generally, this result extends to the ray curvature, being to first order controlled by ζ/c_g, the ratio between $\zeta = \nabla \times \boldsymbol{v}$, the vertical component of the current vorticity, and $c_g = \partial_k \omega_0 = \omega/2\|\boldsymbol{k}\|$, the group velocity. Accordingly, the rays will bend in the direction of decreasing (increasing) current speed. Moreover, a potential velocity field will give little refraction. Yet, a potential velocity field will control the variation of the wave-number magnitude, and thus the group velocity and bending, along the propagation.

To specify the local linear wave propagation, a precise knowledge of the surface currents, local gradients and/or vorticity, thus appears essential. In a realistic numerical setting, Ardhuin et al. (2017) clearly demonstrated that wave energy variations would largely be dominated by the effects of ocean currents at scales of about 10–100 km. From altimeter ocean surface wave energy measurements, Quilfen and Chapron (2019) also showed that mesoscale and sub-mesoscale upper ocean circulation can drive a significant part of the wave variability in the coupled ocean-atmosphere system. Unfortunately, these small-scale currents are not observed and certainly not resolved in operational models. Today, a precise spatio-temporal information is thus largely missing. To overcome these observation difficulties, but to best take into account unresolved small-scale currents, a stochastic framework can be adopted. Such a stochastic model shall then provide means to perform fast simulations and test ensembles of wave-propagation predictions, to best evaluate impacts of underlying near-surface small-scale currents on the evolution of ocean surface swell systems.

2   Random Swell-Rays

To first order in wave steepness, the group velocity $\boldsymbol{v}_g$ is modified by the local velocity of the currents $\boldsymbol{v}$,

$$\frac{d\boldsymbol{x}}{dt} = \boldsymbol{v}_g = \nabla_{\boldsymbol{k}}\omega = \underbrace{\nabla_{\boldsymbol{k}}\omega_0(\boldsymbol{k})}_{\substack{\text{Group velocity} \\ \text{without currents} \\ \text{but changing wave vector}}} + \boldsymbol{v}, \qquad (11)$$

where $\boldsymbol{x}$ is the centroid of a wave group. The ray direction can thus differ from the direction of the wave vector, except in the case of parallel wave and current directions. Unlike depth refraction, the crest alignment does not indicate the wave propagation direction. The coupled wave vector evolution writes

$$\frac{d\mathbf{k}}{dt} = -\boldsymbol{\nabla}\boldsymbol{v}^T \mathbf{k}. \tag{12}$$

Along the propagation ray, velocity gradients induce linear variations. Decelerating currents will shorten waves, and thus reduce the group velocity. The validity of this coupled ray approximation largely depends on the condition $\|\mathbf{k}\|\xi \gg 1$, where ξ is a length scale on which the current field is varying, physically corresponding to the typical eddy size. This condition is well satisfied for wave numbers of interest, of order $\|\mathbf{k}\| \sim 2\pi/250 \, \text{rad.m}^{-1}$, and typical eddy size $\xi \sim 5 \, \text{km}$ or larger. Scattering of the waves by currents can further be assumed to be weak, with $\|\boldsymbol{v}\|$ of order $0.5 \, \text{m/s}$, much smaller than $\|\boldsymbol{v}_g\|$ of order $10 \, \text{m/s}$. Subsequently, each ray will be appreciably deflected, with scattering angle of order $\sim \|\boldsymbol{v}\|/\|\boldsymbol{v}_g\|$ after traveling a typical correlation length $\sim\xi$ along the mean wave vector direction.

To complete the wave field description, the wave action $A(\mathbf{x}, t)$ is considered to be an adiabatic invariant. Wave action is crucial to anticipate wave transformations by currents (White and Fornberg, 1998). This action is the integral of the action spectrum $N(\mathbf{x}, \mathbf{k}, t)$ over all the wave-vectors $\mathbf{k}$:

$$A(\mathbf{x}, t) = \int d\mathbf{k} \, N(\mathbf{x}, \mathbf{k}, t). \tag{13}$$

The wave action spectrum N is the action by unit of surface (unit of $\mathbf{x}$) and by unit of wave-vector surface (unit of $\mathbf{k}$). For linear waves, the wave action spectrum is simply related to the wave energy spectrum E:

$$E(\mathbf{x}, \mathbf{k}, t) = N(\mathbf{x}, \mathbf{k}, t) \, \omega_0(\mathbf{k}). \tag{14}$$

By the Liouville theorem, the $(\mathbf{x}, \mathbf{k})$ space does not contract nor dilate along time[1] Since the dissipation is neglected, the wave action spectrum N is thus conserved (Lavrenov, 2013), i.e.

$$N\left(\mathbf{x}(t_i), \mathbf{k}(t_i), t_i\right) = N\left(\mathbf{x}(t_f), \mathbf{k}(t_f), t_f\right), \tag{15}$$

along the following $(\mathbf{x}, \mathbf{k})$ variable change between initial time t_i and the final time t_f:

$$\begin{pmatrix} \mathbf{x}(t_i) \\ \mathbf{k}(t_i) \end{pmatrix} \mapsto \begin{pmatrix} \mathbf{x}(t_f) \\ \mathbf{k}(t_f) \end{pmatrix}. \tag{16}$$

[1] $\begin{bmatrix} \boldsymbol{\nabla}_x \\ \boldsymbol{\nabla}_k \end{bmatrix} \cdot \left(\frac{d}{dt} \begin{bmatrix} \mathbf{x} \\ \mathbf{k} \end{bmatrix} \right) = \begin{bmatrix} \boldsymbol{\nabla}_x \\ \boldsymbol{\nabla}_k \end{bmatrix} \cdot \left(\begin{bmatrix} \boldsymbol{v} \\ -\boldsymbol{\nabla}_x \boldsymbol{v}^T \mathbf{k} \end{bmatrix} \right) = \boldsymbol{\nabla}_x \cdot \boldsymbol{v} - \boldsymbol{\nabla}_x \cdot \boldsymbol{v} = 0.$

Subsequently, each Fourier mode of a swell wave train can be modified, independently of the others. In absence of source terms, the action spectrum conservation (15) then writes:

$$\frac{dN}{dt} = \partial_t N + v_g \cdot \nabla_x N + \left(-\nabla_x v^T k\right) \cdot \nabla_k N = 0. \tag{17}$$

3   The Time-Decorrelation Assumption

Now, the Eulerian current v is decomposed into a large-scale component $\bar{v}$ and a small-scale unresolved component v':

$$v = \bar{v} + v'. \tag{18}$$

In a stochastic framework, we can work with the Stratonovich notations (Oksendal, 1998; Kunita, 1997). Under Stratonovich calculus rules, expressions become similar to deterministic ones. The Stratonovich dispersion relation is analogous to the deterministic one (6). The method of characteristics is also valid, (11), (12), and (15), with v' defined by $\sigma \circ dB_t/dt$, where dB_t/dt is a spatio-temporal white noise and $\sigma \circ$ denotes a spatial filter which encodes spatial correlations and horizontal incompressibility ($\nabla \cdot \sigma = 0$). For a spatially stationary and isotropic small-scale velocity, the wave characteristic dynamics equations (11), (12) and (15) would then also remain the same with Ito notations (i.e. we can replace $\sigma \circ dB_t$ by σdB_t to derive the evolution). With Ito notations, the action spectrum conservation (17) writes

$$\partial_t N + v_g \cdot \nabla_x N + \left(-\nabla_x v^T k\right) \cdot \nabla_k N = \begin{bmatrix} \nabla_x \\ \nabla_k \end{bmatrix} \cdot \left(D \begin{bmatrix} \nabla_x \\ \nabla_k \end{bmatrix} N\right), \tag{19}$$

where v_g and v include the random small-scale component $v' = \sigma dB_t/dt$, and

$$D = \frac{1}{2dt} \mathbb{E} \left\{ \begin{bmatrix} \sigma dB_t \\ -\nabla_x(\sigma dB_t)^T k \end{bmatrix} \begin{bmatrix} \sigma dB_t \\ -\nabla_x(\sigma dB_t)^T k \end{bmatrix}^T \right\}. \tag{20}$$

Compared to (17), a RHS diffusive term appears, likely acting to increase the initial directional spread of the incident very directional swell components.

Voronovich (1991) and White and Fornberg (1998) discussed the joint random evolution changes of the coupled (x, k), i.e. the location and the wave vector of waves, subject to a random current v. Considering the wave train to undergo slow changes over the typical time to travel through the typical correlation length of the underlying current, the joint time evolution of (x, k) can be approximated to be driven by a diffusion Markov process.

3.1 The Ray Lagrangian Correlation Time

To apply (19), the covariance of the small-scale unresolved component v' – in the wave group frame – is thus to be assessed:

$$\gamma_{v'}^{X_r}(t) = \mathbb{E}\left(v'(t', X_r(t')) \cdot v'(t' + t, X_r(t' + t))\right) = \gamma_{v'}(t, X_r(t'+t) - X_r(t')),$$

(21)

where $\gamma_{v'}$ is the (Eulerian) spatio-temporal covariance of v', assuming statistical homogeneity, and stationarity for v'. Assume a typical isotropic form for this covariance:

$$\gamma_{v'}(t, x) = \gamma\left(\frac{|t|}{\tau_{v'}} + \frac{\|x\|}{l_{v'}}\right),$$

(22)

then,

$$\gamma_{v'}^{X_r}(t) = \gamma\left(\frac{|t|}{\tau_{v'}} + \frac{\|X_r(t'+t) - X_r(t')\|}{l_{v'}}\right) = \gamma\left(\left(\frac{1}{\tau_{v'}} + \frac{\|v_g\|}{l_{v'}}\right)|t| + O(t^2)\right),$$

(23)

for small time increment t. Therefore, $\left(\frac{1}{\tau_{v'}} + \frac{\|v_g\|}{l_{v'}}\right)^{-1}$ is the correlation time of $v'(t, X_r(t))$. The same derivation is valid for $\nabla(v')^T(t, X_r(t))$. Over deep ocean, the swell wave group velocity is $\|v_g^0\| = \|\nabla_k \omega_0\| = \frac{1}{2}\sqrt{\frac{g}{\|k\|}}$, and the along-ray correlation time of the small-scale velocity can be approximated by $l_{v'}/\|v_g^0\|$. The ratio ϵ between this along-ray correlation time and the characteristic time of the wave group properties evolution, will then control the time decorrelation assumption of v':

$$\epsilon = \frac{l_{v'}}{\|v_g^0\|}\|\nabla v^T\|.$$

(24)

Note the Eulerian small-scale velocity v' is not necessarily time uncorrelated. Yet, for small enough ϵ, the Lagrangian small-scale velocity along the ray can be considered time uncorrelated. From the expression of ϵ, such a condition depends upon:

- $\|v_g^0\|$, increasing with the square root of the wave-group wave number. Hence, ϵ decreases with the square root of the wave-group wave-length.
- $l_{v'}$, defined by the separation between large scales $\bar{v}$ and small scales v', e.g. the spatial filtering cutoff of the large-scale velocity $\bar{v}$.
- $\|\nabla v^T\|$ – which is different from $\|\nabla(v')^T\|$ –, related to the overall kinetic energy (KE) and its high-wavenumber spectral slope.

3.2 Ray Absolute Diffusivity

The absolute diffusivity (or Kubo-type formula) usually corresponds, in the so-called diffusive regime, to the variance per unit of time of a fluid particle Lagrangian path $\frac{dX}{dt} = v$. It is approximately equal to the velocity variance times its correlation time. The Eulerian velocity covariance (22) will thus induce an absolute diffusivity

$$a = \int_0^\infty dt\ \gamma_{v'}(t, X(t' + t) - X(t')) \approx \gamma(0)\ \tau_{v'}. \tag{25}$$

Here, a wave group is followed along its propagation, and a ray absolute diffusivity slightly differs from the usual absolute diffusivity to become

$$a^{X_r} = \int_0^\infty dt\ \gamma_{v'}^{X_r}(t) \approx \left(\frac{1}{\tau_{v'}} + \frac{\|v_g\|}{l_{v'}}\right)^{-1} \gamma(0) \approx \frac{l_{v'}}{\|v_g^0\|}\ \gamma(0). \tag{26}$$

In the Fourier space, the current Absolute Diffusivity Spectral Densisty (ADSD) (Resseguier et al., 2020) associated with the wave dynamics is defined by

$$A^{X_r}(k) = \frac{1/k}{\|v_g^0(k^{X_r})\|}\ E_k(k), \tag{27}$$

where k^{X_r} denotes the wave wave-vector, k the current wave number and E_k the current kinetic energy spectra. Accordingly, for noise calibration, we assume A^{X_r} self-similar and we choose a divergence-free spatial filter $\nabla^\perp \psi_\sigma$ such that $v' = \sigma dB_t/dt = \nabla^\perp \breve{\psi}_\sigma \star dB_t/dt$ and $\|\widehat{\sigma dB_t}(k)\|^2/dt = |k\ \breve{\psi}_\sigma(k)|^2 = A_{v'}^{X_r}(k)$.

3.3 A Practical Estimation

To simplify (20), let us consider the solution for an homogeneous and isotropic small-scale velocity $v' = \sigma dB_t/dt = \nabla^\perp \breve{\psi}_\sigma \star dB_t/dt$ and Matérn stream function covariance, $(\breve{\psi}_\sigma * \breve{\psi}_\sigma)$, leading to

$$D = \frac{1}{2dt} \begin{bmatrix} \mathbb{E}\left\{(\sigma dB_t)(\sigma dB_t)^T\right\} & \begin{matrix} 0\ 0 \\ 0\ 0 \end{matrix} \\ \begin{matrix} 0\ 0 \\ 0\ 0 \end{matrix} & \sum_{ij=1}^2 k_i k_j\ \mathbb{E}\left\{(\nabla_x(\sigma dB_t)_i)(\nabla_x(\sigma dB_t)_j)^T\right\} \end{bmatrix}, \tag{28}$$

$$= \begin{bmatrix} \frac{a_0}{2}\mathbb{I}_d & \begin{matrix} 0\ 0 \\ 0\ 0 \end{matrix} \\ \begin{matrix} 0\ 0 \\ 0\ 0 \end{matrix} & \frac{c_{\kappa_M}}{2}\left(kk^T + 3k^\perp\left(k^\perp\right)^T\right) \end{bmatrix}, \tag{29}$$

where $a_0 = \frac{1}{2dt}\mathbb{E}\|\sigma\,dB_t\|^2$ and $c_{\kappa_M} = \frac{1}{8dt}\mathbb{E}\|\nabla_x(\sigma\,dB_t)^T\|^2$ are constants depending on both the correlation length and the spectrum slope of the small-scale velocity. The Ito action spectrum equation (19) then reads:

$$\partial_t N + \boldsymbol{v}_g \cdot \nabla_x N + \left(-\nabla_x \boldsymbol{v}^T \boldsymbol{k}\right) \cdot \nabla_k N$$

$$= \nabla_x \cdot \left(\tfrac{1}{2}a_0 \nabla_x N\right) + \nabla_k \cdot \left(\tfrac{1}{2}c_{\kappa_M}\left[\boldsymbol{k}\boldsymbol{k}^T + 3\boldsymbol{k}^\perp\left(\boldsymbol{k}^\perp\right)^T\right]\nabla_k N\right), \tag{30}$$

$$= \tfrac{1}{2}a_0 \Delta_x N + \tfrac{1}{2}c_{\kappa_M}\frac{1}{\|\boldsymbol{k}\|}\partial_{\|\boldsymbol{k}\|}\left(\|\boldsymbol{k}\|^3 \partial_{\|\boldsymbol{k}\|}N\right) + 3\tfrac{1}{2}c_{\kappa_M}\partial_{\theta_k}^2 N. \tag{31}$$

The ensemble mean then follows:

$$\partial_t \mathbb{E}N + \overline{\boldsymbol{v}}_g \cdot \nabla_x \mathbb{E}N + \left(-\nabla_x \overline{\boldsymbol{v}}^T \boldsymbol{k}\right) \cdot \nabla_k \mathbb{E}N$$

$$= \tfrac{1}{2}a_0 \Delta_x \mathbb{E}N + \tfrac{1}{2}c_{\kappa_M}\frac{1}{\|\boldsymbol{k}\|}\partial_{\|\boldsymbol{k}\|}\left(\|\boldsymbol{k}\|^3 \partial_{\|\boldsymbol{k}\|}\mathbb{E}N\right) + 3\tfrac{1}{2}c_{\kappa_M}\partial_{\theta_k}^2 \mathbb{E}N, \tag{32}$$

This last RHS diffusion term along the ray-direction θ is then reminiscent to Eq. 3.16 in Bôas and Young (2020) and Eq. 36 in Smit and Janssen (2019) derived under the same isotropic and homogeneous turbulence assumptions.

4 Numerical Simulations

To illustrate our purpose, we consider the Surface Quasi-Geostrophic dynamics (Pierrehumbert, 1994; Lapeyre, 2017), abbreviated SQG:

$$(\partial_t + \boldsymbol{v} \cdot \nabla)\left(-\frac{b}{N}\right) = 0 \text{ with } \boldsymbol{v} = \boldsymbol{v}_{\text{SQG}} = -\nabla^\perp(-\Delta)^{-1/2}\left(-\frac{b}{N}\right). \tag{33}$$

Note, real-upper-ocean currents may not strictly follow SQG. Still, after a wind burst, it can be a good approximation at many mid-latitude locations. SQG corresponds to dynamics with extreme locality, i.e a KE spectrum with a shallow slope $-5/3$. Hence, for fixed KE value, a larger current gradient $\|\nabla \boldsymbol{v}^T\|$ is expected. The validity of the time-decorrelation assumption of Sect. 3 will then depend upon the scale separation, defining the correlation length of the unresolved scales.

A reference simulation is obtained at a resolution 512×512 for a 1000-km squared domain, through a pseudo-spectral code (Resseguier et al., 2017, 2020).

Once initialized, the current velocity $\boldsymbol{v}$ is about 0.1 m.s^{-1}.

A swell system enters the southern boundary, propagating to the north. The carrier incident wave has a wave length $\lambda = 250$ m. Its envelope is Gaussian with an isotropic spatial extension of 30λ. Figure 1 illustrates the branched regime

Fig. 1 Swell interacting with a high-resolution (512 × 512) deterministic SQG current. The left panel shows ray trajectories computed by forward advection and superimposed on the current vorticity $\omega = \nabla^\perp \cdot \boldsymbol{v}$. The right panel shows bidirectional wave spectra, computed by backward advection, at 8 locations along a meridional axis (the mean wave propagation direction)

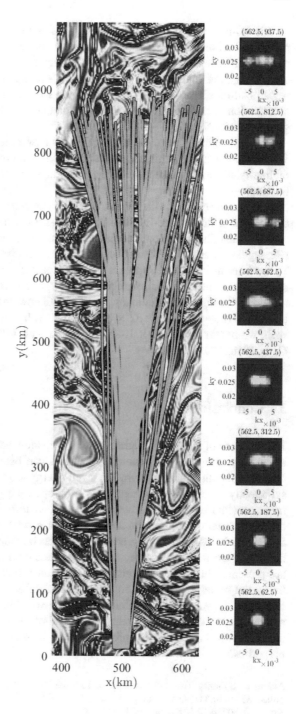

in this homogeneous SQG turbulence. This regime spreads the positions (left panel) and wavevectors (right panel) of the incoming waves. From south to north, spectral diffusion occurs (right panel), in the direction orthogonal (here k_x) to the propagation (here k_y). This accelerates – along the propagation – the zonal wave position spread, to create the branched regime visible in the left panel. This acceleration is explained by the ray equation (11) dominated by the intrinsic wave group velocity $\nabla_k \omega_0 = \frac{\|\nabla_k \omega_0\|}{\|k\|} k$.

To mimic a badly resolved $\overline{v}$, the current v is smoothed at a resolution 32×32. Wave dynamics, using this coarse-scale current, are obtained Fig. 2. The branched regime is strongly weakened, i.e. the spectral small-scale turbulence diffusion is missing.

A stochastic current is then added to this coarse deterministic one. That stochastic component is divergence-free and has a self-similar distribution of energy across spatial scales. Its precise parametrisation is a modification of the ADSD calibration (Resseguier et al., 2020) (see Sect. 3.2). Figure 3 displays the wave simulations. This white-in-time model appears to work for a sufficiently well-resolved large-scale current. Indeed, the decorrelation ratio $\epsilon = (l_{v'}/\|v_g^0\|)\|\nabla v^T\|$ depends on this resolution through $l_{v'}$. Specifically, for this SQG flow, the large-scale current $\overline{v}$ needs to be resolved at least on a 32×32 grid, i.e. with a resolution $l_{v'} = 31.3$ km. As such, we obtain $\epsilon = 3.23 \times 10^{-2}$ (computed with $1/\|\nabla v^T\| = 1.38 \times 10^5$ s and $C_g \simeq 10$ m.s^{-1}).

5   Conclusion

The presence of velocity variations results in random scattering of swell-wave rays. Interactions are weak, but cumulative effects can become significant, to increase the average path length taken by the swell energy to reach an observer. Nowadays, sufficiently precise measurements can then open the possibility to use along-ray measurements to probe the near-surface ocean turbulence. Under a Lagrangian time-decorrelation assumption and using geometrical optics, a practical stochastic framework helps express these scattering effects on the mean swell-action statistics, directly in terms of the KE spectrum of the unresolved surface current field. Results are presented in both Lagrangian and Eulerian forms, where the latter augments the initial radiative transport equation with a diffusive term in directional space. Measured delays in swell arrivals, estimated wave height spectral characteristics and decays, and/or varying directional spread of the swell field shall then be more quantitatively interpreted to infer regional and seasonal upper ocean dynamical properties.

Acknowledgments This work is supported by the R&T CNES R-S19/OT-0003-084, the ERC project 856408-STUOD, the European Space Agency World Ocean Current project (ESA Contract No. 4000130730/20/I-NB), and SCALIAN DS.

Fig. 2 Swell interacting with a low-resolution (32×32) deterministic SQG current. The left panel shows ray trajectories computed by forward advection and superimposed on the low-resolution current vorticity $\overline{\omega} = \nabla^{\perp} \cdot \overline{v}$. The right panel shows bidirectional wave spectra, computed by backward advection, at 8 locations along a meridional axis (the mean wave propagation direction)

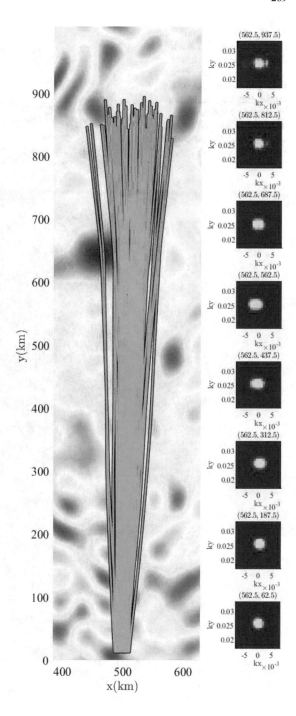

Fig. 3 Swell interacting with a low-resolution (32 × 32) deterministic SQG current plus (one realization of) the time-uncorrelated stochastic model. Ray trajectories are computed by forward advection and superimposed on the low-resolution current vorticity $\overline{\omega} = \nabla^{\perp} \cdot \overline{v}$

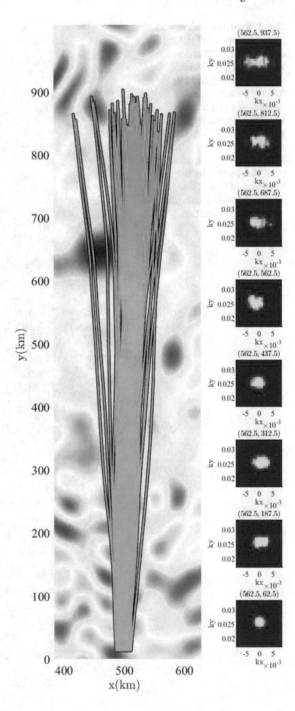

References

Ardhuin F, Chapron B, Collard F (2009) Observation of swell dissipation across oceans. Geophysical Research Letters 36(6)

Ardhuin F, Gille ST, Menemenlis D, Rocha CB, Rascle N, Chapron B, Gula J, Molemaker J (2017) Small-scale open ocean currents have large effects on wind wave heights. Journal of Geophysical Research: Oceans 122(6):4500–4517

Bôas ABV, Young WR (2020) Directional diffusion of surface gravity wave action by ocean macroturbulence. Journal of Fluid Mechanics 890

Collard F, Ardhuin F, Chapron B (2009) Monitoring and analysis of ocean swell fields from space: New methods for routine observations. Journal of Geophysical Research: Oceans 114(C7)

Kunita H (1997) Stochastic flows and stochastic differential equations, vol 24. Cambridge university press

Lapeyre G (2017) Surface quasi-geostrophy. Fluids 2(1):7

Lavrenov I (2013) Wind-waves in oceans: dynamics and numerical simulations. Springer Science & Business Media

Oksendal B (1998) Stochastic differential equations. Spinger-Verlag

Phillips M (1977) The dynamics of the upper ocean. Cambridge University Press

Pierrehumbert R (1994) Tracer microstructure in the large-eddy dominated regime. Chaos, Solitons & Fractals 4(6):1091–1110

Quilfen Y, Chapron B (2019) Ocean surface wave-current signatures from satellite altimeter measurements. Geophysical Research Letters 46(1):253–261

Resseguier V, Mémin E, Chapron B (2017) Geophysical flows under location uncertainty, part II quasi-geostrophy and efficient ensemble spreading. Geophysical & Astrophysical Fluid Dynamics 111(3):177–208

Resseguier V, Pan W, Fox-Kemper B (2020) Data-driven versus self-similar parameterizations for stochastic advection by lie transport and location uncertainty. Nonlinear Processes in Geophysics 27(2):209–234

Smit PB, Janssen TT (2019) Swell propagation through submesoscale turbulence. Journal of Physical Oceanography 49(10):2615–2630

Snodgrass FE, Groves GW, Hasselmann K, Miller GR, Munk WH, Powers WH (1966) Propagation of ocean swell across the pacific. Philos Trans R Soc London, Ser A (249):431–497

Voronovich A (1991) The effect of shortening of waves on random currents. In: Proceedings of nonlinear water waves, Bristol

White BS, Fornberg B (1998) On the chance of freak waves at sea. Journal of fluid mechanics 355:113–138

Modified (Hyper-)Viscosity for Coarse-Resolution Ocean Models

Louis Thiry, Long Li, and Etienne Mémin

Abstract We present a simple parameterization for coarse-resolution ocean models. To replace computationally expensive high-resolution ocean models, we develop a computationally cheap parameterization for coarse-resolution models based solely on the modification of the viscosity term in advection equations. It is meant to reproduce the mean quantities like pressure, velocity, or vorticity computed from a high-resolution reference solution or using observations. We test this new parameterization on a double-gyre quasi-geostrophic model in the eddy-permitting regime. Our results show that the proposed scheme improves significantly the energy statistics and the intrinsic variability on the coarse mesh. This method shall serve as a deterministic basis model for coarse-resolution stochastic parameterizations in future works.

1 Introduction

Ocean general circulation models used at climatic scales are limited for evident computational reasons to too coarse horizontal resolutions to solve correctly ocean mesoscale and sub-mesoscale eddies, even with large computational infrastructure. The horizontal resolution of the most recent climatic ocean models is of the order of the Rossby radius of deformation. These models are hence in the so-called eddy-permitting regime and they can solve partially the mesoscale (i.e. 10–100 km) eddy field. These models however suffer from strong limitations. In particular, they are unable to reproduce accurately large-scale structures such as the eastward turbulent jet in an idealized double-gyre configuration.

Recent parameterizations have shown significant improvements in coarse-resolution models compared to high-resolution reference solutions [2]. However, it remains an important topic of research, as the actual generation of parametrizations

L. Thiry (✉) · L. Li · E. Mémin
INRIA/IRMAR, Rennes, France
e-mail: louis.thiry@inria.fr

© The Author(s) 2023   273
B. Chapron et al. (eds.), *Stochastic Transport in Upper Ocean Dynamics*,
Mathematics of Planet Earth 10, https://doi.org/10.1007/978-3-031-18988-3_17

is not completely able to resolve the effects of the unresolved scales on the large-scale flow structures.

A wide range of subgrid parametrizations relies on eddy viscosity such as Laplacian and biharmonic schemes [16, 10, 4, 3]. It has been shown in [9] that including only these (hyper)viscosity in coarse-resolution models often causes too much dissipation and results in an artificial energy sink at large scales. In general, even eddy-permitting models are not energetic enough and as a result, the long-time average of any coarse model's variable of interest departs completely from the long-time average of high-resolution models subsampled at the same scale. This becomes the main motivation of the present work. In particular, we would like to answer the following question: how can we reduce the excessive resolved kinetic energy loss due to the viscosity while simultaneously ensuring numerical stability?

We propose a simple affine parameterization of (hyper)viscosity. The (bi)laplacian operator $\Delta^p f$ is replaced by $\Delta^p (f - f')$, where f' is a field of same dimension as f that does not depend upon time. We interpret this method as a mathematical regularization technique to guide the solutions towards prior information. We frame f' as the solution of an optimal control problem to reproduce statistics computed from a reference solution or observations. We present a method to solve this optimal control problem.

We test the proposed method with an idealized double-gyre configuration. For that purpose, we release with this article a fast, concise, and CPU-GPU portable Pytorch implementation of a multi-layer quasi-geostrophic model on a rectangular domain. We implement and test our optimization procedure within this setting.

This article is organized as follows: we present in Sect. 2 the double gyre quasi-geostrophic model we use and detail its implementation, we present in Sect. 3 our *modified viscosity* parameterization and we show and discuss numerical results in Sect. 4.

2 Double Gyre Quasi-Geostrophic Model

2.1 Governing Equations

We use the same multi-layer quasi-geostrophic model in a non-periodic rectangular domain as in [6]. Here, we only give a brief review of this system. The quasi-geostrophic pressure and potential vorticity (PV) are stacked in three isopycnal layers. We adopt vector forms to denote the layered pressure and potential vorticity (PV):

$$\mathbf{p} = \begin{bmatrix} p_1 \\ p_2 \\ p_3 \end{bmatrix}, \quad \mathbf{q} = \begin{bmatrix} q_1 \\ q_2 \\ q_3 \end{bmatrix}.$$

The forced and damped quasi-geostrophic (QG) equations can be then written as

$$\partial_t \mathbf{q} = \frac{1}{f_0} J(\mathbf{q}, \mathbf{p}) + f_0 B \mathbf{e} + \frac{1}{f_0}\left(a_2 \Delta - a_4 \Delta^2\right)(\Delta \mathbf{p}), \tag{1}$$

$$\left(\Delta - f_0^2 A\right)\mathbf{p} = f_0 \mathbf{q} - f_0 \beta(y - y_0), \tag{2}$$

where $\Delta = \partial_{xx}^2 + \partial_{yy}^2$ denotes the horizontal Laplacian, Δ^2 the bi-laplacian operator, $J(a, b) = \partial_x a \partial_y b - \partial_x b \partial_y a$ stands for the Jacobi operator, $f_0 + \beta(y - y_0)$ is the Coriolis parameter under beta-plane approximation with the meridional axis center y_0, a_2 and a_4 are the Laplacian and biharmonic viscosity coefficients. Parameters of the configuration are listed in the Tables A.1 and A.2 in Appendix. Besides, the second term on the right-hand side of Eq. (1) represents the external forcing applied on different layers. In this work, we only consider an idealized case in which the ocean basin is driven by a stationary and symmetric wind stress $\tau = (\tau^x, \tau^y)$ on the surface and by a linear Ekman stress at the bottom. In that case, the forcing term can be specified by

$$B = \begin{bmatrix} \frac{1}{H_1} & \frac{-1}{H_1} & 0 & 0 \\ 0 & \frac{1}{H_2} & \frac{-1}{H_2} & 0 \\ 0 & 0 & \frac{1}{H_3} & \frac{-1}{H_3} \end{bmatrix}, \quad \mathbf{e} = \begin{bmatrix} \partial_x \tau^y - \partial_y \tau^x \\ 0 \\ 0 \\ \frac{\delta_{ek}}{2|f_0|} \Delta p_3 \end{bmatrix}, \quad \tau = \tau_0 \begin{bmatrix} -\cos(2\pi y/L_y) \\ 0 \end{bmatrix},$$

where τ_0 is the magnitude of surface wind, H_k is the background thickness of layer k, and δ_{ek} is the bottom Ekman layer thickness. The vertical stratification level of such a model is described by the term $-f_0^2 A \mathbf{p}$ in Eq. (2) with

$$A = \begin{bmatrix} \frac{1}{H_1 g'_{1.5}} & \frac{-1}{H_1 g'_{1.5}} & 0 \\ \frac{-1}{H_2 g'_{1.5}} & \frac{1}{H_2}\left(\frac{1}{g'_{1.5}} + \frac{1}{g'_{2.5}}\right) & \frac{-1}{H_2 g'_{2.5}} \\ 0 & \frac{-1}{H_3 g'_{2.5}} & \frac{1}{H_3 g'_{2.5}} \end{bmatrix},$$

where $g_{k+0.5}$ is the reduced gravity defined across the interface between layers k and $k + 1$. A multi-layered generalization of this model can be found in [5]. Note also that such a multi-layered model can be considered as a vertical discretized approximation of the continuously stratified QG system [17] with $\partial_z(f_0 \partial_z \mathbf{p}/\mathbf{N}^2) \approx -f_0 A \mathbf{p}$ approximated by finite differences, and in which $\mathbf{N}$ denotes the buoyancy (or Brunt-Vaisala) frequency.

2.2 Pytorch Implementation

To facilitate numerical developments and benefit from built-in automatic differenti-ation, we develop a Pytorch [12] implementation of the above-described multilayer QG model.[1] For this purpose, we follow rigorously the strategy of [7]:

1. we use a regular numerical grid with finite differences
2. We solve the PV advection equation (1) on the whole domain except the boundaries. We use a standard 5-point finite difference scheme for the (bi-)Laplacian and the energy-enstrophy conservative Arakawa-Lamb scheme for the Jacobian [1].
3. We apply a vertical change of coordinate to Eq. (2) which becomes a set of three inhomogeneous Helmholtz equations. We solve these equations with the spectral Discrete Sine Transform (DST) method, and we add corresponding homogeneous Helmholtz equation solutions to ensure mass conservation.
4. We update the boundary values of the potential vorticity $\mathbf{q}$ using Eq. (2).

Detailed equations and numerical routine design choices can be found in [7]. We use a Heun–Runge–Kutta 2 time-stepping instead of the Leap-Frog time scheme used by [7].

For sake of numerical efficiency, we follow the recommendation of [14]: we compile computationally demanding routines and simplify finite difference calculations by reducing as much as possible the number of multiplications. We end up with a very concise code (less than 300 lines) that only depends upon Numpy and Pytorch libraries. This implementation will be open-sourced at the time of the publication.

2.3 Eddy-Resolving and Eddy-Permitting Regimes

We consider two spatial settings for our simulations:

1. The eddy-resolving regime, our high-resolution reference with a 5 km resolution.
2. The eddy-permitting regime, our low-resolution setting with a 40 km resolution.

Parameters for these two different regimes are written in Table A.2 in Appendix.

Shevchenko and Berloff [15] studied the resulting flows' differences between these two regimes. The high-resolution eddy-resolving model shows a well-pronounced eastward jet fuelled by mesoscale eddies circulating while the low-resolution eddy-permitting model does not induce a proper eastward jet as shown on Fig. 1. Temporal statistics significantly differ between high- and low-resolution simulations.

---

[1] Available at https://github.com/louity/qgm_pytorch.

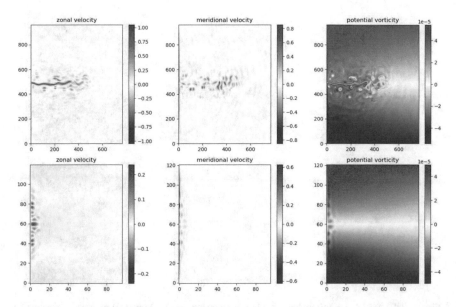

Fig. 1 (Top) high-resolution and (bottom) low-resolution top-layer snapshots after 400 years of integration starting from zero velocity. Velocities are in m s^{-1} and PV in s^{-1}

3 Proposed Modified Viscosity

3.1 Motivation

In both resolutions, we use biharmonic viscosity as in [16, 10, 4, 3] essentially because it is less dissipating at large scales than a Laplacian. Compared to the usual Laplacian viscosity, it preserves large-scale structures. However, hyperviscosity remains much too dissipative in the "eddy-permitting" regime [9]. This too strong dissipation *kills* the eastward jet that is present in the high-resolution and that we expect to see in such a double-gyre quasi-geostrophic model. Figure 2 shows a sequence of snapshots of the low-resolution models where we input a downsampled snapshot of the high-resolution (see Appendix for details on downsampling). After as few as three years, the eastward jet has almost disappeared, showing that the model is too dissipating. Lowering the hyper-viscosity coefficient by a factor of 10 does not solve this problem, and creates spurious gradients in the potential vorticity as shown in Fig. 2. These numerical artifacts are due to a bad representation of the direct enstrophy cascade, causing a piling up of the small-scale vorticity gradients at the cut-off frequency together with aliasing effects.

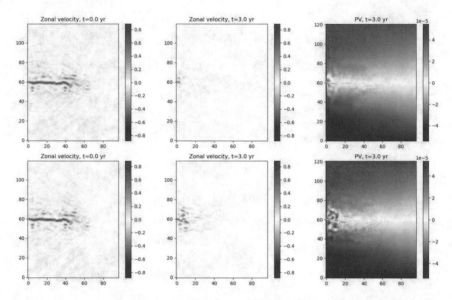

Fig. 2 (left) Initial condition: high-resolution snapshot on the low-resolution grid.(center and right) Zonal velocity and potential vorticity (PV) snapshots after 3 years of integration at low-resolution with Eqs. (1, 2) with (top) standard hyper-viscosity and (bottom) 10 times smaller hyper-viscosity. We can see aliasing effects on potential vorticity snapshots integrated with low hyper-viscosity

3.2 Modified Viscosity

Here we propose a simple affine modification parameterization of hyperviscosity. We add a bias to the term $\Delta \mathbf{p}$ in Eq. (1), which becomes $\Delta \left(\mathbf{p} - \mathbf{p}'\right)$ where $\mathbf{p}'$ is a dimensional field that does not depend upon time. The PV advection equation with hyperviscosity becomes

$$\partial_t \mathbf{q} = \frac{1}{f_0} J(\mathbf{q}, \mathbf{p}) + f_0 B \mathbf{e} + \frac{1}{f_0}\left(a_2 \Delta - a_4 \Delta^2\right)\left(\Delta(\mathbf{p} - \mathbf{p}')\right). \tag{3}$$

The elliptic equation (2) remains unchanged.

The goal of this additional term is to reproduce a relevant time-average pressure field relying on observations or high-resolution solutions. For example the high-resolution average $\overline{\mathbf{p}_{HR}}$ can be downsampled to the targeted coarse grid resolution in $\overline{\mathbf{p}_{HR}} \downarrow$, and we want the average of the modified low-resolution $\overline{\mathbf{p}_{LR}}$ model to be as close as possible to the high-resolution reference $\overline{\mathbf{p}_{HR}} \downarrow$.

We face here an optimal control problem, as the low-resolution average is a function of the control parameter $\mathbf{p}'$. We state it with the following least-square formulation

$$\mathbf{p}'_{\text{opt}} = \underset{\mathbf{p}'}{\operatorname{argmin}} \, \mathcal{F}(\mathbf{p}') \tag{4}$$

$$\mathcal{F}(\mathbf{p}') = \left\| \, \overline{\mathbf{p}_{\text{LR}}}(\mathbf{p}') - \overline{\mathbf{p}_{\text{HR}}} \downarrow \, \right\|^2 \tag{5}$$

This optimization problem is a priori non-convex and we shall not expect to find a global optimum. In the following, we propose a numerical procedure to find a heuristic $\hat{\mathbf{p}}'$ of the optimal solution $\mathbf{p}'_{\text{opt}}$.

Computationally, the implementation of this modified hyperviscosity is simple and computationally cheap. We precompute $\Delta \mathbf{p}'$ and subtract it from $\Delta \mathbf{p}$ at each time-integration step. It increases the integration time of the advection equation (1) by less than 1% on CPUs and GPUs.

3.3 Modified Viscosity Regularization

The continuously stratified QG equations can be rewritten in a variational formulation [8] with a Hamiltonian $\mathcal{J}$ defined as

$$\mathcal{J}(\mathbf{p}) = \frac{1}{2} \int_{\Omega} \frac{1}{f_0} |\nabla \mathbf{p}|^2 + \frac{f_0}{N^2} (\partial_z \mathbf{p})^2.$$

Our model is a discretized version of the continuous stratification. Since we add an external wind forcing term and we use an energy conservative Arakawa advection scheme, we need to add some viscosity or hyperviscosity to dissipate energy. In a variational formulation, these (hyper-)viscous terms become the following penalization

$$\frac{1}{2} \int_{\Omega} a_2 |\Delta \mathbf{p}|^2 + a_4 |\nabla (\Delta \mathbf{p})|^2,$$

added to the Hamiltonian $\mathcal{J}(\mathbf{p})$ to produce a smooth solution. The Gradient norm penalization of Laplacian $\mathbf{p}$ guides the minimization toward solutions of smooth Laplacian. Hyperviscosity corresponds to the Laplacian norm penalization and enforces a solution of minimum Laplacian norm. The parameters a_2 and a_4 quantify the strength of these regularization constraints.

Here, we simply propose to replace it with the following penalization

$$\frac{1}{2} \int_{\Omega} a_2 |\Delta(\mathbf{p} - \mathbf{p}')|^2 + a_4 |\nabla (\Delta(\mathbf{p} - \mathbf{p}'))|^2.$$

We now penalize $(\mathbf{p} - \mathbf{p}')$ instead of $\mathbf{p}$, meaning that we guide the solution to a possibly non-smooth reference $\mathbf{p}'$ that will produce the correct large scale behavior.

3.4 Iterative Procedure

Here we present a method to find a solution to the optimization problem (4). A natural guess for $\mathbf{p}'_{\text{opt}}$ is $\overline{\mathbf{p}_{\text{HR}}} \downarrow$. We solve the equations and compute the average pressure $\overline{\mathbf{p}_{\text{LR}}}$. Results are shown in Fig. 4. It is a good first-guess, but the difference $\overline{\mathbf{p}_{\text{HR}}} \downarrow -\overline{\mathbf{p}_{\text{LR}}}$ is still large.

We propose the following iterative procedure to find a better guess for $\mathbf{p}'_{\text{opt}}$. In the following we assume that we are in low resolution, i.e. $\mathbf{p} = \mathbf{p}_{\text{LR}}$ and $\overline{\mathbf{p}} = \overline{\mathbf{p}_{\text{LR}}}$ unless explicitly written.

- We set $\mathbf{p}'_0$ and we compute the average pressure $\overline{\mathbf{p}}_0$ solving standard equations (1, 2) without modified viscosity.
- Choose $k \in \,]0, 1]$.
- Start with $\mathbf{p}'_1 = \overline{\mathbf{p}_{\text{HR}}} \downarrow$.
- Evolve the ensemble for n years and compute the corresponding average pressure $\overline{\mathbf{p}}_1$ with ensemble average.
- For $n = 1 \ldots$:

  - Set $\mathbf{p}'_{n+1} = \mathbf{p}'_n + k \left(\overline{\mathbf{p}_{\text{HR}}} \downarrow -\overline{\mathbf{p}}_n \right)$.
  - Evolve the ensemble for n years and compute new average pressure $\overline{\mathbf{p}}_{n+1}$.

- return $\mathbf{p}'_n$ and $\overline{\mathbf{p}}_n$

There is no theoretical guarantee that this procedure converges, but we observe in the next section that it converges with the double-gyre QG model that we use.

4 Results and Discussion

4.1 Statistics

We use ensemble averages to compute the statistics. To create ensembles of size N, we start from a zero solution and spin up the models for 100 years with a timestep of 1200 s to reach statistically steady states as in [13]. Then we run the models for 500 years and save 10 snapshots a year to get 5000 snapshots, and we randomly select N snapshots out of these 5000 snapshots. The ensemble averages are simply average over these N ensemble members that we evolve in parallel. Such ensemble averages are denoted with $\overline{\bullet}$ in the following, i.e. the average pressure is denoted by $\overline{\mathbf{p}}$, average velocity by $\overline{u}$, etc.

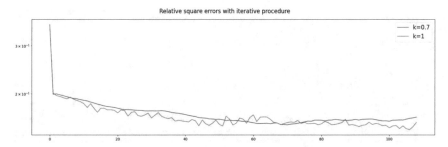

Fig. 3 Evolution of the relative square error $\frac{\|\bar{\mathbf{p}}_n - \overline{\mathbf{p}_{HR}} \downarrow\|^2}{\|\overline{\mathbf{p}_{HR}} \downarrow\|^2}$ w.r.t iterations of the procedure

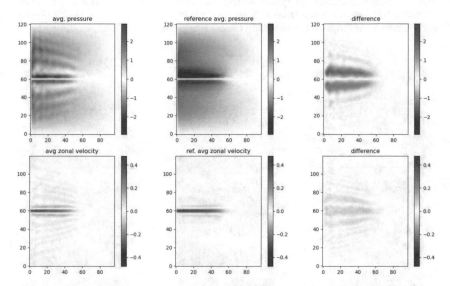

Fig. 4 Top-layer average pressure (top) and velocity (bottom) of (left-to-right) proposed model at low-resolution, reference, and the difference between the two

4.2 Iterative Procedure

We test the iterative procedure described in Sect. 4.2 with the double-gyre model presented in Sect. 2 in the eddy-permitting regime. We use $n = 10$ years to evolve the ensemble after each iterate. We compute the reference pressure average $\bar{\mathbf{p}}_{HR}$ with the same model in the eddy-resolving regime.

Figure 3 shows the relative square error $\|\bar{\mathbf{p}}_n - \overline{\mathbf{p}_{HR}} \downarrow\|^2 / \|\overline{\mathbf{p}_{HR}} \downarrow\|^2$ at iterations of the procedure with $k = 1$ and $k = 0.7$. The procedure converges with $k = 0.7$ and oscillates with $k = 1$.

Figure 4 shows the output average pressure $\bar{\mathbf{p}}_n$ of the iterative procedure, the reference $\bar{\mathbf{p}}_{HR}$ and the difference between the two, as well as for zonal velocity $\mathbf{u}$. Our model can reproduce the eastward jet produced by the high-resolution reference

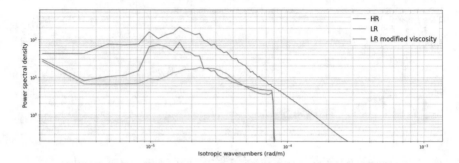

Fig. 5 Top-layer kinetic-energy spectra average with models at high-resolution (HR), at low-resolution (LR) and at low-resolution with proposed modified viscosity. The decreasing slope of the spectrum of the proposed model is much closer to the high-resolution reference

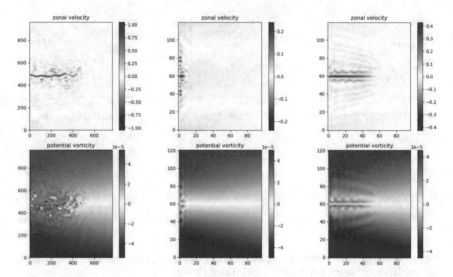

Fig. 6 PV and zonal velocity snapshots form (left-to-right) high-resolution, low-resolution and proposed model at low-resolution

model. Kinetic energy spectra shown on Fig. 5 shows also the improvement of our model compared to low-resolution. Finally, Fig. 6 shows high-resolution and low-resolution snapshots as well a snapshot of the proposed model at low-resolution. Our model effectively produces the eastward jet and a re-circulation zone around it where eddies are created. Artifacts can be also observed on the zonal velocity and potential vorticity on the right of Fig. 6. They can likely be Rossby waves created by the harmonic regularization terms, which remain an artificial constraint, but this needs to be studied further.

5 Conclusion

We presented a simple modified-viscosity scheme for coarse resolution ocean modeling that we derived and tested on a double-gyre multi-layer quasi-geostrophic model. We interpret it as a modified regularization technique that will guide the solution to a reference rather than producing a too smooth solution in the eddy-permitting regime. The technique requires solving an optimization problem, and we presented a procedure to find a good guess for the solutions. We showed that it converges to a reasonable solution that fairly reproduces the input reference.

If this method mimics the average of the high-resolution, it only reproduces partially the variability and higher-order statistics of the high-resolution. We see in Fig. 5 our model's snapshots resemble the averages. In future works, we consider using this method as a deterministic basis for stochastic parameterizations such as Location-Uncertainty [11].

Appendix

Downsampling Procedure

Downsampling the high-resolution solution on a low-resolution grid consists of interpolating the high-resolution (769×961) streamfunction on the low-resolution (97×121) grid. Then we can compute the potential vorticity using Eq. (2). Because of the no-flow constraint, the downsampled streamfunction should be constant on the boundaries and should satisfy a mass conservation constraint [7]. We also want to preserve the frequency information and prevent aliasing.

We use the following procedure:

1. we apply a Gaussian filter and downsample the streamfunction on the domain except on the boundaries.
2. we adding homogeneous solutions of the Elliptic equation (2) to the streamfunction in order to satisfy the mass conservation as in [7].

Parameter Tables

Table A.1 Common parameters for all the models

Parameters	Value	Description
$L_x \times L_y$	(3840×4800) km	Domain size
H_k	$(350, 750, 2900)$ m	Mean layer thickness
g'_k	$(0.025, 0.0125)$ m s^{-2}	Reduced grativity
δ_{ek}	2 m	Bottom Ekman layer thickness
τ_0	2×10^{-5} m^2 s^{-2}	Wind stress magantitude
a_2	0 m^2 s^{-1}	Laplacian viscosity coefficient
f_0	9.375×10^{-5} s^{-1}	Mean Coriolis parameter
β	1.754×10^{-11} (m s)$^{-1}$	Coriolis parameter gradient

Table A.2 Grid-dependent parameters

Grid dimensions	Resolution	Timestep	Hyperviscosity (a_4)
769×961	5 km	600 s	2.0×10^9 m^4 s^{-1}
97×121	40 km	1200 s	5.0×10^{11} m^4 s^{-1}

References

1. Arakawa, Akio and Lamb, Vivian R (1981), A potential enstrophy and energy conserving scheme for the shallow water equations. Monthly Weather Review 109:18–36.
2. Fox-Kemper, B and Bachman, S and Pearson, B and Reckinger, S (2014), Principles and advances in subgrid modelling for Eddy-Rich simulations. CLIVAR Exchanges No. 65, Vol. 19, No. 2.
3. Frisch, Uriel and Kurien, Susan and Pandit, Rahul and Pauls, Walter and Ray, Samriddhi Sankar and Wirth, Achim and Zhu, Jian-Zhou (2008), Hyperviscosity, Galerkin truncation, and bottlenecks in turbulence. Physical Review Letters 101:144501.
4. Griffies, Stephen M and Hallberg, Robert W (2000), Biharmonic friction with a Smagorinsky-like viscosity for use in large-scale eddy-permitting ocean models. Monthly Weather Review 128:2935–2946.
5. Hogg, Andrew Mc C and Dewar, William K and Killworth, Peter D and Blundell, Jeffrey R (2003), A quasi-geostrophic coupled model (Q-GCM). Monthly Weather Review 131:2261–2278.
6. Hogg, Andrew Mc C and Killworth, Peter D and Blundell, Jeffrey R and Dewar, William K (2005), Mechanisms of decadal variability of the wind-driven ocean circulation. Journal of Physical Oceanography 35:512–531.
7. Hogg, AM and Blundell, JR and Dewar, WK and Killworth, PD (2014), Formulation and users' guide for Q-GCM, http://q-gcm.org/downloads/q-gcm-v1.5.0.pdf.

8. Darryl D. Holm, Vladimir Zeitlin (1998), Hamilton's principle for quasigeostrophic motion. Physics of Fluids 10(4):800–806.
9. Jansen, Malte F and Held, Isaac M (2014), Parameterizing subgrid-scale eddy effects using energetically consistent backscatter. Ocean Modelling 80:36–48.
10. Leith, CE (1996), Stochastic models of chaotic systems. Physica D: Nonlinear Phenomena 98:481–491.
11. Mémin, Etienne (2014), Fluid flow dynamics under location uncertainty. Geophysical & Astrophysical Fluid Dynamics 108:119–146.
12. Paszke, Adam and Gross, Sam and Massa, Francisco and Lerer, Adam and Bradbury, James and Chanan, Gregory and Killeen, Trevor and Lin, Zeming and Gimelshein, Natalia and Antiga, Luca and others (2019), Pytorch: An imperative style, high-performance deep learning library. Advances in Neural Information Processing Systems 32:8026–8037.
13. Ryzhov, EA and Kondrashov, D and Agarwal, N and McWilliams, JC and Berloff, P (2020), On data-driven induction of the low-frequency variability in a coarse-resolution ocean model. Ocean Modelling 153:101664
14. Roullet, Guillaume and Gaillard, Tugdual (2021), A Fast Monotone Discretization of the Rotating Shallow Water Equations, ESSOAR/JAMES.
15. Shevchenko, IV and Berloff, PS (2015), Multi-layer quasi-geostrophic ocean dynamics in Eddy-resolving regimes. Ocean Modelling 94:1–14.
16. Smagorinsky J. (1963), General circulation experiments with the primitive equations. Monthly Weather Review 91.3:99–164.
17. Vallis, G. K. (2017), Atmospheric and oceanic fluid dynamics: fundamentals and large-scale circulation. Cambridge University Press.

Primitive Equations Under Location Uncertainty: Analytical Description and Model Development

Francesco L. Tucciarone, Etienne Mémin, and Long Li

Abstract Resolving numerically all the scale interactions of ocean dynamics in a high resolution realistic configuration is today far beyond reach, and only large scale representations can be afforded. In this work, we study a stochastic parameterization of the ocean primitive equations derived within the modelling under location uncertainty framework. First numerical assessments built with the NEMO core's code are provided for a double-gyres configuration.

Keywords Stochastic parametrization · Ocean modelling

1 Introduction

The Ocean covers a major part of Earth's surface and has an important stabilizing effect on the climate. For climatic prediction, accurate likely ensemble forecasts of future ocean states are consequently essential. However, due to an evident computational limitation high resolution simulations are completely unfeasible and only large-scale ocean representations can be handled. To face this difficulty, and the need of generating different likely future scenarios, there has been a growing interest in the geophysical sciences to set up flow models that incorporate in their dynamics noise terms related to uncertainties or errors. In accounting for the actions of unresolved processes in a random way, these stochastic models are in general less diffusive than the classical large-scale deterministic models. The unresolved processes include small-scale turbulence effects, boundary value uncertainties or uncertainties coming either from scale coarsening or from the numerical schemes used. Moreover, compared to classical large-scale deterministic modelling, the additional degree of freedom brought by the stochastic component allows us to devise new intermediate models [4, 3, 6, 7, 8]. The addition of noise in fluid

F. L. Tucciarone (✉) · E. Mémin · L. Li
INRIA Rennes Bretagne Atlantique, IRMAR – UMR CNRS 6625, Rennes, France
e-mail: francesco.tucciarone@inria.fr; etienne.memin@inria.fr; long.li@inria.fr

© The Author(s) 2023
B. Chapron et al. (eds.), *Stochastic Transport in Upper Ocean Dynamics*,
Mathematics of Planet Earth 10, https://doi.org/10.1007/978-3-031-18988-3_18

287

dynamics models cannot be done in a haphazard manner. Ad-hoc choices for model noise can fundamentally perturb the corresponding fluid dynamics models, making them exhibit unrealistic properties [3]. Rigorously justified methodologies for choosing the model noise have recently been introduced by Mémin [1] and Holm [2]. These derivations lead to large classes of stochastic geophysical fluid dynamics models that preserve either energy or circulation, respectively. Such models naturally emerge from a decomposition of the flow velocity field in terms of a smooth component and a time uncorrelated uncertainty random term. This decomposition is reminiscent, in spirit, of the classical Reynolds decomposition, and enables the definition of large-scale representation with a stochastic term representing small-scale effects. The Location Uncertainty (LU) formulation has been found to be more accurate in structuring the large-scale flow [4] and in reproducing long-terms statistics [22] for the barotropic quasi-geostrophic model. It also provides a good trade-off between model error representation and ensemble spread [21, 23] for the rotating shallow water model and the surface quasi-geostrophic model. In this work we explore more specifically a stochastic version of the primitive equations, named primitive equations under Location Uncertainty. The derivation of this model is detailed and first numerical experiments built from the NEMO code are assessed.

2   Location Uncertainty (LU)

In the LU formalism, the Lagrangian displacement $\mathbf{X}_t$ associated to a fluid particle is decomposed as:

$$\mathbf{X}_t(\mathbf{x}) = \mathbf{X}_{t_0}(\mathbf{x}) + \int_0^t \mathbf{v}(\mathbf{X}_s(\mathbf{x}), s)\, \mathrm{d}s + \int_0^t \boldsymbol{\sigma}(\mathbf{X}_s(\mathbf{x}), s)\, \mathrm{d}\mathbf{B}_s, \tag{1}$$

where $\mathbf{X}: \Omega \times \mathbb{R}^+ \to \Omega$ is the fluid flow map, that is the trajectory followed by fluid particles starting at initial map $\mathbf{X}|_{t=0}(\mathbf{x}) = \mathbf{x}_0$ of the bounded domain $\Omega \subset \mathbb{R}^3$. Written in differential form Eq. (1) takes the usual form:

$$\mathrm{d}\mathbf{X}_t(\mathbf{x}_0) = \mathbf{v}(\mathbf{X}_t, t)\, \mathrm{d}t + \boldsymbol{\sigma}(\mathbf{X}_t, t)\, \mathrm{d}\mathbf{B}_t. \tag{2}$$

The first component, $\mathbf{v}(\mathbf{X}_t, t)$, represents the smooth, resolved velocity field of the flow. It corresponds to the integration of the equations of motions, solved on a grid of a given resolution, and it is supposed to be both spatially and temporally correlated. The second term, $\boldsymbol{\sigma}(\mathbf{X}_t, t)\, \mathrm{d}\mathbf{B}_t$, is a stochastic process that assembles the unresolved flow component, uncertainties on the flow and turbulent effects. This stochastic contribution, often referred to as *noise* in the following, is built from the application of an Hilbert-Schmidt kernel integral operator, $\boldsymbol{\sigma}$, to an I_3–cylindrical Wiener process $\mathbf{B}$

$$(\sigma\,(\mathbf{X}_t,t)\,\mathrm{d}\mathbf{B}_t)^i = \int_{\Omega} \breve{\sigma}_{ik}\,(\mathbf{X}_t,\mathbf{y},t)\,\mathrm{d}\mathrm{B}^k_t\,(\mathbf{y})\,\mathrm{d}\mathbf{y},\tag{3}$$

where $\mathbf{B}$ is defined on a filtered probability space $\{\Omega, \mathcal{F}, \mathrm{P}, (\mathcal{F}_t)_t\}$ and $(\mathcal{F}_t)_t$ is the filtration adapted to $\mathbf{B}$. The application of the (integrable) kernel $\breve{\sigma}$ imposes fast/small scales spatial correlation and defines a centered Gaussian process $\sigma\,\mathrm{d}\mathbf{B}_t \sim \mathcal{N}\,(0, \mathbf{Q}\mathrm{d}t)$, with covariance tensor defined as

$$Q_{ij}\,(\mathbf{x},\mathbf{y},t,s) = \mathbb{E}\left[(\sigma\,(\mathbf{x},t)\,\mathrm{d}\mathbf{B}_t)^i\,(\sigma\,(\mathbf{y},s)\,\mathrm{d}\mathbf{B}_s)^j \right]$$

$$= \delta\,(t-s)\,\mathrm{d}t \int_{\Omega} \breve{\sigma}_{ik}\,(\mathbf{x},\mathbf{z},t)\,\breve{\sigma}_{kj}\,(\mathbf{z},\mathbf{y},s)\,\mathrm{d}\mathbf{z}.$$

The strength of the noise is measured by the diagonal components of the covariance tensor per unit of time, i.e. the variance tensor, $\mathbf{a}$, defined as $\mathbf{a}(\mathbf{x},t)\delta(t-t')\mathrm{d}t = \mathbf{Q}(\mathbf{x},\mathbf{x},t,t')$. The variance tensor is symmetric and positive definite at any point $\mathbf{x}$ of the domain. Notably, it has the dimension of a viscosity in $\mathrm{m}^2\mathrm{s}^{-1}$. The covariance operator is self-adjoint, positive definite and compact and admits a convenient spectral decomposition.

In this paper, the noise will always be assumed to be centred, but it can be proven through Girsanov theorem that one can redefine the Lagrangian displacement (2) as

$$\mathrm{d}\mathbf{X}_t\,(\mathbf{x}_0) = [\mathbf{v}\,(\mathbf{X}_t,t) - \boldsymbol{\mu}_t\,(\mathbf{X}_t)]\,\mathrm{d}t + \sigma\,\mathrm{d}\widetilde{\mathbf{B}}_t\,(\mathbf{X}_t),\tag{4}$$

where the Wiener process $\widetilde{\mathbf{B}}_t$ is a centred process under a new probability measure Q drifted by $\boldsymbol{\mu}_t$. Indeed a non centred Wiener process shifted by a random process $(\mathbf{Y}_t)_t$ can be defined as:

$$\widetilde{\mathbf{B}}_t = \mathbf{B}_t + \int_0^t \mathbf{Y}_s\,\mathrm{d}s.\tag{5}$$

Under good properties of $(\mathbf{Y})_t$ ($\mathcal{F}_t$-measurability, almost sure L^2−integrability and Novikov condition) there exists a measure Q such that $(\widetilde{\mathbf{B}}_t)_t$ is a Q− Wiener process With the non centred random process $\widetilde{\mathbf{B}}_t$ we can rewrite the equations with respect to $\widetilde{\mathbf{B}}_t$ as

$$\sigma\,\mathrm{d}\mathbf{B}_t\,(\mathbf{X}_t) = \sigma\,\mathrm{d}\widetilde{\mathbf{B}}_t\,(\mathbf{X}_t) - \sigma\,(\mathbf{X}_t,t)\,\mathbf{Y}_t\,\mathrm{d}t.\tag{6}$$

Denoting $\sigma\,(\mathbf{X}_t,t)\,\mathbf{Y}_t$ as $\boldsymbol{\mu}_t$ one can write the Lagrangian displacement (2) as (4) and under Q the Wiener process $\mathrm{d}\widetilde{\mathbf{B}}_t$ is centred thus the writing of $\mathrm{d}\mathbf{X}_t$ has the same form as (2) but under a new measure. All the arguments provided in the following will hold for this process under Q. The use of a drifted noise $\sigma\,\mathrm{d}\widetilde{\mathbf{B}}_t$ is fundamental when the processes employed to operationally define the noise are not centred, hence displaying a non-zero time average.

3   Stochastic Transport Theorem

The derivation of Eulerian flow dynamics models within the LU formalism relies on a stochastic version of the Reynolds transport theorem (SRTT), introduced in [1], which describes the rate of change of a random scalar q transported by the stochastic flow (2) within a flow volume V_t:

$$d \int_{V_t} q\,(\mathbf{x}, t)\,\mathrm{d}\mathbf{x} = \int_{V_t} \left\{ D_t q + q \nabla \cdot \left[\mathbf{v}^\star\,\mathrm{d}t + \sigma\,\mathrm{d}\mathbf{B}_t \right] \right\} (\mathbf{x}, t)\,\mathrm{d}\mathbf{x}, \tag{7}$$

with the operator

$$D_t q = \mathrm{d}_t q + \left[\mathbf{v}^\star\,\mathrm{d}t + \sigma\,\mathrm{d}\mathbf{B}_t \right] \cdot \nabla q - \frac{1}{2} \nabla \cdot (\mathbf{a}\nabla q)\,\mathrm{d}t, \tag{8}$$

defining the stochastic transport operator. The SRTT is in perfect analogy with the deterministic Reynolds transport theorem (compare with [13] section 5.3), and the various terms can be interpreted physically. Proceeding in order, the first right-hand side term of (8) is the *increment in time* at a fixed location of the process q, that is $\mathrm{d}_t q = q\,(\mathbf{X}_t, t + \mathrm{d}t) - q\,(\mathbf{X}_t, t)$. This contribution plays the role of the partial time derivative for a process that is not time differentiable. The term enclosed in the square brackets is a *stochastic advection displacement*. It involves a time correlated modified advection,

$$\mathbf{v}^\star = \mathbf{v} - \frac{1}{2} \nabla \cdot \mathbf{a} + \sigma^{\mathrm{T}} (\nabla \cdot \sigma), \tag{9}$$

and a fast evolving, time uncorrelated noise $\sigma\,\mathrm{d}\mathbf{B}_t$. The advection by this term of variable q leads to a *multiplicative noise*, which is hence non Gaussian. This type of noise is often denoted as *transport noise* in the literature. The second term of the modified advection is coined as the *Ito-Stokes drift* velocity in [4], $\mathbf{v}_s = \frac{1}{2} \nabla \cdot \mathbf{a}$. It represents an effective transport velocity resulting from statistical effects due to inhomogeneities of the noise term. The last term of the transport operator is a dissipation term that depicts the mixing mechanism due to the unresolved scales. Following [5] one can consider the transport of a characteristic function to introduce an evolution equation for the Jacobian determinant J of the flow:

$$D_t J - J \nabla \cdot \left[\left(\mathbf{v} - \mathbf{v}^s + \sigma^{\mathrm{T}} (\nabla \cdot \sigma) \right) \mathrm{d}t + \sigma\,\mathrm{d}\mathbf{B}_t \right] = 0. \tag{10}$$

This equation provides a clear condition for the stochastic flow to be isochoric:

$$\nabla \cdot \left[\mathbf{v}^\star\,\mathrm{d}t + \sigma\,\mathrm{d}\mathbf{B}_t \right] = 0. \tag{11}$$

4 Boussinesq Equations

Under location uncertainty, a stratified ocean can be modelled with a modified version of Boussinesq equations. The derivation that is outlined here follows almost verbatim the asymptotic derivation given in [12]. First, one applies the SRTT (7) to the density and imposes conservation, that is $\mathrm{d} \int_{V_t} \rho(\mathbf{x}, t)\, \mathrm{d}\mathbf{x} = 0$. Then, assuming that the fluctuations of density are small compared to the mean,

$$\rho(\mathbf{x}, t) = \rho_0 \left[1 + \varepsilon\, \delta \hat{\rho}(t, \mathbf{x})\right], \tag{12}$$

and using ε as an asymptotic ordering parameter to perform an expansion of the conservation of mass, the first order is found to be:

$$\nabla \cdot \left[\mathbf{v}^\star \, \mathrm{d}t + \sigma\, \mathrm{d}\mathbf{B}_t\right] = 0, \tag{13}$$

that can be split in two incompressibility conditions involving both the modified drift velocity $\mathbf{v}^\star$ and the fast scale component $\sigma\, \mathrm{d}\mathbf{B}_t$ thanks to the uniqueness of semi-martingale decomposition [15]. Applying again the SRTT (7) to the momentum reads

$$\rho \mathbb{D}_t \mathbf{v} = -\nabla \left(p - \frac{\mu}{3}\nabla \cdot \mathbf{v}\right) \mathrm{d}t - \nabla \left(\mathrm{d}p_t^\sigma\right) - \rho g \mathbf{e}_3\, \mathrm{d}t, \tag{14}$$

where the right hand side entails pressure forces, compressibility effects [14] and gravitational forces. The compressibility term $\frac{\mu}{3}\nabla \cdot \mathbf{v}$, with μ dynamical viscosity of water, is usually neglected in the deterministic derivation of the Boussinesq model, but in this model is maintained in view of the different incompressibility condition (12), that enforces $\nabla \cdot \mathbf{v} = \nabla \cdot \mathbf{v}_s$. Following classical nondimensionalization procedure [12, 14], characteristic scales are introduced as:

$$\mathbf{x} = L\hat{\mathbf{x}}, \qquad \mathbf{v} = U\hat{\mathbf{v}}, \qquad t = \tau \hat{t}, \qquad p = \frac{\rho_0 U^2}{\epsilon}\hat{p}, \qquad g = \frac{U^2}{\epsilon L}\hat{g}, \tag{15}$$

with $\tau = L/U$ advective time scale. Furthermore, the variance tensor is assumed to scale as $\mathbf{a} = A\hat{\mathbf{a}}$ so that the fast-evolving component $\sigma\, \mathrm{d}\mathbf{B}_t$ and the kernel σ can be scaled as

$$\sigma\, \mathrm{d}\mathbf{B}_t = \sqrt{\frac{AL}{U}}\hat{\sigma}\, \mathrm{d}\hat{\mathbf{B}}_t \quad \text{and} \quad \sigma = \sqrt{A}\hat{\sigma}. \tag{16}$$

In this novel framework a non-dimensional parameter $\Upsilon = UL/A$ is introduced to compare advection and stochastic diffusion terms in the momentum equation. This parameter is termed *stochastic Peclet number*, in perfect similarity with the deterministic advection-diffusion problem [10]. Introducing these variables, following [12], one obtains:

$$\rho_0 \left(1 + \epsilon \delta \hat{\rho}\right) \left\{ d_t \hat{\mathbf{v}} + \left[\left(\hat{\mathbf{v}} - \frac{1}{\gamma} \hat{\mathbf{v}}_s \right) d\hat{t} + \frac{1}{\gamma^{1/2}} \hat{\sigma} d\hat{\mathbf{B}}_t \right] \cdot \hat{\nabla} \hat{\mathbf{v}} \right.$$

$$\left. - \frac{1}{2\gamma} \hat{\nabla} \cdot \left(\hat{\mathbf{a}} \hat{\nabla} \hat{\mathbf{v}} \right) d\hat{t} \right\} = \hat{\nabla} \left(-\frac{\rho_0}{\epsilon} \hat{p} + \frac{1}{\text{Re}\gamma} \frac{1}{3} \hat{\nabla} \cdot \hat{\mathbf{v}}_s \right) d\hat{t} \quad (17)$$

$$- \hat{\nabla} \left(\frac{P^\sigma}{U^2} d\hat{p}_t^\sigma \right) - \rho_0 \left(1 + \epsilon \delta \hat{\rho}\right) \frac{\hat{g}}{\epsilon} \mathbf{e}_3 \, d\hat{t}.$$

Expanding each variable as an asymptotic with ϵ taken as ordering parameter, Eq. (17) provides at lowest order, once dimensional variables are replaced to non-dimensional variables,

$$\nabla p_0 = -\rho_0 g \mathbf{e}_z, \quad p_0(z) = -\rho_0 g z. \quad (18)$$

Decomposing the density into a background constant density and a deviation, corresponds on the pressure variable to a decomposition in terms of a hydrostatic component and a pressure fluctuation. This splitting,

$$\rho(t, \mathbf{x}) = \rho_0 + \rho'(t, \mathbf{x}), \qquad p(t, \mathbf{x}) = p_0 + p'(t, \mathbf{x}), \quad (19)$$

allows the recognition of the first order component of the pressure as the deviation from the hydrostatic pressure p', so that Eq. (17) at first order in dimensional form becomes

$$d_t \mathbf{v} + \left[(\mathbf{v} - \mathbf{v}^s) \, dt + \sigma d\mathbf{B}_t \right] \cdot \nabla \mathbf{v} - \frac{1}{2} \nabla \cdot (\mathbf{a} \nabla \mathbf{v}) \, dt =$$

$$= \nabla \left(-p' + \frac{v}{3} \nabla \cdot \mathbf{v}_s \right) dt - \nabla \left(\frac{dp_t^\sigma}{\rho_0} \right) - \frac{\rho'}{\rho_0} g \mathbf{e}_z \, dt.$$

The splitting (19) also introduces naturally the *buoyancy* $\mathbf{b} = -g\mathbf{e}_3 \rho'(t, \mathbf{x})/\rho_0$ in the equations of motions, representing the upward (or downward) force associated with the density anomaly ρ'. In terms of buoyancy, the momentum equation can be written as

$$D_t \mathbf{v} = \nabla \left(-p' - \frac{dp_t^\sigma}{\rho_0} + \frac{v}{3} \nabla \cdot \mathbf{v}_s \right) dt - \mathbf{b} \, dt. \quad (20)$$

A stochastic transport equation can be written for the buoyancy from mass conservation. However, in this work a tracer transport equation on salinity, S, and temperature, T, is preferred, relating then the buoyancy and the tracers with a buoyancy state equation $b = b(T, S, z)$. The conservation of a given tracer θ is expressed as

$$D_t \theta + \theta \nabla \cdot \left[(\mathbf{v} - \mathbf{v}_s) \, dt + \sigma d\mathbf{B}_t \right] = F^\theta \, dt + D^\theta \, dt, \quad (21)$$

where the variation of tracer quantity is balanced by a forcing term F^θ and a diffusive term D^θ. We note that here these terms are assumed to be regular in time, although additional Brownian terms could be considered to encode intermittent forcing. The resulting system, split into horizontal and vertical equations using the convention $\mathbf{v} = (\mathbf{u}, w)$, is:

Horizontal momentum:

$$D_t\mathbf{u} + f\mathbf{e}_3 \times \left(\mathbf{u}\,dt + \frac{1}{2}\sigma\,d\mathbf{B}_t^{\mathrm{H}}\right) = \nabla_{\mathrm{H}}\left(-p' + \frac{\nu}{3}\nabla\cdot\mathbf{v}\right)dt - \nabla_{\mathrm{H}}dp_t^\sigma \quad (22)$$

Vertical momentum:

$$D_t w = \frac{\partial}{\partial z}\left(-p' + \frac{\nu}{3}\nabla\cdot\mathbf{v}\right)dt - \frac{\partial}{\partial z}dp_t^\sigma + b\,dt \quad (23)$$

Temperature and salinity:

$$D_t T = \kappa_T\,\Delta T\,dt, \quad (24)$$

$$D_t S = \kappa_S\,\Delta S\,dt, \quad (25)$$

Incompressibility:

$$\nabla\cdot\left[\mathbf{v} - \mathbf{v}^s\right] = 0, \qquad \nabla\cdot\sigma\,d\mathbf{B}_t = 0, \quad (26)$$

Equation of state:

$$b = b(T, S, z). \quad (27)$$

Temperature and salinity are introduced as active tracers, as they modify the buoyancy field, and their stochastic evolution is obtained again by application of the SRTT (7), balanced with a diffusion process with diffusivity κ_T and κ_S respectively. The unusual coefficient $1/2$ in the random Coriolis term can be shown to appear naturally from a derivation of the non-inertial acceleration in this stochastic framework, again following the derivation of [12]. Metric terms relative to the rotation of the earth should also be adapted to the stochastic Frenet-Serret formula $d\mathbf{C} = \boldsymbol{\Omega}\,dt \times \mathbf{C}$ in the case of planetary scale simulations. In Eqs. (22) and (23) the *stochastic pressure* is introduced, and corresponds to a zero-mean turbulent pressure related to the small scale velocity component (i.e. noise). It is a martingale term. An operational model referred to as the primitive equations can be obtained through the so-called hydrostatic balance, resulting from neglecting the vertical acceleration terms through a proper scaling of the velocity. In our stochastic setting, the vertical momentum equation reads, after neglecting the large scale acceleration terms and for moderate noise ($\Upsilon \sim \mathcal{O}(1)$ so as the martingale terms related to the vertical velocity component are negligible):

$$-\frac{\partial p'}{\partial z} + b = 0 \quad \text{and} \quad \frac{\partial dp_t^\sigma}{\partial z} = 0, \quad (28)$$

where the bounded variation terms and the martingale terms have been safely separated. The left equation constitutes the usual hydrostatic balance. With the scaling used, the stochastic pressure is constant along depth and is in balance with the stochastic Coriolis component [9, 5]. These two martingale terms can be removed then from the horizontal momentum equation. In this setting the vertical component of the momentum equation becomes a diagnostic component that can be recovered integrating the continuity equation given by (26). In a similar way, the large scale pressure is obtained from the vertical integration of the hydrostatic relation. The scaling parameter Υ can also be related to the ratio between the Mean Kinetic Energy (TKE) when an advective time scale is used, that is

$$\Upsilon = \frac{U^2}{A/\tau} = \frac{1}{\epsilon} \frac{MKE}{TKE} \tag{29}$$

where $\epsilon = \tau_\sigma/\tau$, is the ratio of the fast-scale to the slow-scale correlation times. This ratio can be adapted to the different variables involved (i.e. momentum, temperature or salinity) with a value similar to the inverse of the Schmidt number (ratio of diffusion rates) making hence the noise scaling parameter, Υ, dependant on the variable transported. The parameter Υ appears in dimensional analysis and asymptotic expansions, but plays also a paramount role in the quantification of the strength of the noise.

5 Methods

The experiments are performed with the level-coordinate free-surface primitive equation ocean model NEMO [16]. The domain configuration is a double-gyre configuration consisting of a 45° rotated beta plane centred at $\sim$ 30°N, 3180 km long, 2120 km wide and 4 km deep. The domain is bounded by vertical walls and a flat bottom. The seasonally varying wind and buoyancy forcings induce a strong jet to appear diagonally in the domain, separating a warm sub-tropical gyre from a cold sub-polar gyre. Three experiments were performed: two purely deterministic simulations at different resolutions, 1/27° (R27d) and 1/3° (R3d), and one stochastic simulation at 1/3° (R3LU). Each simulation was run for 10 years with data collected every (and averaged over) 5 days. The focus of this paper is to assess the benefits brought by LU to the coarse simulation, so the parameters of the simulation were chosen following thoroughly [17, 18] (see Table 1 for an overview of their values). In this first study, we restrict ourselves to 3D divergence-free horizontal noise (i.e. with no vertical component). In spectral form the random field and the variance tensor can be written as:

$$\sigma d\mathbf{B}_t = \sum_{i \in \mathbb{N}} \lambda_i^{1/2} \boldsymbol{\varphi}_i(\mathbf{x}) d\beta_t^i, \qquad \mathbf{a} = \sum_{i \in \mathbb{N}} \lambda_i \boldsymbol{\varphi}_i(\mathbf{x}) \boldsymbol{\varphi}_i^{\mathrm{T}}(\mathbf{x}), \tag{30}$$

Table 1 Parameters of the model experiments

	R27d	R3d	R3LU
Horizontal resolution	$1/27°$ (3.9 km)	$1/3°$ (35.3 km)	$1/3°$ (35.3 km)
Horizontal grid points	540×810	60×90	60×90
Vertical levels	30	30	30
Time step	5 min	20 min	20 min
Eddy viscosity	-5×10^{-9} m^{4}s^{-1}	-10^{-12} m^{4}s^{-1}	-10^{-12} m^{4}s^{-1}
Eddy diffusivity	-5×10^{-10} m^{4}s^{-1}	300 m^{2}s^{-1}	300 m^{2}s^{-1}

where $\{\boldsymbol{\varphi}_i(\mathbf{x}), i \in \mathbb{N}\}$ are the orthonormal eigenfunctions of the covariance operator associated to $\{\lambda_i, i \in \mathbb{N}\}$, the (real, positive) eigenvalues ranged in decreasing value order and $\{\beta_t^i, i \in \mathbb{N}\}$ is a set of standard (scalar) Brownian variables. This representation corresponds to the Karhunen-Loeve decomposition [24]. Operationally, the (finite) set of eigenfunctions $\{\boldsymbol{\phi}_i(\mathbf{x}), i \in [1, N]\}$ and of eigenvalues $\{\lambda_i, i \in [1, N]\}$ are computed through a proper orthogonal decomposition (POD) [11] of the temporal fluctuations of the two-dimensional low resolution residual $\mathbf{u}_{\mathrm{LR}}$. This velocity residual is obtained through Gaussian filtering of the high resolution deterministic simulation R27d, $\mathbf{u}_{\mathrm{LR}} = (1 - \mathcal{G}) \mathbf{u}_{\mathrm{HR}}$, with the fluctuations computed through Reynolds decomposition:

$$\mathbf{u}'_{\mathrm{LR}}(\mathbf{x}, t) = \mathbf{u}_{\mathrm{LR}}(\mathbf{x}, t) - \overline{\mathbf{u}_{\mathrm{LR}}(\mathbf{x}, t)}^t = \sum_{i=1}^{N} \boldsymbol{\phi}_i(\mathbf{x}) \alpha_i(t). \tag{31}$$

The POD procedure applied to $\mathbf{u}'_{\mathrm{LR}}(\mathbf{x}, t)$ provides a set $\{\boldsymbol{\phi}_i(\mathbf{x}), i \in [1, N]\}$ of eigenfunctions that are stationary in time and such that

$$\langle \boldsymbol{\phi}_m, \boldsymbol{\phi}_n \rangle = \int_{\Omega} \boldsymbol{\phi}_m^{\mathrm{T}} \boldsymbol{\phi}_n(\mathbf{x}) \, \mathrm{d}\mathbf{x} = \delta_{mn}, \quad \overline{\alpha_m \alpha_n}^t = \lambda_m \delta_{m,n}. \tag{32}$$

The eigenfunctions are used to define the random field and a stationary variance tensor as

$$\sigma \mathrm{d}\mathbf{B}_t(\mathbf{x}) = \sum_{i=1}^{M(z)} \lambda_i^{1/2} \boldsymbol{\phi}_i(\mathbf{x}) \sqrt{\Delta t} \, \mathrm{d}\beta_t^i, \quad \mathbf{a}(\mathbf{x}) = \sum_{i=1}^{M(z)} \lambda_i \Delta t \, \boldsymbol{\phi}_i(\mathbf{x}) \boldsymbol{\phi}_i^{\mathrm{T}}(\mathbf{x}) \tag{33}$$

where $\boldsymbol{\varphi}_i = \boldsymbol{\phi}_i \sqrt{\Delta t}$ and $M(z) \ll N$ chosen to provide at least 85% of the energy of the fluid layer. Due to the constraint posed by Eq. (26) on the noise, incompressibility on the horizontal noise is imposed by applying a Helmoltz-Hodge decomposition [19] on the each snapshot of the horizontal velocity $\mathbf{u}_{\mathrm{LR}}$. Moreover, the set of eigenfunctions $\{\boldsymbol{\phi}_i(\mathbf{x}), i \in [1, N]\}$ is used to construct the drift $\boldsymbol{\mu}_t$ of Eq. (4) in such a way that the distance between $\boldsymbol{\mu}_t$ and $\overline{\mathbf{u}_{\mathrm{LR}}}^t$ is minimized, that is

$$\mu_t = \sum_{i=1}^{N} \phi_i(\mathbf{x}) \, y_t^i \quad \text{with} \quad y_t^i = \arg\min \left\| \overline{\mathbf{u}_{\mathrm{LR}}(\mathbf{x}, t)}^t - \sum_{i=1}^{N} \phi_i(\mathbf{x}) \, y_t^i \right\|_2. \tag{34}$$

Due to the orthogonality of the basis functions the coefficients can be easily recovered as the orthogonal projection $y_t^i = \langle \overline{\mathbf{u}_{\mathrm{LR}}(\mathbf{x}, t)}^t, \phi_i(\mathbf{x}) \rangle$.

6   Results

In this work we focus on the results of a single realisation. From a qualitative point of view, the effect of the coarsening of the resolution can be seen in Figs. 1 and 2, where the leftmost panel represents the result the R27d simulation, the central panel shows the results of the R3d simulation and the rightmost panel shows the R3LU simulation. The first noticeable characteristic of the R27d reference simulation is the presence of a primary jet stream inclined at an almost $-45°$ angle starting at the bottom-left corner and directed towards the centre, and a secondary, smaller jet with the same inclination roughly 80 km above the primary. The presence of both structures is visible in the reference papers [17, 18]. In both figures the comparison between the high resolution and the low resolution deterministic simulation shows a degradation of the information about the jet-streams. Figure 1, depicting the relative vorticity $\overline{\zeta}^{10Y} = \overline{(\partial_x v - \partial_y u)/f}^{10Y}$, shows that the deterministic R3d simulation is incapable of reproducing the primary jet characteristic and its positioning, though showing an increased activity in place of the secondary jet stream. The stochastic R3LU simulation presents instead a intensification of the vortical activity in the

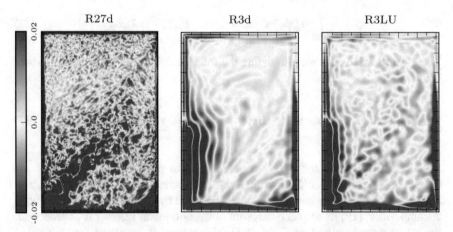

Fig. 1 10-years averaged relative vorticity $\zeta = (\partial_x v - \partial_y u)/f$ at the surface layer of the model for deterministic high-resolution (1/27°, left), for deterministic low resolution (1/3°, middle) and for stochastic low resolution (1/3°, right)

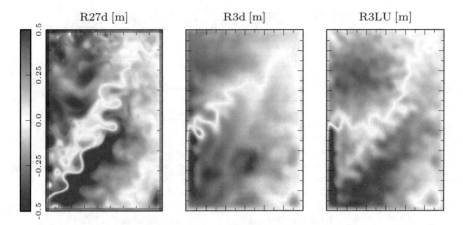

Fig. 2 5-days averaged sea surface height of the model for deterministic high-resolution (1/27°, left), for deterministic low resolution (1/3°, middle) and for stochastic low resolution (1/3°, right)

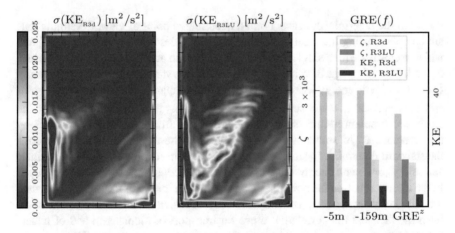

Fig. 3 Left and centre panels, standard deviation of the kinetic energy. The color scale has been adjusted to enhance the differences in the jet region, not considering the highly energetic boundaries where peaks present values as $0.2\,\mathrm{m}^2/\mathrm{s}^2$ for R3d and $0.17\,\mathrm{m}^2/\mathrm{s}^2$ for R3LU. Right panel, the Gaussian relative entropy for relative vorticity, ζ, (cold palette) and kinetic energy (warm palette). The lighter colors represent the deterministic simulation R3d, the darker colors represent the stochastic simulation R3LU. All the statistics are computed over 10 years

regions of the primary and secondary jet. Considering sea surface height, Fig. 2 shows that the best result is obtained by the stochastic simulation that, while not being able to distinguish the primary jet stream by the smaller vortices of the secondary jet, it is capable of reproducing the main behaviour. The left and centre panels of Fig. 3 shows the difference obtained in terms of variance of the kinetic energy in the two coarse simulations, with greater variability obtained with the stochastic model, especially in the area of the jet stream, where a lesser variability is

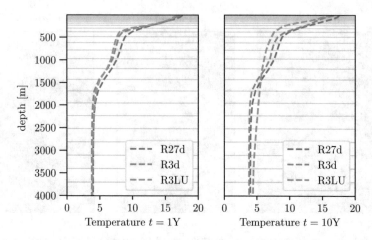

Fig. 4 Vertical profile of temperature after 1 year of simulation (left) and after 10 years (right)

shown in the deterministic case. From a quantitative point of view, the simulations are compared using the Gaussian Relative Entropy described in details in [20] and which measures with a single criterion both the mean and variance reconstructions. In the left panel of Fig. 3, values of the GRE for two variables, the relative vorticity ζ, and the kinetic energy $KE = (u^2 + v^2)/2$ are compared. For two different depths and in a vertical average sense $(\overline{GRE^z})$, the relative entropy is smaller for the stochastic simulation, indicating a smaller distance from the distribution given by the reference R27d simulation. The proposed stochastic model thus outperforms the standard deterministic simulation in terms of both relative entropy and intrinsic variability for kinetic energy and vorticity. This behaviour is observed in every layer. In the tracers equation the noise has been scaled with the aid of the Schmidt number, the ratio between the eddy viscosity and eddy diffusivity. This consideration stems from the fact that the correlation times for transport of momentum and of tracers are not the same, and the difference can be expressed in terms of the Schmidt number. Figure 4 shows the vertical profiles of horizontally-averaged temperature, $\overline{T}^{x,y}(z,t) = \int_A T(x, y, z, t)\,dxdy$, at time $t = 1Y$ and $t = 10Y$ for the three simulations. The vertically averaged temperature shows an increase in mixing of temperature of the stochastic setting with respect to its deterministic counterparts. This process has been observed to be sensible to the noise amplitude and might be caused by the structure of the noise and by the effects of Helmholtz-Hodge decomposition. Further studies to investigate this process with three-dimensional and isopycnal noise are ongoing.

7 Conclusions

The considered stochastic model has been implemented into the NEMO dynamical core. A 3D horizontal, incompressible noise was considered and has been proven to successfully increase the capabilities of a coarse simulation in simulating the dynamical quantities of interest, when corrected with a stochastic drift leading to a change of probability measure. Both the qualitative behaviour of the jet-stream and the quantitative intrinsic variability of the model have been increased. Thermodynamic quantities like temperature and salinity seem to not benefit from this implementation. In future works, more complex non stationary fully 3D noises will be investigated within the same setting.

Acknowledgments The authors acknowledge the support of the ERC EU project 856408-STUOD.

References

1. Mémin, E: Fluid flow dynamics under location uncertainty. Geophysical and Astrophysical Fluid Dynamics 108, 119–197 (2014).
2. Holm, D. D.: Variational principles for stochastic fluid dynamics. Proceedings of the Royal Society A: Mathematical, Physical and Engineering Sciences, 471(20140963), 2015.
3. Chapron, B., Dérian, P., Mémin, E., Resseguier, V.: Large-scale flows under location uncertainty: a consistent stochastic framework. QJRMS, 144(710):251–260, 2018.
4. Bauer, W., Chandramouli, P., Chapron, B., Li, L., Mémin, E.: Deciphering the role of small-scale inhomogeneity on geophysical flow structuration: a stochastic approach. Journal of Physical Oceanography, (2020).
5. Resseguier , V., Mémin, E., Chapron, B.: Geophysical flows under location uncertainty, Part I Random transport and general models Geophysical and Astrophysical Fluid Dynamics 111, 149–176 (2017).
6. Cintolesi, C., Mémin, E.: Stochastic Modelling of Turbulent Flows for Numerical Simulations Fluids 5, (2020).
7. Kadri Harouna, S., Mémin, E.: Stochastic representation of the Reynolds transport theorem: Revisiting large-scale modelling. Computers and Fluids 156, 456–469 (2017).
8. Pinier, B., Mémin, E., Laizet, S., Lewandowski R.: Stochastic flow approach to model the mean velocity profile of wall-bounded flows. Phys. Rev. E, 99(6):063101, 2019.
9. Brecht, R., Li, L., Bauer, W., Mémin, E.: Rotating shallow water flow under location uncertainty with a structure-preserving discretization. J. of Advances in Modelling of Earth Systems, 13(12), 2021.
10. Heinrich, J.C., Huyakorn, P.S., Zienkewicz, O.C., Mitchell, A.R.: An 'Upwind' Finite Element Scheme for Two-dimensional Convective Transport Equation. International Journal for Numerical Methods in Engineering 11, 131–143, (1977).
11. Holmes, P., Lumley, J. L., Berkooz, G.: Turbulence, coherent structures, dynamical systems, and symmetry. Cambridge University Press (1996).
12. Vallis, Geoffrey K.: Atmospheric and Oceanic Fluid Dynamics: Fundamentals and Large-Scale Circulation. Cambridge University Press, 2nd edition (2017).
13. Borisenko, A. I., Tarapov, I. E., Silverman, R. A.: Vector and Tensor Analysis with Applications. Dover Publications (1979).
14. Batchelor, G. K.: An Introduction to Fluid Dynamics. Cambridge University Press (2000)

15. Kunita, H.: Stochastic Flows and Stochastic Differential Equations. Cambridge Studies in Advanced Mathematics (1997)
16. Madec, G., Bourdallé-Badie, R., Chanut, J., Clementi, E., Coward, A.,Ethé, C., Iovino, D., Lea, D., Lévy, C., Lovato, T., Martin, N., Masson, S., Mocavero, S., Rousset, C., Storkey, D., Vancoppenolle, M., Müller, S., Nurser, G., Bell, M., Samson, G.: Nemo ocean engine, Oct. 2019.
17. Lévy, M. , Klein, P., Tréguier, A.-M., Iovino, D. , Madec, G., Masson, S., Takahashi, K.: Modifications of gyre circulation by sub-mesoscale physics. Ocean Modelling, 34(1-2):1–15, 2010.
18. Lévy, M. Resplandy, L., Klein, P., Capet, X., Iovino, D., Ethé, C.: Grid degradation of submesoscale resolving ocean models: Benefits for offline passive tracer transport. Ocean Modelling, 48:1–9, 2012.
19. Denaro, F. M.: On the application of the helmholtz–hodge decomposition in projection methods for the numerical solution of the incompressible Navier–Stokes equations with general boundary conditions. International Journal for Numerical Methods in Fluids, 43:43–69, 09 2003.
20. Grooms, I., Majda, A., Smith, S.: Stochastic superparameterization in a quasi-geostrophic model of the antarctic circumpolar current. Ocean Modelling, 85, 10 2014.
21. Brecht, R. , Li, L., Bauer, W., Mémin, E.: Rotating shallow water flow under location uncertainty with a structure-preserving discretization. Journal of Advances in Modeling Earth Systems, American Geophysical Union, 2021, 13 (12)
22. Bauer, W., Chandramouli, P., Li, L., Mémin, E.: Stochastic representation of mesoscale eddy effects in coarse-resolution barotropic models Ocean Modelling, Elsevier, 2020, 151, pp.1–50.
23. Resseguier, V., Li, L., Jouan, G., Dérian, P., Mémin, E., Chapron, B.: New trends in ensemble forecast strategy: uncertainty quantification for coarse-grid computational fluid dynamics Archives of Computational Methods in Engineering, Springer Verlag, 2021, 28 (1), pp.215–261. https://doi.org/10.1007/s11831-020-09437-x
24. Loeve, M.: Probability theory, volume II. Springer, Graduate Texts in Mathematics (1978).

Bridging Koopman Operator and Time-Series Auto-Correlation Based Hilbert–Schmidt Operator

Yicun Zhen, Bertrand Chapron, and Etienne Mémin

Abstract Given a stationary continuous-time process $f(t)$, the Hilbert–Schmidt operator A_τ can be defined for every finite τ. Let $\lambda_{\tau,i}$ be the eigenvalues of A_τ with descending order. In this article, a Hilbert space $\mathcal{H}_f$ and the (time-shift) continuous one-parameter semigroup of isometries $\mathcal{K}^s$ are defined. Let $\{v_i, i \in \mathbb{N}\}$ be the eigenvectors of $\mathcal{K}^s$ for all $s \geq 0$. Let $f = \sum_{i=1}^{\infty} a_i v_i + f^\perp$ be the orthogonal decomposition with descending $|a_i|$. We prove that $\lim_{\tau \to \infty} \lambda_{\tau,i} = |a_i|^2$. The continuous one-parameter semigroup $\{\mathcal{K}^s : s \geq 0\}$ is equivalent, almost surely, to the classical Koopman one-parameter semigroup defined on $L^2(X, \nu)$, if the dynamical system is ergodic and has invariant measure ν on the phase space X.

Keywords Singular spectrum analysis · Koopman theory · Hilbert–Schmidt theory

1 Introduction

Let $\{f(t) \in \mathbb{C} : t \geq 0\}$ be a continuous time process. We assume that f has zero temporal mean and the lagged moments exist for all $s \geq 0$:

$$\rho(s) := \lim_{T \to \infty} \frac{1}{T} \int_0^T f(t)\bar{f}(t + s)\mathrm{d}t. \tag{1}$$

Define $\rho_{-s} = \bar{\rho}_s$. In [3] the self-adjoint operator A_τ is defined to act on $L^2([0, \tau])$:

Y. Zhen (✉) · B. Chapron
Institut Français de Recherche pour l'Exploitation de la Mer, Plouzané, France
e-mail: zhenyicun@protonmail.com

E. Mémin
INRIA/IRMAR, Rennes, France

B. Chapron et al. (eds.), *Stochastic Transport in Upper Ocean Dynamics*,
Mathematics of Planet Earth 10, https://doi.org/10.1007/978-3-031-18988-3_19

$$(A_\tau g)(t) = \frac{1}{\tau} \int_0^\tau g(s)\rho(t-s)\mathrm{d}s, \tag{2}$$

for every $g \in L^2([0, \tau])$, and for all $t \in [0, \tau]$. When $\rho \in L^2_{\mathrm{loc}}(\mathbb{R})$ and $\rho(s) \neq 0$ for almost all $s \in [0, \tau]$, A_τ is a Hilbert–Schmidt operator. In particular, A_τ is compact and always has a purely punctual spectrum. In other words, the Hilbert space $L^2([0, \tau])$ admits a basis $\{\phi_i \in L^2([0, \tau]) : i \in \mathbb{N}\}$, so that each ϕ_i is an eigenvetor of A_τ. This implies a Karhunen–Loéve type of decomposition. Namely for any $h \in L^2([0, \tau])$, there exists scalars $c_i \in \mathbb{C}$, so that:

$$h(t) = \sum_i c_i \phi(t), \tag{3}$$

for any $t \in [0, \tau]$.

As stated in [3], the singular spectrum analysis (SSA) algorithm is based on the spectral analysis of A_τ. Given a finite sequence of discrete-time measurements: $\{f(n\Delta t) : n = 0, 1, 2, \ldots, N + M, \mathrm{and}(N + M)\Delta t \leq \tau\}$, the $(N + 1) \times (N + 1)$ a discretized version of A_τ can be approximated by:

$$A_\tau \approx C_N := \frac{1}{M+1} H_{NM} H_{NM}^*, \tag{4}$$

where H_{NM} is the trajectory matrix defined by

$$H_{NM} = \begin{pmatrix} f(0) & f(\Delta t) & \cdots & f(M\Delta t) \\ f(\Delta t) & f(2\Delta t) & \cdots & f((M+1)\Delta t) \\ \vdots & & & \\ f(N\Delta t) & f((N+1)\Delta t) & \cdots & f((N+M)\Delta t) \end{pmatrix}, \tag{5}$$

and H_{NM}^* refers to the conjugate transpose of H_{NM}. Matrix H_{NM} can be computed numerically whenever a discrete-time time series is available. Intuitively, for τ large enough and Δt small enough, C_N is a good approximation of A_τ. The SSA method starts with calculating the spectral quantities (i.e. eigenvectors, eigenvalues) of C_N. The spectral quantities of A_τ are the theoretical quantity that the spectral quantities of C_N are supposed to represent.

While in practice the SSA method has been applied successfully to a large variety of time series, in a theoretical purpose, yet with practical consequences, one may ask ourselves what is the relation between A_{τ_1} and A_{τ_2} for different τ_1 and τ_2? And what is the asymptotic behavior of A_τ as $\tau \to \infty$? In what way is the spectral property of A_τ related to intrinsic properties of the dynamical system? These questions are important because for real world data it is often not possible to get finer sampling time Δt. However, longer time series are sometimes available with long enough data. In this article we generalize the idea and tools developed in [4] and apply them to study of A_τ. We shall prove that

$$\lim_{\tau \to \infty} \lambda_{\tau,i} = |a_i|^2, \tag{6}$$

where $\lambda_{\tau,i}$ is the i-th largest eigenvalue of A_τ and a_i is the i-th largest (in modulus) coefficient of some eigenvector v_i (of unit length) of the time-shift operator $\mathcal{K}^s$ (for all $s \geq 0$) in the orthogonal decomposition of f:

$$f = \sum_{i=1}^{\infty} a_i v_i + f^{\perp}, \tag{7}$$

where $f^{\perp}$ denotes the the expression of f in the orthogonal complement of the space spanned by the time-shift operator eigenfunctions. If there are only finitely many i (say only N terms in the summation) in Eq. (7), then we set $a_i = 0$ for $i > N$. The time-shift operator $\mathcal{K}^s$ is closely related to the classical Koopman operator, which is defined to act, as a time-shift operator, on some function space whose domain is the whole phase space of the dynamical system.

In Sect. 2 we present the main result and a brief introduction of the mathematical tools used by the proof of the main result. All the quantities mentioned above are defined rigorously in Sect. 2. The detailed proof of the main result is presented in Sect. 3.

Notes and Comments The main result as well as the techniques and ideas used for the proof are close in spirit to those developed in [4]. However, the Hilbert–Schmidt operator A_τ is defined for continuous time process and the theory developed in [4] does not cover the continuous-time case. The objective of this paper is to confirm that the asymptotic behavior of the Hilbert–Schmidt operator A_τ is well related to Koopman theory.

2 Preliminaries and the Main Result

Let $\{f(t) : t \geq 0\}$ be a continuous-time process.

Assumption 1 *Assume that*

$$\lim_{T \to \infty} \frac{1}{T} \int_0^T f(t)\,dt = 0, \tag{8}$$

and that $\rho(s)$ is well-defined by Eq. (1) for all $s \geq 0$.

For any $s \geq 0$, we use F_s to denote the time series $\{F_s(t) = f(t + s) : t \geq 0\}$. For any two time series $g = \{g(t) : t \geq 0\}$ and $h = \{h(t) : t \geq 0\}$, we define the new time series

$$ag + bh = \{ag(t) + bh(t) : t \geq 0\}, \tag{9}$$

where $a, b \in \mathbb{C}$. We consider the following linear space:

$$\widetilde{\mathcal{H}}_f = \mathrm{Span}_{\mathbb{C}}\{F_s : s \geq 0\}. \tag{10}$$

Each element $h \in \widetilde{\mathcal{H}}_s$ can be written as

$$h = \sum_{i=1}^{n} c_i F_{s_i}, \tag{11}$$

for any $n \geq 1, c_i \in \mathbb{C}, s_i \geq 0$. The existence of $\rho(s)$ allows us to define the following positive semi-definite Hermitian form:

$$\langle h, g \rangle = \lim_{T \to \infty} \frac{1}{T} \int_0^T h(t)\bar{g}(t)\mathrm{d}t. \tag{12}$$

Let $V = \{v \in \widetilde{\mathcal{H}}_f : \langle v, v \rangle = 0\}$. Since the Hermitian form is positive semi-definite, V is a linear subspace of $\widetilde{\mathcal{H}}_f$. And the Hermitian form is strictly positive-definite on the quotient space $\widetilde{\mathcal{H}}_f / V$. Hence it defines an inner product on $\widetilde{\mathcal{H}}_f / V$. We define

$$\mathcal{H}_f := \overline{\widetilde{\mathcal{H}}_f / V} \tag{13}$$

where the closure is taken with respect to the inner product defined above.

We define the operator $\mathcal{K}^s$ on $\widetilde{\mathcal{H}}_f$ for any $s, s_1 \geq 0$:

$$\mathcal{K}^s F_{s_1} = F_{s_1+s}. \tag{14}$$

It is obvious that

$$\langle \mathcal{K}^s h, \mathcal{K}^s g \rangle = \langle h, g \rangle, \tag{15}$$

for any $h, g \in \widetilde{\mathcal{H}}_f$ and any $s \geq 0$. Hence $\mathcal{K}^s$ is well-defined on $\widetilde{\mathcal{H}}_f / V$, and can be further extended to the whole $\mathcal{H}_f$ by continuity. Therefore we obtain a one parameter family of isometric operators $\mathcal{K}^s$ that acts on the Hilbert space $\mathcal{H}_f$. And obviously we have

$$\mathcal{K}^{s_1}\mathcal{K}^{s_2} = \mathcal{K}^{s_1+s_2}. \tag{16}$$

To simplify the notation, we use f to also denote the continuous-time process F_0. We further assume that

Assumption 2

$$\lim_{s \to 0^+} \|\mathcal{K}^s f - f\|_{\mathcal{H}_f} = 0. \tag{17}$$

In other words, Assumption 2 assumes that the curve:

$$\gamma : [0, \infty) \to \mathcal{H}_f$$

$$t \to \mathcal{K}^t f \tag{18}$$

is continuous. Since $\mathcal{H}_f$ is generated by f and $\mathcal{K}^s$ are isometries for all $s \geq 0$, Assumption 2 implies that $\mathcal{K}^s \to I$ in the strong operator topology as $s \to 0^+$. In other words, $\{\mathcal{K}^s : s \geq 0\}$ forms a strongly continuous semigroup of isometries on $\mathcal{H}_f$.

Under Assumption 2, we have the following decomposition theorem (see Theorem 9.3 in [2]).

Theorem 1 *Let $\{\mathcal{K}^s : s \geq 0\}$ be a strongly continuous semigroup of isometries on a Hilbert space $\mathcal{H}$. Then $\mathcal{H}$ has the orthogonal decomposition $\mathcal{H} = \mathcal{H}_U \oplus \mathcal{H}_{NU}$, where $\mathcal{H}_U = \bigcap_{s \geq 0} \mathcal{K}^s \mathcal{H}$, and $\mathcal{H}_{NU}$ is isomorphic to $L^2([0, \infty], \mathcal{H}_0)$ for some Hilbert space $\mathcal{H}_0$. $\mathcal{H}_U$ and $\mathcal{H}_{NU}$ are invariant under $\mathcal{K}^s$ for all $s \geq 0$. The operator $\mathcal{K}^s$ restricted on $\mathcal{H}_U$ is a strongly continuous semigroup of unitary operators. And $\mathcal{K}^s$ restricted to $\mathcal{H}_{NU}$ acts as the unilateral shift operator, i.e. for any $\gamma \in \mathcal{H}_{NU} = L^2([0, \infty], \mathcal{H}_0)$,*

$$(\mathcal{K}^s \gamma)(t) = \gamma(t + s) \in \mathcal{H}_0. \tag{19}$$

Theorem 1 provides us with an useful tool to deal with the completely nonunitary component of $\mathcal{K}^s$. For the unitary component, we have the following spectral representation theorem.

Theorem 2 *Let $\{U(s) : s \geq 0\}$ be a strongly continuous semigroup of unitary operators on a Hilbert space $\mathcal{H}$. Assume that $\mathcal{H}$ can be generated by U and some $f \in \mathcal{H}$. Then there exists a unitary map $\phi : \mathcal{H} \to L^2(\mathbb{R}, d\mu)$ where μ is some positive finite measure on $\mathbb{R}$, such that*

$$(\phi(f))(x) = 1, \tag{20}$$

$$(\phi(\mathcal{K}^s g))(x) = e^{isx}(\phi(g))(x) \tag{21}$$

for all $g \in \mathcal{H}$, $x \in \mathbb{R}$, and $s \geq 0$.

Theorems 1 and 2 suggest the orthogonal decomposition $\mathcal{H}_f = \mathcal{H}_{f,U} \oplus \mathcal{H}_{f,NU} = L^2(\mathbb{R}, d\mu_f) \oplus L^2([0, \infty], \mathcal{H}_{f,0})$. Furthermore, we can write $\mu_f = \mu_{f,d} + \mu_{f,c}$, where $\mu_{f,d}$ is a countable sum of Dirac measures and $\mu_{f,c}$ is continuous with respect to the Lebesgue measure. $\mu_{f,c}$ can be composed both of an absolutely continuous part and a singular continuous part. The decomposition of μ_f suggests the orthogonal decomposition $\mathcal{H}_{f,U} = L^2(\mathbb{R}, d\mu_{f,d}) \oplus L^2(\mathbb{R}, d\mu_{f,c})$. In sum, we have

$$f = f_{NU} + f_d + f_c, \tag{22}$$

where $f_{NU} \in L^2([0, \infty], \mathcal{H}_{f,0})$, $f_d \in L^2(\mathbb{R}, d\mu_{f,d})$, and $f_c \in L^2(\mathbb{R}, d\mu_{f,c})$. Note that these subspaces are pair-wise orthogonal and are all invariant under $\mathcal{K}^s$ for all $s \geq 0$. The support of $\mu_{f,d}$ consists of countably many points. Each point x_i in the support of $\mu_{f,d}$ corresponds to an eigenvector $v_i \in \mathcal{H}_f$ of $\mathcal{K}^s$ for all $s \geq 0$, i.e.

$$(\phi(a_i v_i))(x) = \begin{cases} 1 & \text{if } x = x_i, \\ 0 & \text{otherwise,} \end{cases} \tag{23}$$

and $\mu_{f,d}(\{x_i\}) = |a_i|^2$, where a_i's are the coefficients of the eigenvectors in the following decomposition:

$$f = \sum_i a_i v_i + f_{NU} + f_c. \tag{24}$$

We rearrange the index of v_i so that $|a_1| \geq |a_2| \geq \cdots \geq 0$. In order to make connection with A_τ, we need the following lemmas.

Lemma 1 *For any $\tau > 0$ and any $g \in L^2([0, \tau])$, the following integral*

$$\int_0^\tau g(s)\mathcal{K}^s f \, ds \tag{25}$$

is well-defined and is an element of $\mathcal{H}_f$.

The proof of this and the following lemma use standard argument from mathematical analysis and we leave the proof to the interested readers.

Let

$$\widetilde{\mathcal{H}}_f^{\text{int}} = \left\{ \int_0^\tau g(s)\mathcal{K}^s f \, ds : \tau > 0, g \in L^2([0, \tau]) \right\}. \tag{26}$$

$\widetilde{\mathcal{H}}_f^{\text{int}}$ is a linear subspace of $\mathcal{H}_f$. We have

Lemma 2

$$\overline{\widetilde{\mathcal{H}}_f^{\text{int}}} = \mathcal{H}_f. \tag{27}$$

For simplicity, we use the notation $L_\tau^2 := L^2([0, \tau])$. Given Lemma 1, for any $g_1, g_2 \in L^2([0, \tau])$ and $t \in [0, \tau]$, we define the Hermitian form $\mathbf{A}_\tau : L_\tau^2 \times L_\tau^2 \to \mathbb{C}$:

$$\mathbf{A}_\tau(g_1, g_2) = \frac{1}{\tau}\left\langle \int_0^\tau g_1(t)\mathcal{K}^t f\, dt, \int_0^\tau g_2(s)\mathcal{K}^s f\, ds \right\rangle_{\mathcal{H}_f}. \tag{28}$$

Cauchy-Schwartz inequality implies that

$$|\mathbf{A}_\tau(g_1, g_2)|^2 \le \frac{1}{\tau^2}\left\| \int_0^\tau g_1(s)\mathcal{K}^s f\, ds \right\|_{\mathcal{H}_f}^2 \left\| \int_0^\tau g_2(s)\mathcal{K}^s f\, ds \right\|^2 \tag{29}$$

$$\le \frac{1}{\tau^2}\|g_1\|_{L_\tau^2}^2 \|g_2\|_{L_\tau^2}^2 \|f\|_{\mathcal{H}_f}^4, \tag{30}$$

where $\langle, \rangle_{L_\tau^2}$ refers to the inner product in L_τ^2 and $\langle, \rangle_{\mathcal{H}_f}$ refers to the inner product in $\mathcal{H}_f$. Therefore Riesz representation theorem warrants that there exists a linear bounded operator $A_\tau : L_\tau^2 \to L_\tau^2$ so that $\mathbf{A}_\tau(g_1, g_2) = \langle g_1, A_\tau g_2 \rangle_{L_\tau^2}$. Consequently,

$$(A_\tau g)(t) = \frac{1}{\tau}\left\langle \int_0^\tau g(s)\mathcal{K}^s f\, ds, \mathcal{K}^t f \right\rangle_{\mathcal{H}_f} = \frac{1}{\tau}\int_0^\tau g(s)\rho(t - s)\, ds, \tag{31}$$

which is the same as the definition of A_τ in [3]. Assumption 2 implies that $\rho \in L_{\text{loc}}^2(\mathbb{R})$. This implies that A_τ is a Hilbert–Schmidt operator on L_τ^2. We shall use the following variational description of the eigenvalues.

Proposition 1 (The Min-Max Principle) *Let $\mathcal{H}$ be a Hilbert space and A a Hermitian operator on $\mathcal{H}$. Let $\lambda_1 \ge \lambda_2 \ge \cdots$ be the eigenvalues of A in descending order. Then*

$$\lambda_i = \max_{\substack{\mathcal{M} \subset \mathcal{H} \\ \dim \mathcal{M} = i}} \min_{v \in \mathcal{M}} \frac{\langle v, Av \rangle}{\|v\|^2} \tag{32}$$

Our main result states that,

Theorem 3 (Main Result) *Under Assumptions 1 and 2, we have, for all $i \in \mathbb{N}$*

$$\lim_{\tau \to \infty} \lambda_{\tau, i} = |a_i|^2, \tag{33}$$

where $\lambda_{\tau, i}$ stands for the eigenvalues of A_τ.

The following Proposition [4] demonstrates the correspondence between the eigenfrequencies of the continuous-time time-shift operator and the discrete-time time-shift operator. Please refer to [4] for the notations in the proposition.

Proposition 2 *Let $\{f(X_t) : t \ge 0\}$ be a continuous time process for which ρ_s exists for all $s \ge 0$. Let $\Delta t > 0$ be a time step. Assume that*

$$\lim_{T \to \infty} \frac{1}{T} \int_0^T f(X_t) \bar{f}(X_{t+k\Delta t}) dt$$

$$= \lim_{T \to \infty} \frac{\Delta t}{T} \sum_{\mathbb{N} \ni n=0}^{T/\Delta t} f(X_{n\Delta t}) \bar{f}(X_{(n+k)\Delta t}), \tag{34}$$

for all $k \in \mathbb{N}$. Then $\mathcal{H}_f \hookrightarrow \mathcal{H}_f^{cont}$. Let q be an eigenfrequency of the discrete-time operator $\mathcal{K}^{\Delta t}$, i.e. there exists $h \in \mathcal{H}_f \hookrightarrow \mathcal{H}_f^{cont}$ so that $\mathcal{K}^{\Delta t} h = e^{iq} h$. Then there exists an integer k, and $h_k \in \mathcal{H}_f^{cont}$, so that

$$\mathcal{K}^s h_k = e^{i \frac{q+2k\pi}{\Delta t} s} h_k \tag{35}$$

for all $s \geq 0$.

Remark 1 It is worth to point out that the one-parameter semigroup of isometries $\{\mathcal{K}^s : s \geq 0\}$ is equivalent to the classical Koopman one-parameter semigroup $\{\tilde{\mathcal{K}}^s : s \geq 0\}$ which acts on $L^2(X, dv)$ almost surely (with respect to the initial state of the time series), if the dynamical system is ergodic and has finite invariant measure v on the phase space X. Because if $f \in L^2(X, v)$, then $f\tilde{\mathcal{K}}^s \bar{f} \in L^1(X, dv)$ and Birkhoff ergodic theorem states that $\rho(s) = v(f\tilde{\mathcal{K}}^s \bar{f})$ for almost every initial state $x_0 \in X$. In other words, $\langle f, \mathcal{K}^s f \rangle_{\mathcal{H}_f} = \langle f, \tilde{\mathcal{K}}^s f \rangle_{L^2(X,dv)}$. Note that f is interpreted as a given time series on the left of the equality and interpreted as a function on the right of the equality. This shows that under the assumption that the dynamical system is ergodic and (finite) measure-preserving, there is a natural isometric bijection from $\mathcal{H}_f$ to $L^2(X, dv)$.

For mathematical interests, we present the main result in an abstract mathematical form.

Theorem 4 (Main Result in Mathematical Form) *Let $\mathcal{H}$ be a Hilbert space and $\{\mathcal{K}^s : s \geq 0\}$ a strongly continuous one-parameter semigroup of isometries acting on $\mathcal{H}$. For any $f \in \mathcal{H}$, let $f = \sum_i a_i v_i + f^\perp$, where v_i's are the common eigenvectors of $\mathcal{K}^s$ for all $s \geq 0$, and $f^\perp$ is the component of f that is orthogonal to the eigenspace of $\mathcal{K}^s$ for all $s \geq 0$. Assume that $|a_1| \geq |a_2| \geq \cdots \geq 0$. For any $\tau > 0$, let $A_{f,\tau}$ be the Hermitian operator on $L^2([0, \tau])$, such that for any $g \in L^2([0, \tau])$ and any $t \in [0, \tau]$,*

$$(A_{f,\tau} g)(t) = \frac{1}{\tau} \int_0^\tau g(s) \langle \mathcal{K}^s f, \mathcal{K}^t f \rangle_{\mathcal{H}} ds. \tag{36}$$

Then $A_{f,\tau}$ is a Hilbert–Schmidt operator and hence has purely punctual spectrum. Let $\lambda_{f,\tau,i}$ be the i-th largest eigenvalue of $A_{f,\tau}$. Then we have

$$\lim_{\tau \to \infty} \lambda_{f,\tau,i} = |a_i|^2. \tag{37}$$

3 Proof of the Main Result

For any fixed small $\epsilon \geq 0$, we choose k, so that $\displaystyle\sum_{i=k+1}^{\infty} |a_i|^2 \leq \epsilon$. We have the orthogonal decomposition

$$\begin{aligned}
f &= f_d + f_{NU} + f_c = \sum_{i=1}^{k} a_i v_i + \sum_{i=k+1}^{\infty} a_i v_i + f_{d,k} + f_{NU} + f_c \\
&= f_{d,k} + f_{d,\epsilon} + f_{NU} + f_c,
\end{aligned} \tag{38}$$

where $f_{d,k} \in \mathcal{H}_{f,d,k}$ which is the subspace of $\mathcal{H}_{f,d}$ spanned by $\{v_1, \ldots, v_k\}$, and $f_{d,\epsilon} \in \mathcal{H}_{f,d,\epsilon}$ the subspace spanned by the rest of the eigenvectors, $f_{NU} \in \mathcal{H}_{f,NU}$, and $f_c \in \mathcal{H}_{f,c}$. Note that $\mathcal{H}_{f,d,k}$, $\mathcal{H}_{f,d,\epsilon}$, $\mathcal{H}_{f,NU}$, and $\mathcal{H}_{f,c}$ are pairwise orthogonal and invariant subspaces of $\mathcal{H}_f$. Hence following Eq. (28), for any $g_1, g_2 \in L_\tau^2$,

$$\begin{aligned}
\langle g_1, A_\tau g_2 \rangle_{L_\tau^2} &= \frac{1}{\tau} \Big\langle \int_0^\tau g_1(s) \mathcal{K}^s f \, ds, \int_0^\tau g_2(t) \mathcal{K}^t f \, dt \Big\rangle_{\mathcal{H}_f} \\
&= \frac{1}{\tau} \Big\langle \int_0^\tau g_1(s) \mathcal{K}^s (f_{d,k} + f_{d,\epsilon} + f_c + f_{NU}) ds, \int_0^\tau g_2(t) \mathcal{K}^t (f_{d,k} + f_{d,\epsilon} + f_c + f_{NU}) dt \Big\rangle \\
&= \frac{1}{\tau} \Big\langle \int_0^\tau g_1(s) \mathcal{K}^s f_{d,k} ds, \int_0^\tau g_2(t) \mathcal{K}^t f_{d,k} dt \Big\rangle_{\mathcal{H}_f} \\
&\quad + \frac{1}{\tau} \Big\langle \int_0^\tau g_1(s) \mathcal{K}^s f_{d,\epsilon} ds, \int_0^\tau g_2(t) \mathcal{K}^t f_{d,\epsilon} dt \Big\rangle_{\mathcal{H}_f} \\
&\quad + \frac{1}{\tau} \Big\langle \int_0^\tau g_1(s) \mathcal{K}^s f_c ds, \int_0^\tau g_2(t) \mathcal{K}^t f_c dt \Big\rangle_{\mathcal{H}_f} \\
&\quad + \frac{1}{\tau} \Big\langle \int_0^\tau g_1(s) \mathcal{K}^s f_{NU} ds, \int_0^\tau g_2(t) \mathcal{K}^t f_{NU} dt \Big\rangle_{\mathcal{H}_f} \\
&= \langle g_1, A_{\tau,d,k} g_2 \rangle_{L_\tau^2} + \langle g_1, A_{\tau,d,\epsilon} g_2 \rangle_{L_\tau^2} + \langle g_1, A_{\tau,c} g_2 \rangle_{L_\tau^2} + \langle g_1, A_{\tau,NU} g_2 \rangle_{L_\tau^2},
\end{aligned} \tag{39}$$

in which the definition of $A_{\tau,d,k}$, $A_{\tau,d,\epsilon}$, $A_{\tau,c}$ and $A_{\tau,NU}$ are obvious. It is not hard to show that $A_{\tau,d,k}$, $A_{\tau,d,\epsilon}$, $A_{\tau,c}$ and $A_{\tau,NU}$ all admit eigendecomposition since they are all Hilbert–Schmidt Hermitian operators. Note that the cross product terms all as $\mathcal{H}_{f,d,k}$, $\mathcal{H}_{f,d,\epsilon}$, $\mathcal{H}_{f,c}$ and $\mathcal{H}_{f,NU}$ are pairwise orthogonal and invariant under $\mathcal{K}^s$ for all $s \geq 0$.

Let $\lambda_{\tau_d,k,i}$, $\lambda_{\tau,d,\epsilon,i}$, $\lambda_{\tau,c,i}$, and $\lambda_{\tau,NU,i}$ be the i-th largest eigenvalue of $A_{\tau,d,k}$, $A_{\tau,d,\epsilon}$, $A_{\tau,c}$, $A_{\tau,NU}$ respectively. We will prove the following identities:

Proposition 3

$$\lim_{\tau \to \infty} \lambda_{\tau,d,k,i} = |a_i|^2 \text{ for } i = 1, \ldots, k, \tag{40}$$

$$\lambda_{\tau,d,\epsilon,1} \le \epsilon \text{ for any } \tau > 0, \tag{41}$$

$$\lim_{\tau \to \infty} \lambda_{\tau,c,1} = 0, \tag{42}$$

$$\lim_{\tau \to \infty} \lambda_{\tau,NU,1} = 0. \tag{43}$$

Before we start to prove Eqs. (40)–(43), it is not hard to see that Propositions 1 and 3 directly implies the main result. Indeed, for any fixed n and any $\epsilon > 0$, we can find k so that $n \le k$ and $\sum_{i=k+1}^{\infty} |a_i|^2 \le \epsilon$. Then we find τ large enough so that $\lambda_{\tau,c,1} \le \epsilon$ and $\lambda_{\tau,NU,1} \le \epsilon$. Note that $A_{\tau,d,k}$, $A_{\tau,d,\epsilon}$, $A_{\tau,c}$, and $A_{\tau,NU}$ are all positive semidefinite. Applying the min-max principle we have

$$\lambda_{\tau,n} = \max_{\substack{\mathcal{M} \subset L_\tau^2 \\ \dim \mathcal{M}=n}} \min_{v \in \mathcal{M}} \frac{\langle v, A_\tau \, v \rangle}{\|v\|^2} \tag{44}$$

$$= \max_{\substack{\mathcal{M} \subset L_\tau^2 \\ \dim \mathcal{M}=n}} \min_{v \in \mathcal{M}} \frac{\langle v, A_{\tau,d,k} \, v \rangle + \langle v, A_{\tau,d,\epsilon} \, v \rangle + \langle v, A_{\tau,c} \, v \rangle + \langle v, A_{\tau,NU} \, v \rangle}{\|v\|^2} \tag{45}$$

$$\ge \max_{\substack{\mathcal{M} \subset L_\tau^2 \\ \dim \mathcal{M}=n}} \min_{v \in \mathcal{M}} \frac{\langle v, A_{\tau,d,k} \, v \rangle}{\|v\|^2} = \lambda_{\tau,d,k,n}, \tag{46}$$

and that

$$\lambda_{\tau,n} = \max_{\substack{\mathcal{M} \subset L_\tau^2 \\ \dim \mathcal{M}=n}} \min_{v \in \mathcal{M}} \frac{\langle v, A_\tau \, v \rangle}{\|v\|^2} \tag{47}$$

$$\le \max_{\substack{\mathcal{M} \subset L_\tau^2 \\ \dim \mathcal{M}=n}} \min_{v \in \mathcal{M}} \frac{\langle v, A_{\tau,d,k} \, v \rangle}{\|v\|^2} + 2\epsilon = \lambda_{\tau,d,k,n} + 2\epsilon. \tag{48}$$

Combined with Eq. (40), this implies Theorem 3.

***Proof (Equation* (40))** Recall from Eq. (23) that each eigenvector v_i corresponds to a point x_i in the support of μ_d. For any $g \in L_\tau^2$, Theorem 2 states that $\int_0^\tau g(s)\mathcal{K}^s f_{d,k} \mathrm{d}s$ has the following representation in $L^2(\mathbb{R}, d\mu)$, for any $x \in \mathbb{R}$,

$$\left(\phi\left(\int_0^\tau g(s)\mathcal{K}^s f_{d,k}\,\mathrm{d}s\right)\right)(x) = \begin{cases} \int_0^\tau g(s)e^{isx_j}\,\mathrm{d}s & ,\text{if } x = x_j \text{ for some } j. \\ 0 & ,\text{otherwise.} \end{cases} \tag{49}$$

And

$$\langle g, A_{\tau,d,k}g\rangle_{L_\tau^2} = \frac{1}{\tau}\left\langle\int_0^\tau g(s)\mathcal{K}^s f_{d,k}\,\mathrm{d}s, \int_0^\tau g(t)\mathcal{K}^t f_{d,k}\,\mathrm{d}t\right\rangle_{\mathcal{H}_f} \tag{50}$$

$$= \frac{1}{\tau}\sum_{j=1}^k\left\|\int_0^\tau g(s)e^{isx_j}\,\mathrm{d}s\right\|_{L^2(\mathbb{R},d\mu)}^2 \tag{51}$$

$$= \frac{1}{\tau}\sum_{j=1}^k |a_j|^2\left|\int_0^\tau g(s)e^{isx_j}\,\mathrm{d}s\right|^2. \tag{52}$$

Let $\xi_j \in L_\tau^2$ so that $\xi_j(s) = e^{isx_j}$ for any $s \in [0, \tau]$. Then $\|\xi_j\|_{L_\tau^2}^2 = \tau$ and

$$\langle g, A_{\tau,d,k}\,g\rangle_{L_\tau^2} = \frac{1}{\tau}\sum_{j=1}^k |a_j|^2|\langle\xi_j, g\rangle_{L_\tau^2}|^2 = \sum_{j=1}^k |\langle\frac{a_j\xi_j}{\sqrt{\tau}}, g\rangle_{L_\tau^2}|^2 \tag{53}$$

Let $V_{\tau,k} = \text{Span}_\mathbb{C}\{\frac{a_1\xi_1}{\sqrt{\tau}}, \frac{a_2\xi_2}{\sqrt{\tau}}, \cdots, \frac{a_k\xi_k}{\sqrt{\tau}}\}$. We write $g = g_{\tau,k}+g^\perp$, where $g_{\tau,k} \in V_{\tau,k}$, and $g^\perp \in V_{\tau,k}^\perp$. Then

$$\langle g, A_{\tau,k,d}\,g\rangle_{L_\tau^2} = \sum_{j=1}^k |\langle\frac{a_j\xi_j}{\sqrt{\tau}}, g_{\tau,k}\rangle|_{L_\tau^2}^2. \tag{54}$$

Note that $\dim V_{\tau,k} = k$ for all $\tau > 0$. Direct calculation yields that, for $j \neq \ell$, $\langle\frac{a_j\xi_j}{\sqrt{\tau}}, \frac{a_\ell\xi_\ell}{\sqrt{\tau}}\rangle_{L_\tau^2} = a_j\bar{a}_l\frac{e^{i(x_j-x_\ell)\tau}-1}{i\tau(x_j-x_\ell)} \to 0$ as $\tau \to \infty$. Therefore the distribution of the eigenvalues of $A_{\tau,k,d}$ shall approach to the distribution of the eigenvalues of

$$\begin{pmatrix} |a_1|^2 & 0 & \cdots & 0 \\ 0 & |a_2|^2 & \cdots & 0 \\ \vdots & & & \\ 0 & 0 & \cdots & |a_k|^2 \end{pmatrix} \tag{55}$$

as $\tau \to \infty$. This completes the proof of Eq. (40).

Proof (Equation (41)) Similar to Eq. (53), for any $g \in L_\tau^2$, $\|g\|_{L_\tau^2} = 1$, we have

$$\langle g, A_{\tau,d,\epsilon} \, g \rangle_{L_\tau^2} = \frac{1}{\tau} \sum_{j=k+1}^{\infty} |a_j|^2 |\langle \xi_j, g \rangle_{L_\tau^2}|^2 = \sum_{j=k+1}^{\infty} |\langle \frac{a_j \xi_j}{\sqrt{\tau}}, g \rangle_{L_\tau^2}|^2 \qquad (56)$$

$$\leq \sum_{j=k+1}^{\infty} |a_j|^2 \leq \epsilon. \qquad (57)$$

Then the min-max principle implies that $\lambda_{\tau,k,\epsilon,1} \leq \epsilon$.

Proof (Equation (42)) Following [1] (page 39–41), we first show that

$$\lim_{\tau \to \infty} \frac{1}{\tau} \int_0^\tau |\mu_{f,c}(e^{isx})| ds = 0, \qquad (58)$$

or equivalently

$$\lim_{\tau \to \infty} \frac{1}{\tau} \int_0^\tau |\mu_{f,c}(e^{isx})|^2 ds = 0. \qquad (59)$$

Equation (58) means that the large moments associated to the continuous spectral measure has density zero. For any $\epsilon > 0$, we write $\mu_{f,c} = \mu_{f,c,1} + \mu_{f,c,\epsilon}$, in which $\mu_{f,c,1}$ has compact support, $\mu_{f,c,\epsilon}(\mathbb{R}) < \epsilon$ and $\mu_{f,c,1} \perp \mu_{f,c,\epsilon}$. Denote the support of $\mu_{f,c,1}$ by B_1. Then we have

$$\frac{1}{\tau} \int_0^\tau |\mu_{f,c}(e^{isx})|^2 ds = \frac{1}{\tau} \int_0^\tau |\mu_{f,c,1}(e^{isx})|^2 ds + \frac{1}{\tau} \int_0^\tau |\mu_{f,c,\epsilon}(e^{isx})|^2 ds \qquad (60)$$

$$< \frac{1}{\tau} \int_0^\tau |\int_{\mathbb{R}} e^{isx} d\mu_{f,c,1}(x)|^2 ds + \epsilon \qquad (61)$$

and that

$$\frac{1}{\tau} \int_0^\tau |\mu_{f,c,1}(e^{isx})|^2 ds = \frac{1}{\tau} \int_0^\tau |\int_{\mathbb{R}} e^{isx} d\mu_{f,c,1}(x)|^2 ds$$

$$= \frac{1}{\tau} \int_0^\tau ds \int_{\mathbb{R}} \int_{\mathbb{R}} e^{is(x-y)} d\mu_{f,c,1}(x) d\mu_{f,c,1}(y) \qquad (62)$$

$$= \frac{1}{\tau} \int_{\mathbb{R}} \int_{\mathbb{R}} d\mu_{f,c,1}(x) d\mu_{f,c,1}(y) \int_0^\tau e^{is(x-y)} ds \qquad (63)$$

$$= \frac{1}{\tau} \int_{B_1} \int_{B_1} d\mu_{f,c,1}(x) d\mu_{f,c,1}(y) \int_0^\tau e^{is(x-y)} ds \qquad (64)$$

Note that $|\frac{1}{\tau} \int_0^\tau e^{is(x-y)} ds| \leq 1$ for any $\tau > 0$ and any $x, y \in \mathbb{R}$. And when $x \neq y$

$$1 \geq \left| \frac{1}{\tau} \int_0^\tau e^{is(x-y)} ds \right| = \left| \frac{e^{i\tau(x-y)} - 1}{\tau i (x - y)} \right| \xrightarrow[\tau \to \infty]{} 0. \tag{65}$$

Since $\mu_{f,c,1}$ is continuous, we have that $(\mu_{f,c,1} \times \mu_{f,c,1})(\{(x, y) \in \mathbb{R}^2 : x = y\}) = 0$. Hence, the integral in Eq. (64) boils down to an integral on $\mathbb{R}^2 \setminus \{x = y\}$. Lebesgue's dominated convergence theorem implies that the integral in Eq. (64) converges to 0 as $\tau \to \infty$. Hence $\limsup\limits_{\tau \to \infty} \frac{1}{\tau} \int_0^\tau |\mu_{f,c}(e^{isx})|^2 ds < \epsilon$ for any $\epsilon > 0$. This implies Eq. (59).

For any $g \in L^2(\mathbb{R})$, Theorem 2 implies that

$$\left(\phi \left(\int_0^\tau g(s) \mathcal{K}^s f_{d,c} ds \right) \right)(x) = \int_0^\tau g(s) e^{isx} ds. \tag{66}$$

Therefore

$$\langle g, A_{\tau,c} g \rangle_{L_\tau^2} = \frac{1}{\tau} \left\langle \int_0^\tau g(s) \mathcal{K}^s f_{d,c} ds, \int_0^\tau g(t) \mathcal{K}^t f_{d,c} dt \right\rangle_{\mathcal{H}_f} \tag{67}$$

$$= \frac{1}{\tau} \left\langle \phi \left(\int_0^\tau g(s) \mathcal{K}^s f_{d,c} ds \right), \phi \left(\int_0^\tau g(t) \mathcal{K}^t f_{d,c} dt \right) \right\rangle_{L^2(\mathbb{R}, d\mu_c)} \tag{68}$$

$$= \frac{1}{\tau} \int_{-\infty}^\infty d\mu_{f,c}(x) \int_0^\tau \int_0^\tau g(s) \bar{g}(t) e^{i(s-t)x} ds\, dt \tag{69}$$

$$= \frac{1}{\tau} \int_0^\tau \int_0^\tau g(s) \bar{g}(t) \mu_{f,c}(e^{i(s-t)x}) ds\, dt \tag{70}$$

Hence

$$|\langle g, A_{\tau,c} g \rangle| \leq \frac{1}{\tau} \int_0^\tau \int_0^\tau |g(t)| \cdot |g(s)| \cdot |\mu_{f,c}(e^{i(s-t)x})| dt\, ds \tag{71}$$

$$= \frac{1}{\tau} \iint_{0 \leq s \leq t \leq \tau} |g(t)| \cdot |g(s)| \cdot |\mu_{f,c}(e^{i(s-t)x})| dt\, ds \tag{72}$$

$$+ \frac{1}{\tau} \iint_{0 \leq t \leq s \leq \tau} |g(t)| \cdot |g(s)| \cdot |\mu_{f,c}(e^{i(s-t)x})| dt\, ds \tag{73}$$

$$= \frac{2}{\tau} \int_0^\tau |g(t)| \int_t^\tau |g(s)| \cdot |\mu_{f,c}(e^{i(s-t)x})| dt\, ds \tag{74}$$

$$= \frac{2}{\tau} \int_0^\tau |g(t)| \int_0^{\tau-t} |g(t+s)| \cdot |\mu_{f,c}(e^{isx})| ds\, dt \tag{75}$$

$$\leq \frac{2}{\tau} \int_0^\tau \int_0^{\tau-t} \frac{1}{2} (|g(t)|^2 + |g(t+s)|^2) |\mu_{f,c}(e^{isx})| ds\, dt \tag{76}$$

$$= \frac{1}{\tau} \int_0^\tau |\mu_{f,c}(e^{isx})| \int_0^{\tau-s} (|g(t)|^2 + |g(t+s)|^2) \, ds \, dt \tag{77}$$

$$\leq \frac{1}{\tau} \int_0^\tau 2|\mu_{f,c}(e^{isx})| \cdot \|g\|_{L_\tau^2}^2 \, ds \tag{78}$$

Therefore

$$\lambda_{\tau,c,1} = \max_{g \in L_\tau^2} \frac{\langle g, A_{\tau,c} g \rangle}{\|g\|_{L_\tau^2}} \to 0, \tag{79}$$

as $\tau \to \infty$. This completes the proof of Eq. (42).

Proof (Equation (43)) Recall that $\mathcal{H}_{f,NU} \cong L^2([0, +\infty], \mathcal{H}_0)$. Hence f_{NU} can be represented as a curve from $[0, \infty]$ to $\mathcal{H}_0$. We denote this curve by γ. Without ambiguity, we do not distinguish between γ and f_{NU}. Hence for each $t \geq 0$, $\gamma(t) \in \mathcal{H}_0$. And $\|\gamma\|_{\mathcal{H}_{f,NU}}^2 = \int_0^\infty \|\gamma(t)\|_{\mathcal{H}_0}^2 \, dt$. Recall that $(\mathcal{K}^s \gamma)(t) = \gamma(t+s)$. We set $\gamma(t) = 0$ for all $t < 0$. Hence for any $g \in L_\tau^2$,

$$\langle g, A_{\tau,NU} g \rangle_{L_\tau^2} = \frac{1}{\tau} \left\langle \int_0^\tau g(s_1) \mathcal{K}^{s_1} \gamma \, ds_1, \int_0^\tau g(s_2) \mathcal{K}^{s_2} \gamma \, ds_2 \right\rangle_{\mathcal{H}_{f,NU}} \tag{80}$$

$$= \frac{1}{\tau} \int_0^\infty \int_0^\tau \int_0^\tau \bar{g}(s_2) g(s_1) \langle \gamma(t+s_1), \gamma(t+s_2) \rangle_{\mathcal{H}_0} \, ds_1 \, ds_2 \, dt \tag{81}$$

$$= \frac{1}{\tau} \int_0^\tau \int_0^\tau \bar{g}(s_2) g(s_1) \int_0^\infty \langle \gamma(t+s_1), \gamma(t+s_2) \rangle_{\mathcal{H}_0} \, dt \, ds_1 \, ds_2 \tag{82}$$

We first show the following identity:

$$\lim_{s \to \infty} \langle \gamma, \mathcal{K}^s \gamma \rangle_{\mathcal{H}_{f,NU}} = \lim_{s \to \infty} \int_0^\infty \langle \gamma(t), \gamma(t+s) \rangle_{\mathcal{H}_0} \, dt = 0. \tag{83}$$

To prove Eq. (83), without loss of generality we assume that $\|\gamma\|_{\mathcal{H}_{f,NU}} = 1$. For any $\epsilon > 0$, there exists N_ϵ, so that $\int_0^{N_\epsilon} \|\gamma(t)\|_{\mathcal{H}_0}^2 \, dt > 1 - \epsilon$. This means that $\int_{N_\epsilon}^\infty \|\gamma(t)\|^2 \, dt < \epsilon$. Therefore for any $s \geq N_\epsilon$,

$$\left| \int_0^\infty \langle \gamma(t), \gamma(t+s) \rangle_{\mathcal{H}_0} \, dt \right|^2 \leq \left| \int_0^\infty \|\gamma(t)\|_{\mathcal{H}_0}^2 \, dt \right|^2 \cdot \left| \int_{N_\epsilon}^\infty \|\gamma(t)\|_{\mathcal{H}_0}^2 \, dt \right|^2 < \epsilon^2. \tag{84}$$

This proves Eq. (83). Now we continue with Eq. (82):

$$\langle g, A_{\tau,NU} g \rangle_{L_\tau^2} \leq \left| \frac{2}{\tau} \int_0^\tau \int_{s_1}^\tau \bar{g}(s_2) g(s_1) \langle \mathcal{K}^{s_1} \gamma, \mathcal{K}^{s_2} \gamma \rangle_{\mathcal{H}_{f,NU}} \, ds_1 \, ds_2 \right| \tag{85}$$

$$\leq \left| \frac{2}{\tau} \int_0^\tau \int_0^{\tau - s_1} g(s_1) \bar{g}(s_1 + s) \langle \gamma, \mathcal{K}^s \gamma \rangle_{\mathcal{H}_{f,NU}} \, \mathrm{d}s_1 \, \mathrm{d}s \right| \tag{86}$$

For any $\epsilon > 0$, find M_ϵ, so that for any $|\langle \gamma, \mathcal{K}^s \gamma \rangle| < \epsilon$ for any $s > M_\epsilon$. Now for any $\tau > M_\epsilon / \epsilon$ and any $\|g\|_{L_\tau^2} = 1$, we have

$$\langle g, A_{\tau, NU} g \rangle_{L_\tau^2} \tag{87}$$

$$\leq \frac{2}{\tau} \int_0^\tau \int_0^{M_\epsilon} |g(s_1)| \cdot |g(s_1 + s)| \cdot |\langle \gamma, \mathcal{K}^s \gamma \rangle_{\mathcal{H}_{f,NU}}| \mathrm{d}s_1 \mathrm{d}s + \tag{88}$$

$$\frac{2}{\tau} \int_0^\tau \int_{M_\epsilon}^{\tau - s_1} |g(s_1)| \cdot |g(s_1 + s)| \cdot |\langle \gamma, \mathcal{K}^s \gamma \rangle_{\mathcal{H}_{f,NU}}| \mathrm{d}s_1 \mathrm{d}s \tag{89}$$

$$\leq \frac{1}{\tau} \int_0^\tau \int_0^{M_\epsilon} (|g(s_1)|^2 + |g(s_1 + s)|^2) |\langle \gamma, \mathcal{K}^s \gamma \rangle_{\mathcal{H}_{f,NU}}| \mathrm{d}s_1 \mathrm{d}s + \tag{90}$$

$$\frac{1}{\tau} \int_0^\tau \int_{M_\epsilon}^{\tau - s_1} (|g(s_1)|^2 + |g(s_1 + s)|^2) |\langle \gamma, \mathcal{K}^s \gamma \rangle_{\mathcal{H}_{f,NU}}| \mathrm{d}s_1 \mathrm{d}s \tag{91}$$

$$\leq \frac{1}{\tau} \int_0^\tau \int_0^{M_\epsilon} |g(s_1)|^2 \mathrm{d}s_1 \mathrm{d}s + \frac{1}{\tau} \int_0^\tau \int_0^{M_\epsilon} |g(s_1 + s)|^2 \cdot |\langle \gamma, \mathcal{K}^s \gamma \rangle_{\mathcal{H}_{f,NU}}| \tag{92}$$

$$\mathrm{d}s_1 \mathrm{d}s + \frac{1}{\tau} \int_0^\tau \int_{M_\epsilon}^{\tau - s_1} \epsilon(|g(s_1)|^2 + |g(s_1 + s)|^2) \mathrm{d}s_1 \mathrm{d}s \tag{93}$$

$$\leq \frac{M_\epsilon}{\tau} + \frac{M_\epsilon}{\tau} + 2 \frac{\epsilon}{\tau} (\tau - M_\epsilon) \leq 4\epsilon. \tag{94}$$

Therefore for $\tau > M_\epsilon / \epsilon$,

$$\lambda_{\tau, NU, 1} = \max_{\substack{g \in L_\tau^2 \\ \|g\| = 1}} \langle g, A_{\tau, NU} g \rangle \leq 4\epsilon. \tag{95}$$

This completes the proof of Eq. (43).

References

1. Paul R. Halmos. *Lectures on Ergodic Theory*. Chelsea Publishing Company, New York, N.Y, 1956.
2. Béla Szőkefalvi-Nagy, Ciprian Foias, Hari Bercovici, and László Kérchy. *Harmonic Analysis of Operators on Hilbert Space*. Springer Science & Business Media, 2010.
3. Robert Vautard and Michael Ghil. Singular spectrum analysis in nonlinear dynamics, with applications to paleoclimatic time series. *Physica D: Nonlinear Phenomena*, 35:395–424, 1989.
4. Yicun Zhen, Bertrand Chapron, Etienne Mémin, and Lin Peng. Eigenvalues of autocovariance matrix: A practical method to identify the koopman eigenfrequencies. *arXiv*, 2021.

Index

B
Boulvard, Pierre-Marie, 57

C
Chapron, Bertrand, 211, 223, 259, 301
Collard, Fabrice, 211
Crisan, Dan, 19, 43

D
Debussche, Arnaud, 15
Dinvay, Evgueni, 27
Dufée, Benjamin, 43

F
Fablet, Ronan, 211
Fiorini, Camilla, 57
Flandoli, Franco, 69

G
Goodair, Daniel, 87

H
Hascoet, Erwan, 259
Holm, Darryl D., 109
Hug, Berenger, 15
Hu, Ruiao, 109, 135

L
Lang, Oana, 159
Li, Long, 57, 179, 273, 287

L
Lobbe, Alexander, 195
Luongo, Eliseo, 69

M
Mémin, Etienne, 15, 43, 57, 179, 273, 287, 301
Mensah, Prince Romeo, 1

O
Ouala, Said, 211

P
Pan, Wei, 159
Patching, Stuart, 135
Platzer, Paul, 223

R
Reich, Sebastian, 237
Resseguier, Valentin, 259

S
Street, Oliver D., 109

T
Tandeo, Pierre, 211, 223
Thiry, Louis, 273
Tissot, Gilles, 179
Tucciarone, Francesco L., 287

Z
Zhen, Yicun, 301

© The Author(s) 2023
B. Chapron et al. (eds.), *Stochastic Transport in Upper Ocean Dynamics*,
Mathematics of Planet Earth 10, https://doi.org/10.1007/978-3-031-18988-3